BLASTING AGENT

IRRITAN
6

D1532401

FLAMMABLE GAS
2

FLAMMABLE SOLID
4

FLAMMABLE LIQUID
3

CHLORINE
2

9

HARMFUL
STOW AWAY FROM
FOODSTUFFS
6

OXYGEN
2

OXIDIZER
5.1

ORGANIC PEROXIDE
5.2

Emergency Responder Training Manual for the Hazardous Materials Technician

Emergency Responder Training Manual for the Hazardous Materials Technician

CENTER FOR LABOR EDUCATION AND RESEARCH

Edited by

Lori P. Andrews, P.E.

Contributing Authors

Lori P. Andrews, P.E.
Lynn M. Artz, M.D., M.P.H.
John E. Backensto, C.H.M.I.
W. Donald Fattig, Ph.D.
Barbara M. Hilyer, M.S., M.S.P.H.
James C. Hilyer, Ed.D., M.P.H.
Ralph Johnson, Ph.D.
Charles H. Morton, B.S.
Kenneth W. Oldfield, M.S.P.H.
Higdon C. Roberts, Jr., Ph.D.
D. Alan Veasey, M.A. Ed.

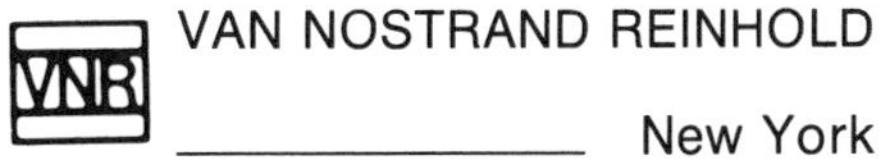

VAN NOSTRAND REINHOLD
New York

Library of Congress Catalog Card Number 91-45222
ISBN 0-442-00877-5

I(T)P Van Nostrand Reinhold is an International Thomson Publishing company.
ITP logo is a trademark under license.

Printed in the United States of America

Van Nostrand Reinhold
115 Fifth Avenue
New York, NY 10003

International Thomson Publishing GmbH
Königswinterer Str. 418
53227 Bonn
Germany

International Thomson Publishing
Berkshire House,168-173
High Holborn, London WC1V 7AA
England

International Thomson Publishing Asia
221 Henderson Bldg. #05-1
Singapore 0315

Thomas Nelson Australia
102 Dodds Street
South Melbourne 3205
Victoria, Australia

International Thomson Publishing Japan
Kyowa Building, 3F
2-2-1 Hirakawacho
Chiyoda-ku, Tokyo 102
Japan

Nelson Canada
1120 Birchmount Road
Scarborough, Ontario
M1K 5G4, Canada

16 15 14 13 12 11 10 9 8 7 6 5 4 3

Library of Congress Cataloging-in-Publication Data

Emergency responder training manual for the hazardous materials technician / Center for Labor Education and Research : edited by Lori P. Andrews.

p. cm.

Includes bibliographical references and index.

ISBN 0-442-00877-5

1. Hazardous substances—Accidents—Handbooks, manuals, etc.

I. Andrews, Lori P. II. University of Alabama at Birmingham.
Center for Labor Education and Research. 91-45222
T55.3.H3E446 1992 CIP

Contents

Preface xvii

Acknowledgments xix

1. BOOK INTRODUCTION AND OVERVIEW 1

How to Use the Book 2
Hazardous Materials: A Risky Business 3
Objectives 4
Defining a Hazardous Materials Emergency 4
Dangers in HAZMAT Emergencies 5
Chemicals and Human Health 5
Chemicals and Fire 9
Chemicals in the Environment 10
References 10

2. FEDERAL OCCUPATIONAL SAFETY AND HEALTH REGULATIONS APPLICABLE TO EMERGENCY RESPONSE PERSONNEL 11

Objectives 12
The Role of OSHA in Regulating Occupational Safety and Health 12
Background and Creation of OSHA 13
Responsibilities of OSHA 13
OSHA Safety and Health Standards 13
The Scope of OSHA Authority 14
Coverage of Employees of Private Industry and the Federal Government 14
Coverage Under State Occupational Safety and Health Plans 14
Origin and Scope of OSHA Regulations Pertaining to Hazardous Waste Operations and Emergency Response 15
Background and Development of HAZWOPER 15
Purpose and Applicability 16

Applicability to Emergency Response Operations 16
Provisions of 29 CFR 1910.120 Applicable To Emergency Response Operations 17
Emergency Response Plan 17
Emergency Response Chain of Command and Personnel Roles 18
Emergency Response Procedures 20
Training 20
Medical Surveillance and Consultation 24
Personal Protective Equipment 25
Postemergency Response Operations 26
Summary 27
References 27

3. INCIDENT COMMAND SYSTEM 28

Objectives 29
Basics of the ICS 29
Components of the ICS 30
Major Functional Areas of the ICS 32
Interaction of the Functional Areas of the ICS 33
Command of the Incident Response 34
Single and Unified Command Structures 34
Command Staff Responsibilities 36
The Operations Section 38
Structure of the Operations Sections 38
Basic Considerations for Operations 39
The Planning Section 40
The Logistics Section 41
The Finance Section 41
Summary 42
References 42

4. HAZARDOUS MATERIALS TERMINOLOGY 43

Objectives 44
Terminology for Materials Which Pose Hazards 44
Hazardous Chemicals 44
Hazardous Materials 44
Hazardous Substances 45
Hazardous Wastes 45
Terminology To Describe Hazard Classification 45
Terminology for Items Used to Communicate Hazard Information 46
Labels and Markings on Containers 49

Shipping Papers 53
Placards 54
Summary 57
References 58

5. HAZARD AND RISK ASSESSMENT 59

Objectives 59
Decision Making in Emergencies 60
Pre-incident Hazard Assessment 63
Materials Present On Site 64
Contingency Plans 64
Confinement Systems 64
Maps 65
Incident Perimeter Survey 66
Where is the Spill? 66
Where is the Spill Headed? 66
What is the Indentity of the Leaking Material? 66
What are the Hazards to Responders? 67
Entry and On-site Survey 67
Monitor the Air 67
Observe Containers 67
Observe Released Material 71
Researching Identified Materials 71
Printed Materials 71
NIOSH *Pocket Guide* 81
Chris *Manual* U.S. Coast Guard 92
Telephone Hotlines 92
Computer Networks and Data Bases 92
Predicting Dispersal of Hazardous Materials 93
Solids 93
Liquids 93
Gases and Vapors 96
Summary 99
References 99

6. SITE CONTROL 101

Objectives 102
Objectives of Site Control 102
Enforcement of Site Control 103
Isolation 104

Access Control 104
Evacuation and Protection in Place 105
Zoning 106
Hot Zone 108
Warm Zone 111
Cold Zone 111
Access Control 112
Other Zoning Considerations 112
Communications 113
Summary 115
References 115

7. SOPs AND TERMINATION 116

Objectives 117
Standard Operating Procedures 117
Considerations 117
Response Activities 119
Operating Guides—Generic Plans 119
Typical Standard Operating Procedures 120
Termintion Procedures 122
Other Types of Termination Procedures 123
Specific Procedures 124
Job Descriptions 124
References 124
Appendix 1: Chemical Spill Response Procedures 125
Appendix 2: Emergency Operation Codes 126
Appendix 3: SOPs for HAZ-1 128

8. SARA TITLE III—STATE ERPs 134

Objectives 134
History 135
Relevant Provisions of SARA TITLE III 137
Subtitle A (including Sections 301–305) 137
State and Local ERPs 144
Major Components 144
Reference Sources 152
Hotline Telephone Numbers 152
Manuals Widely Used by Response Personnel 153
References 156

9. PERSONAL PROTECTIVE EQUIPMENT 158

Objectives 158
Respirators 159
Respiratory Protection Requirements 159
Classification of Respiratory Protective Equipment 159
Selection of Respiratory Protective Equipment 165
Importance of Respirator Fit and Fit Testing 174
Respirator Protection Factor 178
Chemical Protective Clothing and Accessories 179
Selection of CPC 179
Types of CPC and Accessories 190
Classification of Protective Ensembles 192
EPA Levels of Protection 192
NFPA Certification Standards for Chemical Protective Garments 203
Safe Use of PPE 209
PPE Program Requirements 210
Training in PPE Use 211
Factors Limiting Safe Work Mission Duration 211
Personal Factors Affecting Respirator Use 212
Donning Pipe 213
Inspection of PPE 213
In-Use Monitoring of PPE 215
Doffing PPE 215
Storage of PPE 215
Reuse of CPC 216
Maintenance of PPE 216
Heat Stress and Other Physiological Factors 217
Summary 218
References 219

10. DECONTAMINATING PERSONNEL AND EQUIPMENT 220

Objectives 220
Definition of and Justification for Decontamination 221
Methods of Decontamination 221
Physical Methods 221
Chemical Solutions 225
Level of Decontamination Required 227
Setting up a Decontamination Area 228
Location 228
Personnel Decontamination Line 229

Decontamination of Tools, Equipment, and Vehicles 235
Emergency Medicine and Decontamination 236
Management of the Decontamination Area 239
Orderly Cleaning and Doffing 239
Protection of Decontamination Personnel 240
Containment of Liquids 241
Post-incident Management 242
Summary 242
References 243
Appendix 1: EPA-Suggested Decontamination Lines 244

11. BASIC HAZARDOUS MATERIALS CONTROL 250

Objectives 251
The Role of Hazard and Risk Assessment and Decision Making in Hazardous Materials Control 252
Types of Releases 252
Other Considerations 253
Procedures and Considerations for Basic Hazardous Materials Control 253
Controlling Land Releases 254
Controlling Water Releases 260
Controlling Air Releases 264
Collection Techniques 268
Pumping and Vacuum Collection 268
Sorbent Collection 268
Other Techniques and Considerations 271
Summary 272
References 273

12. ADVANCED HAZARDOUS MATERIALS CONTROL 274

Objectives 275
Assessment and Decision Making for Containment Operations 275
Containment-Related Planning and Decision Making 276
Assessment Considerations for Selection of Containment Procedures 276
Equipment, Supplies, and Tools used in Containment 278
Plugs, Patches, and Related Items 279
Adhesives, Sealants, and Gaskets 281
Tools 281
Other Containment Related Items 281

Containment Procedures 285
General Procedures 285
Procedures for Small Containers 286
Procedures for Large Containers 291
Procedures for Plumbing Leaks 293
Procedures for Pressurized Cylinders 297
Summary 299
References 300
Appendix 1: Grounding and Bonding Flammable Liquid Containers 301
Appendix 2: Overpacking Damaged Containers 306

13. THE BEHAVIOR OF CHEMICALS 312

Objective 312
Atoms 313
Structure of the Atom 314
Combinations of Atoms 314
Families of Chemicals 316
Names of Chemicals 318
Symbols 319
Prefixes and Suffixes 319
Organic Chemicals 320
Properties of Chemicals 323
Boiling Point 324
Vapor Pressure 325
Vapor Density 325
Solubility 325
Specific Gravity 326
Flash Point 326
Explosive Limits and Range 327
Hydrogen Ion Concentration (pH) 327
Appearance and Odor 327
Physical State 327
Solids 328
Liquids 328
Gases 328
Changes in Physical State 329
Chemical Reactions 330
Water-Reactive Materials 330
Air-Reactive Materials 330
Oxidizers 331
Unstable Materials 332
Incompatible Materials 332
Toxic Combustion Products 333

Radioactivity 336
Recognition 336
Response 337
Sources of Chemical Information 337
Reference Documents for Use in an Emergency 337
Reference Documents for Desk Use 337
References 338

14. HUMAN HEALTH EFFECTS OF HAZARDOUS MATERIALS 339

Objectives 339
Toxicity 340
Setting Safe Human Exposure Limits 341
Where Do the Numbers Come From? 341
Extrapolating Study Data to Human Exposures Limits 347
Making Words Out of Alphabet Soup 349
Routes of Entry Into the Body 355
Contact with the Body Surface 356
Inhalation into the Respiratory System 357
Ingestion into the Digestive System 359
Assessing Human Exposure to Chemicals 363
Recognizing Symptoms 363
Medical Surveillance and Monitoring 363
Summarizing the Steps for a Safe (Nontoxic) Response 364
References 364

15. AIR SURVEILLANCE 366

Objectives 367
Standard Operating Procedures 367
Surveillance Strategies 368
Perimeter and Background Survey 369
Initial Entry 370
Periodic Monitoring 372
Termination Monitoring 373
Air Monitoring Versus Air Sampling 373
Direct-Reading Instruments 374
Components of DRIs 374
Commonly used Instruments 376
Gas Chromatography 381
Relative Response 383
Calibration 383
Inherent Safety 385

Colorimetric Indicators 386
Operation of Colorimetric Indicator Tubes 386
Limitations of Colorimetric Indicator Tubes 387
Air Sampling 388
Sample Period 389
Sampling Systems 389
Sampling Media 391
Laboratory Analysis 392
Using Air Sampling Data 392
Selecting Air Surveillance Equipment 393
General Considerations 393
Planning Purchases 395
Summary 397
References 397

16. COMPUTER USE IN EMERGENCY RESPONSE 398

Objectives 398
Risk Assessment 399
Operations, Data Management, and Reporting 400
Inventory Control 400
Site Location Maps 400
Reporting 401
Employee Safety 402
Event Tracking and Recording 403
Process Tracking 403
Emergency Response Training 403
CAMEO 3.0 404
ALOHA 408
First Aid and Cardiopulmonary Resuscitation 408
ARCHIE 408
General Bulletin Board Systems 409
Evaluating Emergency Response Software 409
Computer Hardware: Technical Considerations 410
References 411
Appendix: Sources of Emergency Response Computer Software Mentioned in This Chapter 412

17. MENTAL STRESS IN EMERGENCY RESPONSE 415

Objectives 415
Defining Stress 415

Stress at Work 416
Technological Change 417
Social Change 418
Political Change 419
Acute and Chronic Stressors 421
Symptoms of Stress 421
Reaction to Stressful Situations 422
Shaping up to Manage a Stressful Job 422
Relaxation 422
Exercise 423
Structure 423
Meditation 423
Biofeedback 424
Reducing Stressors on the Job 424
References 425

18. TEMPERATURE STRESS IN EMERGENCY RESPONSE 427

Objectives 427
Heat Stress 428
Sources of Heat 428
Thermoregulation in Hot Environments 429
Heat-Induced Illness 429
Prevention of Heat-Induced Illness 431
Cold Stress 436
Conditions Leading to Injury 436
Thermoregulation in Cold Environments 437
Harmful Effects of Cold Stress 439
Prevention of Cold Injury 440
Summary 441
References 442

19. PHYSICAL FITNESS FOR EMERGENCY RESPONSERS 443

Objectives 444
Health Benefits of Physical Activity and Physical Fitness 444
Defining Physical Fitness 446
Functional Physical Fitness 446
Job-Related Fitness 446
Health-Related Fitness 448
Components of Health-Related Physical Fitness 448
Cardiorespiratory Endurance 448
Body Composition 453

Joint Flexibility 457
Muscle Endurance 460
Muscle Strength 462
Physical Fitness Training 465
Specificity of Training 465
The Overload Principle 465
Planning a Good Workout 467
Time Allotment for Exercise 468
Physical Abilities Testing as a Selection Method for Emergency Responders 469
A Comprehensive Health and Fitness Program for Emergency Responders 470
Equipment for Physical Fitness Training 470
Summary 471
References 471

20. EXERCISING THE EMERGENCY RESPONSE PLAN 473

Objectives 473
Benefits of Conducting Exercises 474
Levels of Exercise 474
Orientation Seminar 475
Tabletop Exercise 478
Functional Exercise 480
Full-Scale Exercise 482
The Exercise Process 484
Development 484
Conducting the Exercise 485
Evaluation 487
Follow-Up 488
Summary 489
References 489

Index 491

Related Titles from Van Nostrand Reinhold:

Quick Selection Guide to Chemical Protective Clothing, Second Edition
by Krister Forsberg and S.Z. Mansdorf

Firefighter's Hazardous Material Reference Book and Index, Second Edition
by Daniel J. Davis, Julie Davis, and Grant T. Christianson

Fighting Fires with Foam
by Steven P. Woodworth and John A. Frank

A Practical Guide to Chemical Spill Response
by John W. Hosty with Patricia Foster

Rapid Guide to Hazardous Chemicals in the Workplace, Third Edition
by Richard J. Lewis, Sr.

HM-181 and HM-126F:
A Compliance Guide for DOT's New Hazmat Transportation Regulations
by the Environmental Resource Center

A Comprehensive Guide to the Hazardous Properties of Chemical Substances
by Pradyot Patnaik

Sax's Dangerous Properties of Industrial Materials, Eighth Edition
by Richard J. Lewis, Sr.

Hazardous Chemicals Desk Reference, Third Edition
by Richard J. Lewis, Sr.

Hawley's Condensed Chemical Dictionary, Twelfth Edition
by Richard J. Lewis, Sr.

The Dose Makes the Poison, Second Edition:
A Plain-Language Guide to Toxicology
by M. Alice Ottoboni

Preface

The Center for Labor Education and Research (CLEAR) has been involved in training workers in occupational safety and health since its inception in 1972. This particular aspect of our labor education effort has remained a constant and growing part of our program up to the present. In the early years we worked exclusively within the State of Alabama. However, in 1979 we were funded under the Department of Labor's New Directions grant program, and this enabled us to increase our staff and expand our efforts throughout the South.

Following our 5 years with the New Directions program, we were designated the official, exclusive workers' occupational safety and health training program for the state-funded asbestos removal projects. We coupled this work with several contracts from the Department of Labor under which we developed a series of films, slide shows, and videos to help train Occupational Safety and Health Administration (OSHA) inspectors who would be conducting hazardous and toxic waste site investigations.

In 1987, the Center and the School of Public Health, Deep South Educational Resource Center for Occupational Safety and Health, were selected as one of the 11 National Institute of Environmental Health Science grantees in the Superfund cleanup program. We are responsible for training workers in all the Southern states. As this book, our second under this focus, goes to press, we are in the fourth year of the projected 5-year grant period.

Throughout the education and training efforts outlined above, we have had to develop or adapt a wide range of both teaching techniques and student learning materials. The diversity of our student clientele, and the rapidly changing and expanding field of occupational safety and health, have required ongoing evaluation, review, and restructuring of our classroom and field work. The Deep South Educational Resource Center has assisted us by providing a significant amount of this effort for the grant. This book is the second one which has been produced through our general experience and learning in the broad aspects of occupational safety and health training and our particular expertise acquired in the area of hazardous and toxic waste, cleanups and emergency responses.

"Worker," to us, is a generic term. We interpret it broadly to mean anyone who, in the normal course of his or her job, may be exposed to hazardous or toxic substances. Whether you are a remediation site worker, a public hazmat responder, or a member of an industrial hazmat team, there are fundamental commonalities to the potential risks of exposure. There are also specific occupational or job-related risks. This book is designed to be a comprehensive yet basic guide to responding safely to hazardous material incidents.

HIGDON C. ROBERTS, JR., Ph.D.
Director, Center for Labor Education and Research

Acknowledgments

The efforts of the entire CLEAR staff are greatly appreciated. Special thanks go to Karen Blackwood, who prepared the manuscript throughout its many revisions, and to Judith A. Jones for development of the graphics.

Emergency Responder Training Manual for the Hazardous Materials Technician

1

Book Introduction and Overview

Lori P. Andrews, P.E., and
Barbara M. Hilyer M.S. M.S.P.H.

Training for responders to hazardous materials (hazmat) incidents is required by the U.S. Occupational Safety and Health Administration (OSHA) and recommended by the National Fire Protection Association (NFPA). It is obvious that hazmat responders put themselves at risk as they accomplish their jobs. According to William J. Keffer of the U.S. Environment Protection Agency (EPA), there are 63,000 hazardous chemicals in use, 48,500 of which are industrial chemicals. In 1985, 6,928 hazardous materials incidents involving chemicals other than fuel oil were reported to EPA; approximately 75 percent of these occurred in-plant at fixed facilities and 25 percent occurred in transit.

Thus this text, the second product of the Hazardous Waste Worker Training Project at the University of Alabama at Birmingham at the Center for Labor Education and Research (CLEAR), represents the work of the teaching staff in the development of training for the first three levels of training for hazardous materials emergency responders under OSHA Law 29 CFR 1910.120. The text is directed to those individuals who may respond to a hazardous materials, chemicals, substances, or wastes emergency at a fixed facility setting or in a transportation incident. The three levels of training discussed in this book are:

- First responder awareness level (Part I)
- First responder operation level (Part II)
- Hazardous material technician (Part III)

HOW TO USE THE BOOK

If this text is utilized as a training manual, the following hours are suggested for courses:

- Awareness level (Chapter 1–8): 9 hours
- Operations level (Chapter 1–11): 25 hours
- Hazardous materials technician (Chapter 1–15): 45 hours

Please note that the courses are cumulative for each successive level. For instance, if individuals require training to the operations level, then they would take the first 25 hours of training, which includes the awareness level (9 hours) and specialized training at the operations level (an additional 16 hours). This module style format provides flexibility in training several categories of students at one time. One limitation of this program is that the training for hazardous material technicians requires courses at all three responder levels to meet the OSHA standard. For instance, if an individual has received training to the operations level and his job classification changes to hazardous material technician, then not only will the remaining 20 hours suggested need to be completed to achieve the 45 hr total but also some additional hours of review. The review hours should be dependent on the time elapsed between courses; the longer the interval, the longer or more detailed the review.

The remaining chapters of the text deal with issues that are important to all emergency responders and can be used as reference materials. With the computer revolution, many software programs have been developed to provide the fast, accurate data that are crucial in an incident response. Chapter 16 familiarizes the responder with the capability of such programs. Chapter 17–19 deal with the extraordinary stresses that each responder undergoes in these situations. Finally, Chapter 20 discusses the implementation of an emergency response plan by establishing simulated incidents at each level (from tabletop to full-blown, multiagency incidents).

The text and the specified hours can be used to complete the training at the intended level of compliance for OSHA, EPA, and NFPA standards. Extensive equipment is required for each course level. Each chapter discusses the equipment that should be included in the actual training program. Examples include personal protective equipment, direct-read equipment, computers, reference libraries, patching and plugging materials, tools, air bags, and booms. Based on the experiences of the CLEAR staff, at least 45 percent of class time should be allocated to hands-on training.

Although written to serve an academic tool, this book can also be used as a practical guide for response activities. Each chapter includes an actual incident response that illustrates the practical importance of that chapter's topic or issues. These incidents were supplied to the CLEAR staff from the personal experiences of public and private emergency responders. The individuals interviewed have been referenced in the summary of the incident unless otherwise requested.

The remainder of this chapter deals with the dangers of hazardous materials as a group, and provides information about potential outcomes and hazards resulting from a chemical release.

HAZARDOUS MATERIALS: A RISKY BUSINESS

Allen Knifpher, of the Birmingham-Jefferson County, Alabama, Emergency Management Agency, describes one incident in which a cargo aircraft being serviced by a contractor caught fire. Two hazmat teams were

FIGURE 1-1. An operations-level responder uses a neoprene mat to keep a liquid release out of a manhole entering the sewer system. (Barbara Hilyer)

present at the scene when explosives on the aircraft detonated. Three responders were killed, and 18 were hospitalized with serious injuries.

The first responder at the awareness or operations level will probably be the first person to encounter such a hazardous material release. This individual has the responsibility to recognize the release as hazardous, to inform designated responders, and, if he is trained at the operations level, to stop the release from spreading or contaminating soil or water (Figure 1-1). As illustrated by the incident described emergency response provider an environment of many risks and potential risks. The remainder of this chapter deals with the overview of these risks.

OBJECTIVES

- Provide information about the characteristics of a chemical which make it hazardous
- Develop awareness of the ways in which chemicals cause damage to humans and the environment, and lead to fire and explosion
- Create a healthy respect for the potential dangers which may follow a chemical release
- Provide reassurance that proper training and careful response actions will minimize the risk to humans and the environment

DEFINING A HAZARDOUS MATERIALS EMERGENCY

The rule that requires that hazmat responders take a training course (29 CFR 1910.120) defines a hazardous materials emergency as an occurrence that results, or is likely to result, in an uncontrolled release of a hazardous substance.

Hazardous materials, hazardous chemicals, and hazardous substances are all slightly different by definition, and the terms may be confusing. A good definition for hazmat responders to use is the one put forth by the EPA: A hazardous material is one which, due to its concentration, quantity, or chemical or physical properties, may cause or significantly contribute to an increase in mortality, or to an increase in serious, irreversible, or incapacitating reversible illness, or pose a substantial present or potential hazard to human health or the environment when improperly managed.

The EPA lists specific chemicals which it considers hazardous. In addition to listed chemicals, EPA defines criteria to determine if unlisted chemicals are hazardous substances. A chemical on one of the lists or meeting EPA criteria for ignitability, corrosivity, reactivity, or toxicity is

TABLE 1-1. A Portion of the EPA Hazardous Substances List

Common Name	Synonyms
Acetaldehyde	Ethanal, ethyl aldehyde, acetic aldehyde
Acetic acid	Glacial acetic acid, vinegar acid
Acetic anhydride	Acetic oxide, acetyl oxide
Acetone cyanohydrin	2-Methyllactonitrile, alpha-hydroxyisobutyronitrile
Acetyl bromide	—
Acetyl chloride	—
Acrolein	2-Propenal, acrylic aldehyde, acrylaldehyde, acraldehyde

legally a hazardous substance and subject to hazmat regulations following a release; a chemical not listed by EPA and not meeting the criteria is not. Table 1-1 shows a portion of the EPA hazardous substances list.

As an emergency responder, you are concerned about, and would consider hazardous, any chemical or other substance that might injure you and other people, burn or explode, damage other parts of the facility where it has been released, or cause harm to living things in air, streams, lakes, or soil.

Ludwig Benner, writing for the Federal Emergency Management Agency, defined a hazardous material as a substance that "jumps out of its container at you when something goes wrong, and hurts or harms the things it touches."

DANGERS IN HAZMAT EMERGENCIES

Chemicals and Human Health

Even though human bodies consist of the same atoms of carbon, hydrogen, oxygen, calcium, nitrogen, and other elements that dangerous chemicals are made of, the atoms in the chemicals are combined with each other in different ways which may seriously harm us (Figure 1-2). For example, carbon, oxygen, and hydrogen in our bodies combine to form the sugar glucose, which our cells use for energy; in industrial chemicals, these same atoms may be combined into many different hazardous chemicals, such as acrolein. A small quantity of hydrochloric acid is produced in the stomach, where the lining cells are resistant to it, but hydrochloric acid spilled on the skin or inhaled into the lungs may cause a serious burn.

A great deal of information regarding chemical hazards to humans is available. It has been gathered by two basic means: laboratory studies of the effects of chemicals on animals and studies of people who became sick after exposure to chemicals.

FIGURE 1-2. Hazardous chemicals can cause minor or serious health problems. (Barbara Hilyer)

For a chemical to do harm, it must first get into or onto the body. There are three primary routes of entry, and each of them can be protected against chemical invasion. The physical state of the material—solid, liquid, or gas—determines its main route of entry (Table 1-2). The routes of entry are discussed below.

Inhalation of a Gas, Vapor, Dust, Mist, or Fume

Substances in tiny particles, droplets, or gases in the air may enter the body as you breathe. These may irritate and damage part of the respiratory system and, if they are able to cross the lining of the lungs into the bloodstream, will be carried throughout the body to harm the cells of various organs and cause damage far from the point of entry.

TABLE 1-2. Routes of Entry of Hazardous Materials into the Human Body

Route of Entry	Physical State
Contact with skin and eyes	Liquid, gas, vapor, solid particles
Inhalation by the respiratory system	Gas, vapor, fumes, solid particles
Ingestion into the digestive system	Liquid, solid, inhaled particles

Ingestion of Materials on Hands, Foods, or Beverages

Solid and liquid materials may be ingested, as may dusts and vapors which dissolve or condense on skin or food. Even some inhaled materials are pushed up out of the throat, by the movement of protective hair-like cells, into the mouth, where they are swallowed. Ingestion, like inhalation, may be followed by movement of chemicals across the digestive tract lining into the bloodstream and by transport to other organs.

Contact with Skin or Eyes

The skin is the largest organ of the body and is exposed to a great deal of chemical attack. Damage to the skin by corrosive chemicals, or absorption into underlying blood vessels for transport throughout the body, are hazards which are preventable by protecting skin from exposure to chemicals. Materials that contact eyes can be very damaging, especially since the eyes are always wet and some chemicals readily dissolve there.

Damage to the Body

A number of factors determine how much damage the body will suffer from a chemical exposure; the most important factors are the nature of the chemical and the amount to which the person is exposed. Generally speaking, the greater the dose the greater the harm, but some chemicals are far more toxic than others and are harmful in smaller doses. OSHA has set permissible exposure limits (PELs) for chemicals based on scientific determination of the dose likely to cause harmful effects.

The kinds of toxic effects that may be produced by chemical exposure are many and varied. These will now be discussed.

Cell Poisoning

Some chemicals damage or destroy individual cells. The ultimate result of this process depends on which cells have been poisoned. Different chemicals are toxic to different cells and cause damage in different organs of the body, as shown in Table 1-3.

TABLE 1-3. Toxic Chemicals and Their Sites of Damage

Chemical Name	Sites of Damage
Asbestos	Lungs
Hydrochloric acid	Respiratory system, skin, eyes
Lead	Gastrointestinal tract, kidneys, central nervous system
Toluene	Central nervous system, liver, kidneys, skin

Irritation and Tissue Destruction

Some chemicals irritate or destroy the tissues at the surface where they are contacted. These tissues may be skin, eyes, lung linings, or digestive tract lining. Chemicals that have these effects are corrosives, either strong acids like hydrogen chloride or strong bases like caustic soda. Skin or internal organ linings may be mildly irritated or damaged so severely that they cannot recover.

Reproductive or Fetal Damage

Some chemicals, like certain pesticides and solvents, damage sperm or eggs or lower the sex drive. Others, such as alcohol and some polychlorinated biphenyls (PCBs), may damage an unborn baby if the pregnant woman is exposed, resulting in fetal death, congenital deformity in the newborn, or failure of the baby to grow and thrive. Some of these effects are so long-lasting that they can even influence a child's achievement in school.

Cancer

"Cancer" is a catchall term used to describe a number of diseases in which cells multiply in an uncontrolled manner. A number of chemicals are known or suspected to cause human cancers, but the determination of the cause is clouded by the mechanisms involved and the length of time between exposure and disease. Some cancers may not develop for as long as 40 years after exposure. Table 1-4 lists some cancer-causing chemicals.

Chemical Combinations

Chemicals which combine outside the body can undergo chemical reaction. This occurs when two incompatible chemicals, for example hexane and hydrogen peroxide, mix and react. The reaction can release toxic

TABLE 1-4. Chemicals Known or Suspected to Cause Human Cancer

Acetaldehyde	Beryllium
Acrylonitrile	Cadmium dust
Aldrin	Dibromochloropropane
Aniline	Ethylene oxide
Arsenic	Methylene chloride
Asbestos	Nitropropane
Benzene	Tetrachloroethane
Benzidine	Vinyl chloride

products that are different from the two reacting chemicals. Reaction may lead to splattering and cause skin exposure.

Chemicals that enter the body at the same time, when one is exposed to two or more of them in the same incident, can cause unexpected problems. In some cases, the body cannot detoxify either chemical as well as it could have detoxified one of them in a single exposure. The results may be more harmful than those of exposure to either of the chemicals alone.

Chemicals and Fire

In addition to generating toxic gases, chemical reactions may be dangerous in other ways. An especially hazardous reaction is one that rapidly generates an explosive gas, such as hydrogen, or one in which as chemical generates oxygen gas that supports the burning of a flammable or combustible substance.

A chemical that releases oxygen when it reacts with another chemical is called an "oxidizer." Since oxygen is one of the three factors necessary to support a fire, as seen in the fire triangle in Figure 1-3, oxidizers which spill in the presence of ignitable materials increase a fire hazard. Identifi-

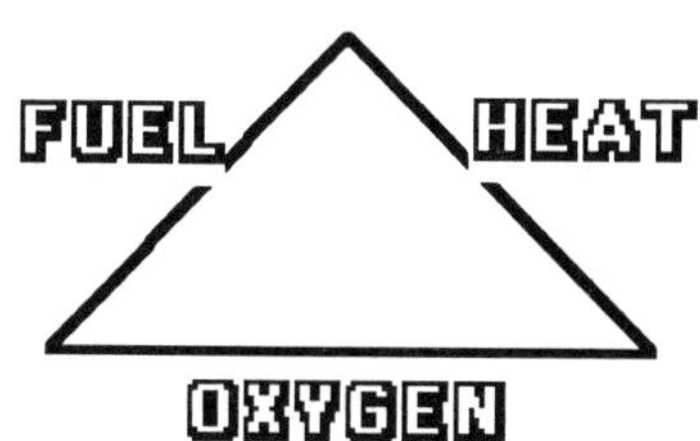

FIGURE 1-3. The fire triangle: all three sides are necessary for combustion.

cation of an oxidizer can be made based on any one of several clues: a label or placard showing the Department of Transportation's oxidizer hazard class; the name of the chemical including "per" (true of many, but not all, oxidizers); or a notation on the Material Safety Data Sheet or other information source that the chemical is incompatible with organic or combustible materials.

Some chemicals are described as being "volatile" because they evaporate from the liquid to the vapor state when released from their containers. The vapors may be lighter than air and drift away, but some chemical vapors are heavier than air and will remain in the area of the release. These heavy vapors can even drift along the ground downhill to an ignition source. If the vapors are flammable, as many are, they will ignite. Heavy vapors can be identified in reference materials by looking up the vapor density; if it is greater than 1.0, the vapors are heavier than air.

Chemicals in the Environment

One of the major goals of confinement actions by an emergency responder is to keep a hazardous material out of nearby soil and waterways. The immediate advantage of preventing environmental contamination is to comply with the law and avoid having to deal with, and be fined by, the EPA; the long-term advantage is to keep hazardous materials from harming fish, birds, wildlife, and plants. Humans who live or drink downstream or downgrade are also at risk from an environmental release.

Every attempt should be made to keep hazardous materials out of air, water, and soil. Once chemicals have spread into any of these places, it is difficult, often impossible, to recover them.

Hazardous materials can be bad news, as you can see. The good news is that emergency responders who are prepared and protected can help to minimize, and perhaps prevent, risks to humans and the environment.

REFERENCES

Henry, M.F., editor. 1989. *Hazardous Materials Response Handbook.* Quincy, Mass.: National Fire Protection Association.

Klaassen, C.D., M.O. Amdur, and J. Doull, editors. 1986. *Casarett and Doull's Toxicology*, 3rd ed. New York: Macmillan Publishing Co.

U.S. Department of Health and Human Services. 1990. *NIOSH Pocket Guide to Chemical Hazards*. Washington, DC: U.S. Government Printing Office.

U.S. Department of Labor. Title 29 Part 1910. Washington, DC: U.S. Government Printing Office. 1990.

U.S. Environmental Protection Agency. Title 40. Washington, DC: U.S. Government Printing Office. 1990.

2

Federal Occupational Safety and Health Regulations Applicable to Emergency Response Personnel

D. Alan Veasey, M.A. Ed.

A worker at a hospital in Troy, New York, failed to close the valve on a cylinder of ethylene oxide before disconnecting the cylinder for replacement. As a result, ethylene oxide (a flammable, toxic, and irritating gas used for sterilizing surgical instruments) began to escape into the room. The worker immediately left the room and sounded the alarm. As a result of the release, officials were forced to evacuate the building. The release was stopped by two hazmat team members who entered the room wearing full protective clothing and closed the cylinder valve. Fortunately, the hazmat team had recently practiced for just such an incident and was operating under an emergency response plan which included specific information on the hazards of ethylene oxide and a storage map showing the exact location of the release.

In recent times, incidents such as the ethylene oxide release at the hospital have resulted in significant loss of life, personal injury, property damage, and environmental damage. Many such incidents were aggravated by lack of an adequate emergency response capability. This problem was addressed at the federal level by passage of the Superfund Amendment and Reauthorization Act (SARA) of 1986. In passing Title I of SARA, Congress addressed concerns for the safety and health of personnel involved in hazardous substance emergency response operations. To that end, Congress directed OSHA and EPA to develop regulations ensuring that adequate protective provisions (such as the training and protective equipment utilized in the hospital incident) are provided to response personnel. This chapter is intended to acquaint the trainee with these federally mandated protective provisions.

OBJECTIVES

- Understand the role of OSHA in regulating occupational safety and health
- Be aware of the purpose and scope of OSHA regulations pertaining to hazardous waste operations and emergency response (29 CFR 1910.120)
- Know and understand the general provisions of OSHA Standard 29 CFR 1910.120 applicable to emergency response operations
- Be familiar with specific regulations applicable to topics such as emergency response plans, medical surveillance programs, and personal protective equipment
- Be aware of general response roles and training requirements for the following levels of certification:
  - First responder awareness level
  - First responder operations level
  - Hazardous materials technician
  - Hazardous materials specialist
  - On-scene incident commander

THE ROLE OF OSHA IN REGULATING OCCUPATIONAL SAFETY AND HEALTH

Background and Creation of OSHA

Prior to the passage of the Occupational Safety and Health Act (OSHAct) of 1970, no uniform federal safety and health regulations existed. State regulations varied widely, and enforcement proceedings against violators of existing regulations were almost nonexistent. As a result of unsafe and unhealthy workplace conditions, unacceptably large numbers of American workers were experiencing illness, injury, or death. The OSHAct was passed by the U.S. Congress as a means of addressing this problem.

The OSHAct was intended to ensure safe and healthful conditions in the American workplace. The act requires that employers take steps to protect employees from recognized workplace hazards that are likely to cause illness or injury. If practical, recognized hazards should be completely eliminated from the workplace, such as through the use of engineering controls. If elimination of a hazard is not practical, the employer must provide other measures, such as personal protective equipment (PPE) to protect employees. The act also requires that employees comply with all applicable rules, regulations, and standards pertaining to occupational safety and health.

A number of organizations came into existence as a result of the OSHAct. One of them, the Occupational Safety and Health Administration, is of particular importance and will be referred to frequently in this text.

Responsibilities of OSHA

OSHA was created within the Department of Labor to act as the primary guardian of worker safety and health. As such, OSHA was given the authority to develop and implement workplace safety and health standards.

OSHA is also responsible for enforcing compliance with standards. To this end, OSHA has the authority to conduct inspections, issue citations, and levy fines.

OSHA SAFETY AND HEALTH STANDARDS

OSHA safety and health standards are legally enforceable sets of industry-specific regulations intended to address concerns for the safety and

TABLE 2-1. Examples of Specific Safety Topics and Applicable OSHA Standards

Safety Topics	Applicable Standards: Labor (29 CFR 1910)		Applicable Standards: Construction (29 CFR 1926)	
Ventilation	Subpart G	1910.94	Subpart D	1926.57
Noise		1910.95		1926.52
Ionizing radiation		1910.96		1926.53
Hazardous materials	Subpart H			
Personal Protective Equipment				
General	Subpart I	1910.132	Subpart E	
Eye/face		1910.133		1926.102
Hearing		1910.95		1926.101
Respiratory		1910.134		1926.103
Head		1910.135		1926.100
Foot		1910.136		
Fire protection	Subpart L		Subpart F	
Materials handling and storage	Subpart N		Subpart H	
Electrical	Subpart S		Subpart K	
Toxic/hazardous substances	Subpart Z			
Trenching and excavation			Subpart P	

health of workers. These standards are developed and revised on a constant basis. The opportunity to comment on proposed new standards or revisions is extended to employers, employees, and all other interested parties. OSHA standards will be referenced throughout this text.

Numerous safety standards have been developed by OSHA to protect the safety and health of workers involved in construction activities and general labor. Examples of specific safety topics and applicable standards are shown in Table 2-1. Standards are referenced according to their location within the Code of Federal Regulations (CFR). For example, regulations limiting employee exposure to noise are contained in part 1910.95 of Title 29 of the CFR. From the standpoint of emergency response to releases of hazardous substances, the most significant OSHA standard is 29 CFR 1910.120. This standard will be described at length in the following section.

THE SCOPE OF OSHA AUTHORITY

Coverage of Employees of Private Industry and the Federal Government

OSHA's authority to enforce workplace safety and health standards as originally established by the OSHAct covered all private sector workplaces. In 1980, protection under OSHA standards was extended to employees of the federal government. However, OSHA currently has no authority to regulate the safety and health of employees of state and local governments.

Coverage Under State Occupational Safety and Health Plans

Individual states may elect to assume authority from OSHA for regulating occupational safety and health in the workplace. States which have taken advantage of this option are shown in Figure 2-1. State occupational safety and health regulatory programs must be documented in an OSHA-approved written state plan. The protective provisions incorporated into state plans must be at least as stringent as equivalent provisions of OSHA.

In order to receive OSHA approval, states are required to extend protection under the state plan to all state and local government employees, in addition to private sector employees, within the jurisdiction of the state. Thus, public-sector employees, such as law enforcement and fire service personnel, must be provided occupational safety and health pro-

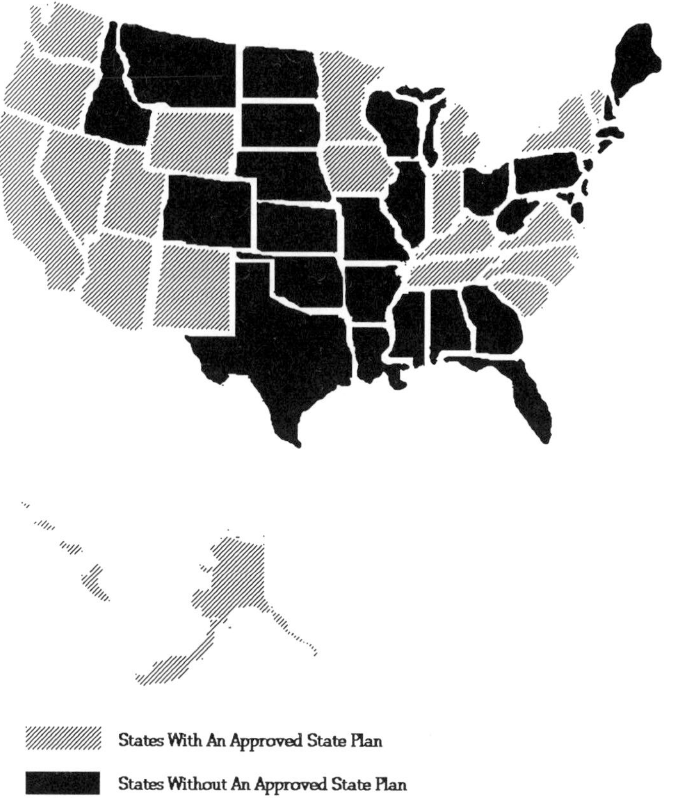

FIGURE 2-1. States having OSHA-approved occupational safety and health plans as of April 1991.

tection, at least equivalent to that required by federal OSHA, in these states.

ORIGIN AND SCOPE OF OSHA REGULATIONS PERTAINING TO HAZARDOUS WASTE OPERATIONS AND EMERGENCY RESPONSE

Background and Development of HAZWOPER

With the passage of SARA in 1986, the federal government addressed a variety of concerns related to hazardous waste operations and preparedness for emergency response to accidental hazardous substance releases.

One major concern was the safety and health of personnel involved in these hazardous operations. Through SARA Title I, OSHA was directed to develop and put into force a standard designed specifically to protect the safety and health of workers engaged in these types of operations. Under the authority of SARA Title I, OSHA developed and implemented 29 CFR 1910.120, entitled "Hazardous Waste Operations and Emergency Response" and sometimes referred to as "HAZWOPER."

Purpose and Applicability

OSHA regulations contained in 29 CFR 1910.120 are intended to protect personnel engaging in various activities involved in hazardous waste and emergency response operations. The general types of operations covered by the standard include the following:

- Cleanup of uncontrolled hazardous waste sites
- Work at RCRA TSD facilities (i.e., facilities involved in treating, storing, and disposing of hazardous wastes in accordance with the requirements of the Resource Conservation and Recovery Act)
- Hazardous substance emergency response operations

Applicability to Emergency Response Operations

Given the scope of this text, coverage of 29 CFR 1910.120 will focus solely on the requirements and provisions of paragraph q. This is the part of the standard which applies to emergency response operations at locations such as industrial facilities, chemical plants, and transportation corridors. The provisions of the standard included here apply to emergency response activities intended to stop or prevent an accidental release of a hazardous substance. These requirements apply in all situations involving a response effort by employees from outside the immediate release area (or other designated responders) to an occurrence which results, or is likely to result, in an uncontrolled release of a hazardous substance. If a release is small enough to be handled using personnel and equipment routinely located in the immediate area of the release, the event is treated as an incidental release, rather than an emergency, and the standard does not apply.

It should be noted that employers are exempt from the requirements of 29 CFR 1910.120 (q) if the following conditions are met:

- An emergency action plan in accordance with 29 CFR 1910.38 (a) has been developed and implemented by the employer in order to ensure the safety of employees should an incident occur.

- In the event of an emergency release of a hazardous substance, all facility employees will be evacuated from the affected area.
- An offsite organization (such as the local fire service agency) will be called in to perform all hazardous substance emergency response operations.
- No facility employees will be allowed to assist directly in a response operation.

In states without an OSHA-approved state occupational safety and health plan (see Figure 2-1), OSHA protection does not extend to employees of state and local governmental agencies. For this reason, employees of agencies such as law enforcement and fire service organizations in these states are not directly covered under 29 CFR 1910.120. In order to close this loophole, EPA was directed under SARA Title I to issue regulations covering these employees. To that end, EPA developed a standard (40 CFR 311) which incorporates 29 CFR 1910.120 in its entirety by reference. The EPA standard also covers unpaid state and local employees, such as members of volunteer fire service organizations. Thus, the provisions of 29 CFR 1910.120 (q), as described in the following section, are applicable to personnel engaging in emergency responses to hazardous substance releases in both the private and public employment sectors.

PROVISIONS OF 29 CFR 1910.120 APPLICABLE TO EMERGENCY RESPONSE OPERATIONS

Emergency Response Plan

An Emergency Response Plan (ERP) must be developed and implemented by the employer prior to the commencement of emergency response operations. This plan must be adequate for any emergencies which could reasonably be expected to occur on site. The plan must be in writing and available to employees, employee representatives, and OSHA personnel. This plan must address, as a minimum, the following topics:

- Preemergency planning and coordination with outside parties
- Personnel roles, lines of authority, training, and communication
- Emergency recognition and prevention
- Safe distances and places of refuge
- Site security and control
- Evacuation routes and procedures
- Decontamination

- Emergency medical treatment and first aid
- Emergency alerting and response procedures
- Critique of response and follow-up
- Personnel protective equipment and emergency equipment

Emergency response organizations may use the local and/or state ERPs as part of their ERP, if applicable, to avoid duplication. Emergency response topics which are properly addressed by existing plans required under SARA Title III need not be duplicated for compliance with 29 CFR 1910.120.

Emergency Response Chain of Command and Personnel Roles

The Incident Command System

Under 29 CFR 1910.120, the Incident Command System (ICS) is required to be used in managing hazardous material incident responses (see Chapter 3). The senior emergency response official responding to an emergency will become the individual in charge of the site-specific ICS. This individual will control and coordinate emergency response activities and communications. Command will be passed up a preestablished line of authority as personnel or officials having greater emergency response seniority arrive.

The individual in charge of the ICS shall identify the hazards present and address the following considerations:

- Site analysis
- Use of engineering controls
- Maximum exposure limits
- Hazardous substance handling procedures
- Use of any new technologies

The individual in charge of the ICS shall implement appropriate emergency operations based on incident-specific conditions. This person must ensure that the personal protective equipment worn is appropriate for the hazards present.

Designated Safety Official

The person in charge of the ICS shall designate a safety official who is knowledgeable in the operations involved. The safety official shall identify and evaluate the hazards and provide direction with respect to the safety of the operations involved.

The safety official shall have the authority to alter, suspend, or terminate any activities which, according to his judgment, involve an immediately dangerous to life and health (IDLH) condition and/or an imminent danger condition. The safety official shall immediately inform the person in charge of the ICS of any actions needed to lessen the hazards involved.

Skilled Support Personnel

Skilled support personnel are skilled in operating equipment such as earth movers, excavators, and cranes, which may be needed temporarily for immediate emergency support work which cannot be reasonably performed by the employer's regular response personnel. Skilled support personnel may be exposed to hazards during emergency response operations but are not required to be trained, as the employer's regular response personnel are. However, skilled support personnel must be briefed before becoming involved in an emergency response. The initial briefing must include:

- Use of PPE
- Chemical hazards involved
- Duties to be performed

Skilled support personnel are entitled to all safety and health precautions provided to the employer's regular responders except training.

Specialist Employees

Specialist employees work with specific hazardous substances in the course of their regular job duties and are knowledgeable about the hazards of those substances. These employees may be called on to provide assistance to the individual in charge of the response to a hazardous substance release. For example, if two chemicals are accidentally mixed in an incident, a chemist may be called upon to predict the reactions which will occur. No specific training requirements for specialist employees are listed in the standard. However, specialist employees are required to receive training appropriate for their area of specialization, or to demonstrate competence in their area of specialization, annually.

Specific Personnel Roles

The standard divides emergency response roles into several categories (see Figure 2-2). Responders must be trained for the specific roles which they are expected to perform in the event of an emergency. Specific emergency response roles are described in the section on training.

Emergency Response Procedures

Operations in hazardous areas must be performed in groups of two or more, using the buddy system. The individual in charge of the ICS shall not allow anyone except personnel actively performing emergency operations to occupy any area of potential exposure during an emergency.

During emergency operations, backup personnel shall remain on standby with appropriate equipment in case the primary responders require assistance or rescue. Advance first aid support personnel shall stand by, with medical equipment and transportation available.

After the termination of emergency operations, the person in charge of the ICS shall implement appropriate decontamination procedures. All personnel and equipment must be fully decontaminated before leaving the operations area.

Training

Training provided to each responder must be specific to that responder's intended role within the emergency response organization (Figure 2-2). This training must be completed before the employee is allowed to participate in a response operation. Training for emergency response will be provided according to the categories that will now be discussed.

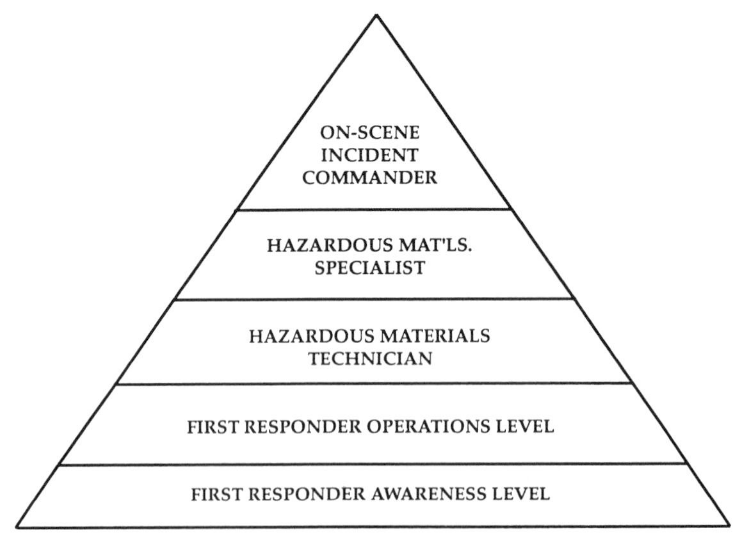

FIGURE 2-2. Emergency response roles/certification levels under 29 CFR 1910.120.

First Responder Awareness Level

First responders at the awareness level are personnel who are likely to discover or witness a hazardous substance emergency. The most important duty of these personnel is to make proper notification in order to begin the emergency response sequence. The response role of these personnel should involve no potential for exposure to hazards related to an incident. First responders at the awareness level must be sufficiently trained and/or experienced to demonstrate competence in the following areas:

- Understanding of what hazardous materials are and the risks associated with them
- Understanding of the potential outcomes associated with an emergency involving hazardous materials
- Ability to recognize hazardous materials present in an emergency
- Ability to identify hazardous materials involved in an emergency, if possible
- Understanding of the role of first responder awareness level personnel in the employer's emergency response plan, including site security and control
- Ability to use the Department of Transportation's (DOTs) *Emergency Response Guidebook*
- Ability to realize that additional resources are needed for the response operation and to make appropriate notifications to the emergency response communication center

First Responder Operations Level

First responders at the operations level are personnel who are involved in an initial response for the purpose of protecting people, property, and the environment from hazardous substances. These personnel are trained to respond defensively, instead of actually trying to stop the release at the source. This type of response involves working at a distance from the point of release to control the released material, keep it from spreading, and prevent exposures. First responders at the operations level must have awareness-level competence and must receive at least 8 hours of training, or have sufficient experience to demonstrate competence, in the following areas:

- Knowledge of basic hazard and risk assessment techniques
- Ability to select and use PPE provided to the first responder operations level
- Understanding of basic hazardous materials terms

- Ability to perform basic control, containment, and/or confinement operations within the capabilities of available resources and PPE
- Ability to implement basic decontamination procedures
- Understanding of relevant standard operating procedures and termination procedures

Hazardous Materials Technician

Hazardous materials technicians are personnel who respond to an emergency in order to stop or prevent a release of hazardous substances. These personnel assume an offensive emergency response role as they approach the point of release and plug, patch, or otherwise stop the release at the source. Hazardous materials technicians must receive at least 24 hours of training incorporating all objectives required for first responder operations level training. In addition, hazardous materials technicians must be able to demonstrate competence in the following areas:

- Knowledge required to implement the employer's ERP
- Knowledge in the use of field survey instruments and equipment to identify, classify, and verify known and unknown materials
- Ability to function within an assigned role in the ICS
- Knowledge required to select and use proper specialized chemical PPE provided for emergency response
- Understanding of hazard and risk assessment techniques
- Ability to perform advance control, containment, and/or confinement operations within the capabilities of available resources and PPE
- Ability to understand and implement decontamination procedures
- Understanding of termination procedures
- Understanding of basic chemical and toxicological terminology and behavior

Hazardous Materials Specialist

Hazardous materials specialists are personnel who respond with, and provide support for, hazardous materials technicians. Their duties parallel those of the hazardous materials technicians but require a more specific knowledge of the hazardous substances involved. The hazardous materials specialist also provides site liaison with federal, state, and/or local government authorities. Hazardous materials specialists must receive at least 24 hours of training incorporating all objectives required for hazardous materials technician training. In addition, the hazardous materials specialist must be able to demonstrate competence in the following areas:

- Knowledge required to implement the local ERP
- Understanding of the use of advanced survey instruments and equipment in the classification, identification, and verification of known and unknown materials
- Knowledge of the state ERP
- Ability to select and use proper specialized chemical PPE provided for emergency response
- Understanding of in-depth hazard and risk assessment techniques
- Ability to perform specialized control, containment, and/or confinement operations within the capabilities of the available resources and PPE
- Ability to determine and implement decontamination procedures
- Understanding of chemical, radiological, and toxicological terminology and behavior

On-Scene Incident Commander

On-scene incident commanders are personnel who will assume control of the incident scene beyond the first responder awareness level. Incident commanders will receive a minimum of 24 hours of training incorporating all objectives required for first responder operations level training. In addition, the on-scene incident commander must be able to demonstrate competence in the following areas:

- Knowledge and ability required to implement the employer's ICS
- Knowledge required to implement the employer's ERP
- Knowledge and understanding of the hazards and risks associated with working in PPE
- Knowledge required to implement the local ERP
- Knowledge of the state ERP and the Federal Regional Response Team
- Knowledge and understanding of the importance of decontamination procedures

Other Considerations for Training

Personnel who receive training as required by the standard must be provided annual refresher training, or else must demonstrate competence in the areas covered by training on an annual basis.

Under 29 CFR 1910.120 it is the employer's responsibility to document that the employee has received training as required for the employee's emergency response role. For first responders at the awareness and operations levels, training requirements can be considered satisfied by an employee's previous experience, provided that the employee is able objectively to demonstrate competence in the areas required to be covered by training. If a statement of competence is made, the employer must keep a record of the methodology used to demonstrate competence.

Medical Surveillance and Consultation

OSHA 29 CFR 1910.120 requires that all hazardous materials technicians who are members of organized and designated hazmat teams and all hazardous materials specialists receive a baseline physical examination and be placed under a program of medical surveillance. The standard does not require that first responders at the awareness and operations levels be provided coverage under a formal medical monitoring program. However, these employees must be provided with medical examinations if they are injured as a result of exposure during an emergency incident or experience symptoms which may be related to exposure. Minimum requirements for medical surveillance programs are presented here.

The standard includes specific requirements for medical examinations performed under the medical surveillance program. These medical examinations must be conducted

- Before assignment to a hazmat team
- At least annually during hazmat team membership unless the attending physician believes that some longer interval, not to exceed 2 years, is appropriate
- After overexposure or the appearance of potentially exposure-related symptoms
- Whenever deemed necessary by the physician
- At the time of reassignment to an area or job which does not require medical surveillance
- At the time of termination

No termination or reassignment examination is required if an employee has received a complete examination within 6 months of the time of termination or job reassignment and has had no significant exposures or potentially exposure-related symptoms since the exam.

Medical examinations must be performed at no cost or loss of pay to the employee, at a reasonable time and place, and by or under the direct supervision of a licensed physician.

The specific content or focus of the medical examination must be determined by the examining physician based on conditions expected to be encountered in the course of response operations. Exams shall include a complete or updated medical and work history and shall focus on any symptoms which may be related to chemical exposure. Fitness for duty under site conditions (such as the use of required PPE under expected temperature extremes) should be emphasized.

Under 29 CFR 1910.120, an employee covered by the medical surveillance program is entitled to receive a written physician's opinion. The results of specific exams and tests will be included if requested by the employee. The opinion must state any medical conditions which require treatment or place the employee at greater risk due to expected hazards and duties. Any recommended work-assignment limitations will also be included in the written physician's opinion.

The standard mandates confidentiality of medical examination results. Therefore, specific findings unrelated to occupational exposure cannot be revealed by the examining physician to the employer.

Personal Protective Equipment

The standard requires that personal protective equipment (PPE) be used as required to protect responders from hazards encountered during emergency operations. In addition to conventional safety hazards, employees must be protected from exposure to hazardous substances in excess of applicable exposure limits (see Chapter 9).

Selection and Use of PPE

Selection of PPE must be based on incident-specific conditions and updated as those conditions change or additional information is generated about the incident. A written PPE program is required and must incorporate, at a minimum, the following topics:

- Selection of PPE
- Use and limitations of PPE
- Work mission duration
- Maintenance protocols
- Storage procedures
- Decontamination and disposal procedures
- Training and proper fitting
- Donning and doffing
- Inspection procedures

- Limitations during temperature extremes
- Program evaluation

Specific Requirements for PPE

Emergency response personnel exposed to hazardous substances representing a potential inhalation hazard shall wear positive-pressure self-contained breathing apparatus (SCBA) until the individual in charge of the ICS determines through air monitoring that a decreased level of respiratory protection is appropriate. When deemed necessary, approved SCBAs may be used with approved cylinders from other SCBAs, provided that the cylinders have the same capacity and pressure rating. All SCBA cylinders must meet applicable DOT and National Institute of Occupational Safety and Health (NIOSH) requirements.

Personnel must be provided with one of the following methods of respiratory protection for work in IDLH atmospheres:

- Positive-pressure SCBA fitted with a full facepiece
- Positive-pressure air-line respirator fitted with a full facepiece and an escape air supply

For operations involving hazards to the skin which may result in an IDLH situation, totally encapsulating chemical-protective (TECP) suits must be used. These suits must be capable of maintaining a positive internal pressure and resistant to inward gas leakage. Methods to be used in testing TECP suits are described in Appendix A of the standard.

Postemergency Response Operations

After the completion of emergency response operations, it may be necessary to remove hazardous substances, health hazards, and/or contaminated materials (such as contaminated soil) from the site. If so, employers engaging in such activities must comply with all requirements of the standard applicable to hazardous waste site cleanup operations. However, if cleanup is performed on plant property using plant employees, the standard doesn't apply. In this case, the employees involved should receive training as required in

- 29 CFR 1910.38(a): Emergency Action Plans
- 29 CFR 1910.134: Respiratory Protection
- 29 CFR 1910.1200: Hazard Communication

Also, any additional safety and health training required by the tasks the employees are required to perform, such as use of PPE and decontamination, shall be provided.

All equipment used in postemergency cleanup must be in serviceable condition. This equipment must be thoroughly inspected before being used.

SUMMARY

Responding to hazardous substance emergencies is inherently dangerous. However, the hazards involved can be minimized if the personnel required to respond are provided with appropriate training and equipment, and if appropriate procedures are followed during response operations. In addressing the hazards inherent in emergency response operations, the federal government has created regulations requiring the employer to provide for the safety and health of response personnel. These regulations are intended to provide a fundamental right of safety to emergency responders. It is therefore important for everyone involved in emergency response to be familiar with these regulations in order to ensure that they are not violated.

REFERENCES

U.S. Department of Labor, 1985. *All About OSHA*. OSHA Publication No. 2056. Washington, DC: U.S. Government Printing Office.

U.S. Department of Labor, 1990. *Hazardous Waste and Emergency Response*. OSHA Publication No. 3114. Washington, DC: U.S. Government Printing Office.

U.S. Department of Labor. Title 29 Part 1910. Washington, DC: U.S. Government Printing Office.

U.S. Department of Labor Title 29 Part 1926. Washington, DC: U.S. Government Printing Office.

U.S. Department of Transportation. 1990. *Emergency Response Guidebook*. Washington, DC: U.S. Government Printing Office.

U.S. Environmental Protection Agency, 1991. *Hazardous Waste Operations and Emergency Response: General Information and Comparison*. EPA Publication No. 9285.2-09FS. Washington, DC: U.S. Government Printing Office.

U.S. Environmental Protection Agency, 1991. *Hazardous Waste Operations and Emergency Response: RCRA TSD and Emergency Response Without Regard to Location*. EPA Publication NO. 9285.2-07FS. Washington, DC: U.S. Government Printing Office.

3

Incident Command System

D. Alan Veasey, M.A. Ed., and
Charles H. Morton, B.S.

A tractor-trailer rig transporting a mixed load of hazardous materials overturned on a major highway in northern Alabama. The hazardous materials involved included methyl ethyl ketone, an alkaline corrosive material, and an additive used in treating swimming pools. As a result of the accident, the three materials were being released, intermixing freely, and entering a storm drain which led to a major waterway approximately a quarter of a mile away. The initial response was by law enforcement and volunteer fire service personnel who lacked adequate training in hazardous materials emergency response. These personnel made a direct approach to the wrecked vehicle without performing any sizeup or hazard and risk assessment. No site control procedures were initiated, and vehicular traffic was allowed to continue through the area. The hazardous materials continued to flow freely from the location of the accident. This condition continued for some time, until personnel with adequate hazmat training arrived and took charge of the scene. The Incident Command System (ICS) was immediately established and utilized to organize the response operation. Site control procedures were initiated, and hazard and risk assessment was conducted. The highway was closed and some nearby residents were evacuated. Procedures were quickly undertaken to stop the spread of the released materials. Response team members entered the trailer utilizing appropriate PPE, and terminated the release at the source. Cleanup procedures were then initiated. One responder familiar with the incident stated that the establishment of the ICS had the effect of bringing order out of chaos.

As this incident indicates, a successful response to a hazardous materials incident requires the allocation and direction of available resources so as to bring the incident under control. Available resources may include various types of personnel, equipment, and supplies. When responding to large or complex incidents, large numbers of response personnel may be required to engage simultaneously in a variety of activities at different locations in order to mitigate the incident. Needless environmental damage, property damage, and even loss of life can result from poor organization of a response operation. For this reason, it is critical that some managerial approach or concept be used to coordinate the activities of emergency responders in order to maximize the efficiency of the response operation. A managerial concept which has been used successfully in filling this need is the ICS. This chapter is intended to provide the knowledge and understanding needed to function effectively in an assigned role within the ICS during an emergency response operation.

OBJECTIVES

- Develop a basic understanding of the ICS concept
- Know the essential components of the ICS
- Know the major functional areas of the ICS and how they are organized within the system
- Understand generally how the ICS works
- Be familiar with the roles of incident commander, designated safety official, and other personnel roles within the ICS
- Be able to function in an assigned role within the ICS (assuming that all additional required training is provided)

BASICS OF THE ICS

The ICS was developed by federal fire protection agencies. It is designed to be used in managing response operations for all types of emergencies and is generally considered the most effective way to manage a hazardous materials incident response.

For this reason, OSHA and EPA have developed regulations that require the use of the ICS at hazardous materials incidents (see Chapter 2).

The ICS is applicable to users throughout the country. With its basic common elements, it is simple enough for a small organization at a small incident, or it can be expanded to coordinate the activities of several organizations at a major incident. It can be used in incidents that require (1) single-jurisdiction/single-agency involvement, (2) single-jurisdiction/multiagency involvement, or (3) multijurisdiction/multiagency involvement.

In order to minimize confusion and conflict, the ICS is designed so that an individual responder working within the system receives orders from only one person and reports to only one person. The system is intended to control all actions taken in response to an incident. The ICS does not allow freelancing. "Freelancing" refers to an individual or a group doing what they want, without the permission and knowledge of the incident commander or a member of his staff. Discipline is a very important aspect of the ICS which helps to promote efficiency, teamwork, and safety.

Components of the ICS

The ICS consists of eight essential components. Interaction of these components is vital to the proper performance of the ICS. The ICS components are defined and characterized below.

Comprehensive Resource Management

In order to bring a hazardous materials incident under control, it is important that all available resources be utilized in an effective manner. These resources may include personnel, equipment, and supplies required to perform a response operation. The ICS provides centralized control and coordination of these resources to mount an efficient response.

Modular Organization

The structure of the ICS is modular, and is intended to expand or unfold from the top down based on the managerial needs of an incident. This provides the flexibility required to respond to both major and minor incidents.

Common Terminology

It is important that terminology be utilized consistently by all personnel required to operate within the ICS. For this reason, standardized terminology has been developed for personnel roles, equipment, and facilities used in emergency response operations. Terms such as "operations section," "hot zone," "decontamination", and "command post" should therefore have the same meaning for all personnel involved in a response operation. This is especially important given the potential for multiple agencies to be involved in an incident.

Unified Command Structure

For major incidents requiring a multiagency response, all involved agencies must be provided input into the processes of establishing response objectives, selecting strategies, and planning for tactical operations. This is the purpose of a unified command structure.

Manageable Span of Control

As a general rule, a person in a managerial role during a response operation can efficiently direct and supervise the activities of three to seven subordinates. Because of this, a span of control of five responders is used as a general rule of thumb in establishing the ICS. If a reasonable span of control is exceeded, the managerial skills of response personnel can be overwhelmed, leading to a breakdown of control within the system. Thus, as the ICS unfolds or ''builds down'' and greater numbers of response personnel become involved in an operation, managerial responsibilities must be delegated to various response personnel according to the emergency response chain of command. This will ensure that no individual's span of control is exceeded so that centralized control and coordination of all activities under the ICS can be maintained.

Consolidated Action Plans

An efficient incident response requires a well-defined plan of action. Action plans must cover the tactical and supporting activities required to perform the hazmat operation. It is important that all personnel and agencies involved in an incident response operate under action plans which are consolidated so that duplication of activities and contradictory actions are avoided.

Integrated Communications

Clear and concise communications are vital for effective response operations. Communications systems required for incident management may include radio, telephone (both on and off site), and public address equipment. Other items such as cellular telephones and facsimile (FAX) machines can also be used. The ICS must provide a means for controlling and coordinating incident communications to form an integrated communications system. Such a communications system may involve fixed centers, mobile and hand-held units, and personnel trained to operate the equipment.

It is important that communications be well integrated so that all personnel can clearly understand them. This is especially important for multiagency responses, because the use of nonstandardized codes can produce a breakdown of communication. One simple approach to this problem is to use plain English rather than ''10 codes'' for all communications during multiagency response operations. Specific communication procedures are described in Chapter 6.

Designated Incident Facilities

Various types of facilities may be established at the scene of an incident in order to facilitate the response. The specific type and location of these facilities will vary according to the specific requirements of an incident. Two commonly used incident facilities are the command post and staging areas (see Chapter 6).

Major Functional Areas of the ICS

The ICS can be divided into five major functional areas: command, operations, logistics, planning, and finance (Figure 3-1). The activities within each functional area are performed by a separate section of the ICS. It is important for all response personnel to understand the types of duties performed under each of the functional areas and the way in which the functional sections interact within the system as a whole.

Command is responsible for managing the response operation. This involves assessing the situation, establishing strategic goals, and ordering and allocating personnel, equipment, and other resources as required to achieve those goals. The incident commander (IC) is the official in charge of the command section and is thereby ultimately responsible for management of the entire response operation, including the safety of response personnel. If required by the size or complexity of the incident response, the IC may be assisted by command staff personnel (as described below) in managing the incident.

Operations directs all tactical operations undertaken to bring a hazardous materials emergency under control. This involves a direct response at the hands-on level, such as performing procedures intended to control, confine, and contain hazardous material releases.

Logistics provides equipment, supplies, and services needed to support the response operation. This may require that various items, ranging from earth moving equipment to food for responders, be procured and transported to the incident scene.

Planning is responsible for collecting and evaluating information related to an incident and for developing action plans and alternative plans.

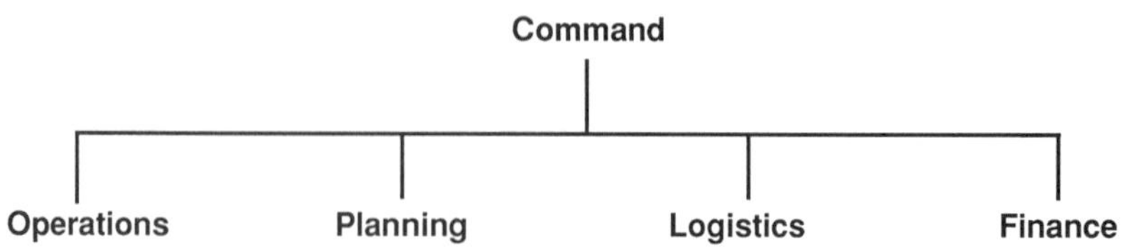

FIGURE 3-1. Major functional areas of the ICS.

Information and advice provided by the planning section are utilized by command in decision making.

Finance is involved with the financial aspects of an incident response. Financial concerns include supplies used during a response operation, purchasing of materials required, timekeeping for response personnel, and other monetary matters.

Interaction of the Functional Areas of the ICS

Command must be clearly established at the beginning of all emergency response operations. The other major functional areas may or may not be formally delineated, depending on the specifics of the incident.

For example, assume that the first-arriving team of responders has reached the scene of a minor hazmat incident. In accordance with the requirements of 29 CFR 1910.120 (see Chapter 2), the senior official of the team assumes the role of on-scene IC and the team is able immediately to stabilize the incident. Due to the simple nature of this incident, the acting on-scene IC is able to manage all functional areas of the ICS by carrying out all duties related to operations, logistics, planning, and finance required to bring the incident under control.

In contrast to the previous example, the response to a major incident may require that the actions of many teams of responders be controlled and coordinated over a long period of time. The modular organization of the ICS allows the managerial system to shrink or expand as needed so that all personnel involved in managing the system are able to maintain a reasonable span of control.

For example, assume that the first-arriving industrial response team has reached the scene of a major hazmat incident. The senior official of the team assumes the role of on-scene IC, begins to assess the scene, and immediately calls for additional help. As personnel having greater seniority arrive, command will be relinquished by the acting IC and passed up a preestablished line of authority, as specified in the emergency response chain of command. As progressively larger numbers of personnel become involved in the response operation, it will be necessary to delineate formally the functional sections subordinate to command and to assign someone to manage each of the sections. Thus, for a major incident, management of the major functional areas of the ICS may require the involvement of a planning section chief, an operations section chief, a logistics section chief, and a finance section chief in addition to the IC (see Figure 3-2). Furthermore, the IC may require direct assistance from command staff personnel, as described below, in carrying out his duties.

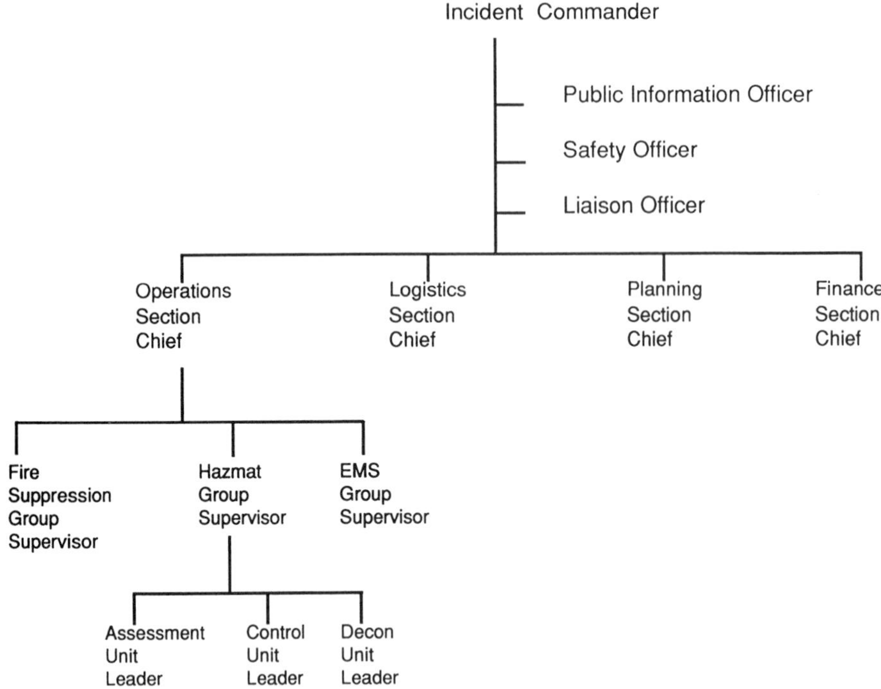

FIGURE 3-2. Hazardous materials ICS managerial roles.

As additional personnel become involved in a response operation, the ICS must expand or build down as required to maintain an appropriate span of control. Thus, group supervisors may be deployed by an operations chief. Likewise, unit leaders can be established to perform as directed by a group supervisor (Figure 3-2). As response personnel begin to bring the incident under control and fewer personnel are needed for the operation, the ICS can be reduced progressively by reversing the process of expansion.

COMMAND OF THE INCIDENT RESPONSE

Single and Unified Command Structures

Command may be single or unified, depending on the incident. If the incident is located in one jurisdiction and only one agency is involved in the response operation, the single command system will be used. In the single command system, one designated official will act as IC. For example, assume that a release occurs at a fixed facility. The facility hazmat team responds under the command of the designated on-scene IC and brings the incident under control, preventing any hazardous chemicals from exiting the property. This is an example of response by a single organization (the in-house hazmat team) utilizing the single command system.

As another example, assume that a release occurs at a fixed facility which has taken advantage of an exclusion from the emergency response planning requirements of 29 CFR 1910.120(q)(1). The local fire department is notified and, in keeping with the conditions of the exclusion, employees are evacuated from the workplace and are not allowed to assist in the response operation. The local fire department responds, with a designated member of the department (usually the fire chief or ranking fire officer) acting as IC, and rapidly brings the incident under control, without harm to the environment or threat to off-site population. This is another example of a single agency (the local fire department) utilizing the single command system.

The unified command system is used when an incident involves more than one jurisdiction and/or more than one agency participates in the response. For example, assume that a major release occurs at a fixed facility located near the boundary line separating two municipalities. The release represents a potential threat to off-site populations in both municipalities and to the environment. A long-term response operation is required involving the facility hazmat team, the local fire departments from both municipalities, the county emergency management agency, and the state environmental regulatory agency. In the unified command system, the individuals designated by their respective jurisdictions/agencies must make joint decisions concerning objectives, strategy, and priorities.

In most public-sector hazardous material incident response operations, the local fire department will assume command and the ranking fire officer will be the IC. As the incident unfolds, it is not uncommon to change from a single to a unified command system as various agencies are notified and begin to arrive at the incident site. As representatives of environmental agencies and other regulatory agencies with legal authority arrive, a unified command system can develop, or the arriving agencies may agree to continue under a single command system, with members of the agencies acting as staff advisors to the IC. The designated IC may be from the fire service, law enforcement, emergency management, or some other agency or organization involved in the response operation.

A clearly understood, preestablished chain of command is vital for all personnel and agencies which become involved in a response operation. For example, assume that a major incident is underway at a fixed industrial facility. The highest-ranking member of the facility emergency response organization has assumed the role of on-scene IC and has initiated the response operation in accordance with the site ERP. The local fire department has been summoned and is en route to the scene. Due to the nature of the incident, the fire department's involvement will be vital in bringing the incident under control. In this type of response, there is no time for confusion or conflict regarding who will be in command after the arrival of the public sector responders. Some states have laws to prevent this type of conflict by mandating that the ranking fire official at the scene of any emergency will be in command. These potential conflicts must be resolved during preemergency planning before incidents occur. To this end, the private sector response organization must work through agencies such as the local emergency planning committee to interface its emergency response chain of command with those of applicable emergency response agencies from the public sector. The ICS allows for development in either the single or unified command system. However, it is vital that all agencies and organizations agree on the evolution of the system and work together for successful termination of the incident.

Command Staff Responsibilities

An important part of the ICS for major incidents is the command staff. This is probably the most overlooked part of the ICS, and in many ways it can be the most important part. The command staff consists of the information officer, safety officer, and liaison officer (see Figure 3-2).

Public Information Officer

The public information officer's function is to control the location of the press and release accurate information concerning the incident as it is cleared by the IC. To this end, the information officer should establish a press area, preferably in a location where the press can safely see what is happening without jeopardizing the safety or effectiveness of the operation. From this location, the information officer can meet with the press on a regular basis and give information and updates as needed.

Under no circumstances should the press be allowed to roam at will during the emergency phase of the incident. The information officer must be aware that some reporters will try to do this. Members of the press should be made to understand that, unless they remain at the press area, their safety is in question and that irresponsible actions on their part may jeopardize the safety of the operation as a whole.

The information officer should maintain a unit log of his activities during the course of the incident.

Safety Officer

The safety officer is another member of the command staff. His role is to ensure that safety procedures are established and carried out throughout the incident response. The safety officer's specific functions are to

- Obtain briefing from the IC
- Identify hazardous situations related to the incident
- Participate in planning meetings
- Review incident action plans
- Identify potentially unsafe situations
- Exercise emergency authority to stop and prevent unsafe acts
- Investigate accidents that occurred within the incident area
- Review and approve the medical plan
- Maintain a unit log

The safety officer's importance cannot be overstated, as this person is responsible for the safety of all personnel involved with the incident response and the safety of the incident itself. The safety officer is the only person who can countermand an order given by the IC or the operations officer. The safety officer should not hesitate to do so if, in his opinion, an operation is unsafe. However, any time an order is countermanded by the safety officer, the IC must be notified immediately.

In larger incidents, the safety officer may be aided by a safety staff, such as a medical officer, sector safety officers, and a decon safety officer (who is often also the medical officer). This staffing allows the safety officer to remain at the command post (CP) with the IC during planning sessions to offer safety suggestions. When a safety staff is developed, its members have the same authority to terminate unsafe operations as the safety officer, but also carry the same responsibility to notify the IC immediately upon such action.

It has long been a goal of emergency services to be part of the solution rather than the problem. The safety officer and his staff will play an important role in achieving this goal when they are assigned in a timely manner and perform properly.

Liaison Officer

Another member of the command staff is the liaison officer. His responsibility is to coordinate the involvement of agencies such as fire, law enforcement, Red Cross, public works, and cleanup contractors, who may become involved in an incident response. The specific functions of the liaison officer are to

- Obtain a briefing from the IC
- Provide a point of contact for assisting/cooperating agency representatives
- Identify the representatives from each agency, including their communication links and locations
- Respond to requests from incident personnel for interorganizational contacts
- Monitor incident operations to identify current or potential interorganizational problems
- Maintain a unit log

Larger incidents involving several agencies will require a liaison officer because the IC simply cannot keep track of which agencies are involved at various stages of the operation. A competent and capable liaison officer will be able to advise the IC of which agency can perform specific tasks and notify those agencies of their assigned tasks. The liaison officer is therefore an invaluable asset to the IC during the course of larger-scale response operations.

THE OPERATIONS SECTION

In small operations the IC may choose to manage the operations section directly. However, for larger incidents, the IC will need to appoint an operations chief to oversee the operations section (Figure 3-2).

Structure of the Operations Section

The size of the response operation will dictate the structure of the operations section. For a small hazmat spill confined to a small area (by the nature of the material, the geographical area of the spill, the amount of material, or a combination of these), operations may be very small, with the hazmat group utilizing an entry unit and a decon unit. For larger spills, the operations section may need to be subdivided into *divisions* and *groups*. Divisions are geographical subdivisions and groups are functional subdivisions.

Division or Sectoring at the Incident Scene

Divisions are created to divide an incident into geographical areas of operation. For example, with a hazmat spill in a warehouse, we might want to establish a front division of operations and a rear division of operations. A synonym for "division" is the term "sector." Using the above example, there would be a front sector and a rear sector.

It is common practice in emergency operations to use sectoring. However, it is very important for the operations officer to appoint a supervisor for each sector created. It should be noted that some emergency response organizations also use the term "sector" as a synonym for "group," as defined in the following section.

Operational Groups

Groups are established to divide the incident into functional areas of operation. The most common groups at a hazmat incident are

- Hazmat
- Fire suppression
- Emergency medical service (EMS)

The hazmat group will be subdivided into assessment, control, and decon units (see Figure 3-2). The fire suppression group will concentrate on preventing or controlling a fire at the incident scene. The EMS group is responsible for the health and safety of emergency responders working at the incident and may be involved in rescue operations. This will depend on the level of training and PPE available to the EMS personnel.

Basic Considerations for Operations

Control of Staging

Some ICs choose to have logistics handle staging, while others recommend that operations perform this task. If staging is being handled by the operations section instead of the logistics section or the IC, the operations chief must establish a staging area and appoint a staging officer.

Communication and Interaction with the IC

There must be clear communication between the IC and the operations chief at all times. In most cases, the operations chief will be located at the CP. This makes communication much easier and allows the operations chief to have first-hand knowledge of what is happening in planning and logistics.

Transfer of Information during Shift Changes for Prolonged Operations

If the incident is prolonged, assistants must be appointed to keep operations running while the operations chief sleeps. The most important aspects of the prolonged operation are good record keeping and proper transfer of information as shifts change. Each time someone assumes a job, there is a significant chance for information to be lost or at least not passed along. The officers in the ICS must continuously guard against this loss of information.

THE PLANNING SECTION

Planning must include information gathered both before and during an incident, since preplanning is a critical aspect of emergency response. The site-specific ERP should include general provisions for safely terminating any hazmat emergency which can reasonably be anticipated. The plan can then be updated or fine-tuned based on information gathered during assessment of the actual incident.

At the scene of any operation, every effort should be made to gather as much useful information as possible before mitigation operations begin. This information-gathering and decision-making process begins as emergency response units approach the incident location and proceeds until the operation is completed. For example, placards, labels, container types, and release locations may be identified as the hazard area is surveyed by the initial emergency responders. Based on this, information can be obtained from Material Safety Data Sheets, various reference documents, and agencies such as Chemtrec. Computer data can be provided by modem and FAX at the incident scene in some instances.

For incidents involving transport vehicles, shipping papers are obtained as soon as possible, sometimes by the initial response team and sometimes by the recon team (if entry into a hazardous area using PPE is required). As the incident unfolds, information from other agencies, such as the shipper or the manufacturer of the product involved, may be helpful. A detailed discussion of information gathering and decision making is provided in Chapter 5.

Assigning a planning section chief in the early stages of an incident is of the utmost importance. As information is gathered, the planning section chief (and his staff in major incidents) can review the information and help the IC and the operations officer with decision making.

Armed with as much information as possible beforehand, recon, rescue, confinement, and containment operations are undertaken. During these operations, especially recon, more information is gathered and additional decisions concerning the incident must be made. The IC must depend on his staff to constantly gather and review information and to offer advice on decision making. Thus, staff meetings are held almost constantly during a hazmat incident.

With a constant flow of information and continuous decision making, the IC guides the response to its conclusion. The ultimate responsibility for each decision rests with the IC.

THE LOGISTICS SECTION

As decisions are made and instructions are issued to the operations section, personnel, equipment, supplies, and services must be available so that operations can function as required. Logistics provides the supplies, services, and facilities required to support the entire response operation. Logistical needs may include

- Operational support, such as supplying breathing air, heavy equipment, fuels, sorbent materials, and other items needed for the response operation
- Facilities, such as the CP, required at the scene to support the operation
- Communications, ensuring that required communication equipment is available and sufficient for the needs of the operation
- Rest and rehabilitation for personnel involved in the operation

This requires a logistics section headed by a logistics chief answering to the IC and coordinating with the operations officer. The logistics section chief is the person who must be able to get anything at any time, as dictated by the situation, the IC, and the operations officer.

All reserve personnel, equipment, and supplies should be located at a staging area. As personnel, equipment, and supplies arrive, they should be directed to report to staging; and as the operations section needs additional personnel, equipment, and supplies, the operations officer should request them from staging. The logistics officer and his staff should log personnel, equipment, and supplies in and out of the staging area and should coordinate this logging activity with the safety officer and his staff. This is especially important for accountability of personnel.

At the CP, constant communication between the IC, operations section chief, safety officer, planning section chief, and logistics section chief must take place. The logistics chief must know the quantity, quality, and type of personnel, equipment, and supplies that will be needed and should be provided this information well in advance if possible.

THE FINANCE SECTION

Another major functional section of the ICS is the finance section. Nothing involving a hazardous materials incident response is free except advice. Hazmat operations cost money, and someone must pay. Even if the responsible party is on the scene and accepts the responsibility for payment, someone must keep a record of what is being used so that a statement can be prepared. This information may prove vital when claims are filed. For major response operations, the IC should appoint a finance section chief with the responsibility of putting a price on everything that is expended in the response operation.

SUMMARY

Although the ICS is developed around a simple resource management concept, the system has proven very effective in coordinating and controlling emergency response operations. The flexibility of the system makes it equally applicable to all types and sizes of hazardous materials incidents. However, in order to function effectively within an assigned ICS role, all response personnel must have a working knowledge of the ICS as a whole, as well as of the specific duties they are to perform within the system. For this reason, the ICS is a vital area of training for a hazardous material incident response.

REFERENCES

Fire Protection Publications. 1983. *Incident Command System*. Stillwater, OK: Fire Protection Publications.

Henry, M.F., ed. 1989. *Hazardous Materials Response Handbook*. Quincy, MA: National Fire Protection Association.

Noll, G.G., M.S. Hildebrand, and J.G. Yvorra. 1988. *Hazardous Materials: Managing the Incident*. Stillwater, OK: Fire Protection Publications.

U.S. Department of Labor. Title 29 Part 1910. Washington, DC: U.S. Government Printing Office.

4

Hazardous Materials Terminology

Barbara M. Hilyer, M.S., M.S.P.H., and
D. Alan Veasey, M.A. Ed.

The driver of a tractor-trailer was climbing a long hill approaching the heavily populated residential outskirts of a large metropolitan area when he discovered a fire in the cab of his truck. He pulled onto the shoulder of the interstate highway and was quickly joined by police and hazmat-trained firefighters. There were no placards on the truck, and since responders were unable to retrieve the shipping papers from the cab due to the fire, the driver was the only source of information about the contents of the trailer. The driver was carrying a load for a landscaping company and stated that the material was urea nitrate. Urea nitrate is classified as a flammable solid under certain conditions and as a high explosive (Explosive A) if it is shipped dry, or wet with more than 25 pounds in one outside package. The road was closed; a large residential area along both sides of the highway was evacuated, with extensive manpower and expense; and attempts were made to contact the generator. The original shipper was finally contacted, and FAXed the information about the product—urea. Urea is a soluble source of nitrogen used as fertilizer; it is not flammable, not a hazardous material, and not regulated during transport. Correct terminology would have gone a long way to prevent trouble and cost in this incident.

When an emergency responder detects or confirms an accidental chemical release, he should look for evidence which may help him determine the nature of the material from a safe distance. Before any decisions are made or any response is initiated, the released material must be identified, if possible, or at least categorized. This chapter will provide information regarding the words, symbols, and other visible markings a responder can use in identifying and evaluating the chemicals involved in a hazmat incident.

OBJECTIVES

- Learn the terminology used to categorize materials which may pose chemical hazards
- Understand the terminology describing the hazards posed by released chemicals
- Know how to utilize hazard classification systems and their required labels, markings, and placards

TERMINOLOGY FOR MATERIALS WHICH POSE HAZARDS

Several different federal agencies regulate chemicals which may cause damage to people, property, or the environment if they are released during an accident. Each agency uses different terms to define and describe hazardous agents. The agencies whose terminology is most likely to be encountered by emergency responders are OSHA, DOT, and EPA.

Hazardous Chemicals

Hazardous chemicals, as defined by OSHA, are chemicals that are hazardous to people in the workplace or to the community if released. A Material Safety Data Sheet for each hazardous chemical used in a workplace must be prepared by the employer or requested from the manufacturer, and kept on file in a location accessible to workers and members of the community upon their request. The employer has the responsibility to instruct workers about the chemical hazards with which they work.

Hazardous Materials

Hazardous materials are those materials that may present a danger during shipment by truck, rail, air, or water, as determined by the U.S. Secretary of Transportation. If a material listed in the DOT Hazardous Materials Table is transported in one package in a quantity equal to or greater than the listed reportable quantity for that material, the shipment is subject to DOT regulation.

DOT has been the primary regulatory agency for hazardous materials in interstate transport since the enactment of the Transportation Safety Act in 1974. The Hazardous Materials Transportation Act of 1975 gave DOT the authority to impose stiff financial penalties for violations. In 49

CFR, DOT regulations are spelled out which establish criteria for packaging, labeling, placarding, shipping papers, and the training and responsibilities of transportation personnel. The Hazardous Material Regulations (HMR; 49 CFR Parts 171–180) were comprehensively revised by the Hazardous Materials Transportation Uniform Standards Act of 1990.

Hazardous materials are called "dangerous goods" in Canada and are regulated in much the same way by the Canadian government. An agreement between Canadian and American agencies makes regulations similar so that materials shipped between these countries can comply with both sets of regulations.

Hazardous Substances

Hazardous substances are substances determined by EPA to present a danger to the environment. If a hazardous substance is spilled or otherwise released into the air, ground, or water in excess of EPA's listed reportable quantity (RQ), the release must be reported to EPA.

EPA regulations are set forth in 40 CFR; however, since Congress instructed EPA to coordinate closely with DOT, the agency relied heavily on that department's regulations already in effect, and 49 CFR was expanded to include EPA's list of hazardous substances. A small part of this list is provided in Table 1-1 in Chapter 1.

Hazardous Wastes

Hazardous wastes are regulated by EPA under the Resource Conservation and Recovery Act (RCRA) of 1976. Hazardous wastes can be broadly defined as hazardous substances that have no commercial value. RCRA authorized EPA to institute a program of hazardous waste management initiating "cradle-to-grave" tracking of hazardous wastes.

TERMINOLOGY TO DESCRIBE HAZARD CLASSIFICATION

There are several hazardous chemicals classification systems, each written by a different agency.

EPA classifies substances and wastes based on whether the materials are ignitable, corrosive, reactive, or toxic. Materials which are not specifically listed in the Hazardous Substances Table are still considered hazardous if they meet the definition of one of these hazard classes. Table 4-1 shows EPA hazard classes and how they are defined.

DOT classifies hazardous materials according to the types of hazards the materials pose, but the categories are defined somewhat differently

TABLE 4-1. EPA Hazard Categories

Hazard Category	Definition
Ignitable	Burnable solids, liquids, and gases
Corrosive	Having a pH of 2 or below, or of 12.5 or above
Reactive	Explodes or reacts violently with air or water
Toxic	Poisonous to living organisms

than those used by EPA in classifying hazardous substances. DOT's categories are called "hazard classes" and are shown in Table 4-2. These classes should be studied by responders who may be called to hazmat transportation incidents, or to spills around a truck or rail loading area at an industrial facility.

There are also International Maritime Organization (IMO) regulations, applicable to the transport of hazardous materials between countries when any portion of the transportation involves carriers on water. The classification system is called the "International Maritime Dangerous Goods (IMDG) Code"; it is also sometimes referred to as the "United Nations system," since it is based on UN recommendations.

The newly promulgated DOT regulations of 1990, which were written partly to bring the U.S. system more closely into compliance with the international system, make greater use of the IMO hazard classes. A comparison of DOT and IMO classifications is presented in Table 4-3. The new regulations are effective October 1, 1991, but compliance was authorized on January 1, 1991. The transition period for coming into compliance varies for different sections; for the classification and hazard communication sections, which include labels and placards, the final compliance date was stated as October 1, 1993, but may be delayed. During the transition period, both sets of specifications for labels and placards will be legal and will be seen on containers and vehicles.

TERMINOLOGY FOR ITEMS USED TO COMMUNICATE HAZARD INFORMATION

The following items are instruments of hazard communication which might be encountered by an emergency response team in a hazardous materials incident.

TABLE 4-2. DOT Hazard Classes

Hazard Class	Definition
Blasting Agent	Designated for blasting; little probability of initiating an explosion
Class A Explosive	Functions by detonation
Class B Explosive	Functions by rapid combustion
Class C Explosive	Manufactured articles containing restricted quantities of Classes A or B
Combustible Liquid	Flash point 100°–200°F
Compressed Gas	Absolute pressure exceeding 40 psi at 70°F, 104 psi at 130°F, or flammable and 40 psi at 100°F
a. Flammable Gas	Flammable or explosive
b. Nonflammable Gas	Nonflammable, nonexplosive; hazardous due to being under pressure
Corrosive Material	Visible destruction or irreversible damage to skin; severe corrosion of steel
Cryogenic Liquid	Gas converted to liquid at extremely cold temperatures (below −130°F)
Etiologic Agent	May cause disease in humans
Flammable Liquid	Flash point less than 100°F
Flammable Solid	Likely to cause fire by absorption of water or spontaneous chemical change
Irritating Material	Gives off dangerous or intensely irritating fumes or vapors when exposed to air or fire
Organic Peroxide	Derivative of hydrogen peroxide in which hydrogen has been replaced by an organic
ORM	Other Regulated Materials that do not meet hazard class definitions but are regulated
A	Can cause extreme annoyance or discomfort to passengers and crew if leaked
B	Can cause damage to vehicle if leaked
C	Unsuitable for shipment unless properly prepared
D	Consumer commodity which presents limited hazard
E	Certain hazardous substances and wastes
Oxidizer	Yields oxygen readily in chemical reactions
Poison A	Specifically listed gases or liquids of such a nature that a small quantity of the gas or vapor from the liquid is dangerous to life
Poison B	So toxic to humans as to create a hazard
Radioactive	Spontaneously emits radiation capable of damaging living tissue
I	Emits <0.5 mrem/hr at package surface
II	Emits 0.5–50.0 mrem/hr at package surface
III	Emits >50.0 mrem/hr at package surface
Toxic by Inhalation	Gives off toxic vapors at normal temperatures

TABLE 4-3. Comparison of IMO and DOT Hazard Classes

IMO Class, Division, and General Description	DOT Hazardous Material Class
Class 1—Explosives	
1.1 Mass explosion hazard	Explosive A
1.2 Projection hazard, but no mass explosion hazard	Explosive A or B
1.3 Fire hazard with minor projection or blast hazard	Explosive B
1.4 No significant hazard	Explosive C
1.5 Very insensitive substance	Blasting Agent
Class 2—Gases	
2.1 Flammable gases	Flammable Gas
2.2 Nonflammable gases	Nonflammable Gas
2.3 Poison gases	Poison A
Class 3—Flammable liquids	
3.1 Low flash point (<0°F)	Flammable Liquid
3.2 Intermediate flash point (0°–73°F)	Flammable Liquid
3.3 High flash point (73°–141°F)	Combustible Liquid
Class 4—Flammable solids or substances	
4.1 Flammable solids	Flammable Solid
4.2 Substances liable to spontaneous combustion	Flammable Solid Flammable Liquid (pyroforic)
4.3 Substances which emit flammable gases when wet	Flammable Solid (with Dangerous When Wet label)
Class 5—Oxidizing substances	
5.1 Oxidizing substances	Oxidizer
5.2 Organic peroxides	Organic Peroxide
Class 6—Poisonous or infectious substances	
6.1 Poisonous substances	Poison B
6.2 Infectious substances	Etiologic Agent
Class 7—Radioactive materials	Radioactive Material
Class 8—Corrosives	Corrosive Material
Class 9—Miscellaneous dangerous substances	Other Regulated Materials (ORM) except ORM-D, which was retained

Labels and Markings on Containers

The DOT regulations include extensive rules for labeling containers. The containers are also specified in the regulations and may be cans, boxes, carboys, cylinders, drums, or other containers that are designed to DOT specifications to enclose hazardous materials safely. Hazard class labels and other labels and markings on containers can be extremely valuable to emergency responders seeking to identify leaking materials.

Hazard Class Labels

Hazard class labels show the DOT hazard class of the contained material. Each hazard class label has a unique design and provides information in several different ways. A label includes the stated hazard class, a symbol of the hazard class, is color coded to the hazard class, and shows the IMO hazard class number. Before the DOT hazmat regulations were amended, the IMO number was optional but was seen on most labels. Under the new regulations, this number is required and is larger than on previous labels (Figure 4-1).

Mixtures in a Container

When two or more chemicals are mixed inside one container, they may form a new chemical which has a name. If not, the proper shipping name

FIGURE 4-1. Corrosive label.

is based on the properties of the mixture; for example, the proper shipping name of two flammable liquids which have been combined, but do not form a new named chemical, is "Flammable Liquid, N.O.S." where N.O.S. stands for "not otherwise specified." The regulations require that the two components in a mixture which contribute most greatly to the hazard must be identified by name on the label.

Identification Numbers

Each hazardous material regulated by DOT has a four-digit identification number. These numbers are listed in the Hazardous Materials Table, found in 49 CFR 172.101. In most cases, the number designates only one chemical. You will find these numbers on containers. Table 4-4 shows a part of the Hazardous Materials Table.

Other Labels and Markings

Other markings on a container provide important information in an emergency. The container must be marked with the name of the material it contains, and the name must be a proper technical one, not a trade name.

TABLE 4-4. Examples of Old and New DOT Classification, Labeling, and Placarding Regulations From the 172.101 Hazardous Materials Table

Old table: legal until further notice

Hazardous Materials Descriptions and Proper Shipping Name	Hazard Class	ID Number	Label	Placard
Acetyl bromide	Corrosive Material	UN1716	Corrosive	Corrosive
Acetyl chloride	Flammable Liquid	UN1717	Flammable Liquid	Flammable
Acetylene	Flammable Gas	UN1001	Flammable Gas	Flammable Gas
Acetylene (liquid)	Forbidden			

New table: legal after January 1, 1991; required after date to be announced

Hazardous Materials Descriptions and Proper Shipping Name	Hazard Class or Division	ID Number	Label	Placard
Acetyl bromide	8	UN1716	Corrosive	Corrosive
Acetyl chloride	3	UN1717	Flammable Liquid and Corrosive	Flammable Liquid
Acetylene, dissolved	2.1	UN1001	Flammable Gas	Flammable Gas
Acetylene (liquefied)	Forbidden			

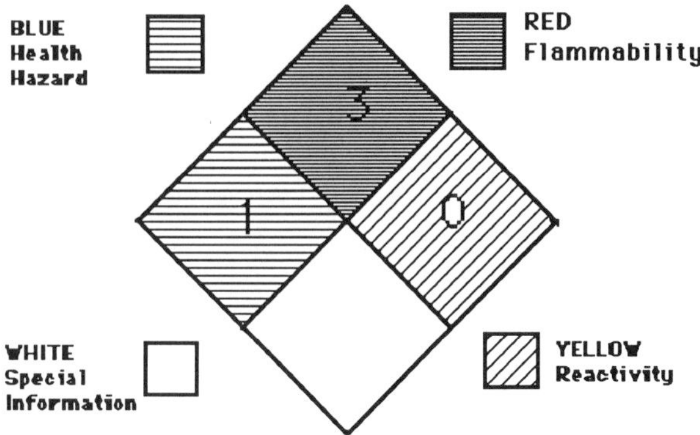

FIGURE 4-2. NFPA label for acetone.

NFPA 704 System

The NFPA's 704 System is used at many fixed facilities to indicate hazardous materials. Figure 4-2 shows an NFPA label. It is a diamond divided into four smaller diamonds, each representing a type of hazard. Health hazards are indicated in blue at the far left, flammability is shown at the top in the red diamond, reactivity with other chemicals appears on the right in yellow, and the lower white diamond leaves room for other important information. Ratings range from 0 to 4, with 4 indicating the highest hazard level in each category.

Health Hazards (Blue):

- 4 Materials too dangerous to health to expose firefighters in SCBA and turnout gear, which is not adequate protective clothing.
- 3 Materials extremely hazardous to health, but areas may be entered with extreme care. SCBA should be worn and no skin exposed.
- 2 Materials hazardous to health, but areas may be entered with full-face SCBA.
- 1 Materials only slightly hazardous to health. It may be desirable to wear SCBA.
- 0 Materials which on exposure under fire conditions would offer no hazard beyond that of ordinary combustible material.

Flammability (Red):

- 4 Very flammable gases or very volatile flammable liquids.
- 3 Materials which can be ignited under almost all normal temperature conditions.

2 Materials which must be moderately heated before ignition will occur.
1 Materials which must be preheated before ignition can occur.
0 Materials which will not burn.

Reactivity (Stability) (Yellow):
It should be noted that this category does not describe chemical reactions that occur between two mixed chemicals, but rather the stability of the material by itself.

4 Materials which are readily capable of detonation, explosive decomposition, or explosive reaction at normal temperatures and pressures.
3 Materials which are capable of detonation, explosive decomposition, or explosive reaction and which require a strong initiating source, or which must be heated under confinement before initiation.
2 Materials which are normally unstable and readily undergo violent chemical change but do not detonate. This category includes materials which undergo chemical change, with rapid release of energy at normal temperatures and pressures, or which can undergo violent chemical change at elevated temperatures.
1 Materials which are normally stable but may become unstable at elevated temperatures and pressures, or which may react with water, with some release of energy but not violently.
0 Materials which are normally stable even under fire exposure conditions and which are not reactive with water.

Other Information (White):
The diamond includes symbols like the ones for water reactivity or radioactivity, or the letters "OX" indicating that the material is an oxidizer.

Pesticide Label
Pesticides are regulated by the Federal Insecticide, Fungicide, and Rodenticide Act (FIFRA) and are labeled under an EPA terminology system different from DOT or NFPA labeling systems (Figure 4-3). Pesticides are generally defined as chemicals which are designed to kill some living organism, like an insect, weed, or fungus, which is considered to be a pest. Any chemical which kills other living things will most likely have some bad health effects on humans; some of them are seriously toxic in small amounts. A signal word on the label indicates the degree of toxicity of the contents: "Danger" is used for the most toxic pesticides, "Warn-

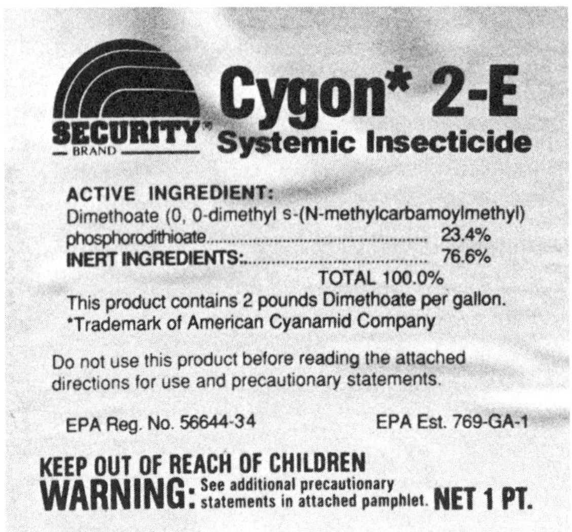

SECURITY BRAND
Cygon* 2-E
Systemic Insecticide
ACTIVE INGREDIENT:
Dimethoate (0, 0-dimethyl s-(N-methylcarbamoylmethyl)
phosphorodithioate........ 23.4%
INERT INGREDIENTS:........ 76.6%
TOTAL 100.0%
This product contains 2 pounds Dimethoate per gallon.
*Trademark of American Cyanamid Company
Do not use this product before reading the attached
directions for use and precautionary statements.
EPA Reg. No. 56644-34
EPA Est. 769-GA-1
KEEP OUT OF REACH OF CHILDREN
WARNING: See additional precautionary statements in attached pamphlet.
NET 1 PT.

FIGURE 4-3. Pesticide label.

ing'' for less toxic ones, and ''Caution'' indicates a relatively minor degree of toxicity.

Pesticide labels show the trade name (which is not useful for looking up hazard information), the EPA registration number (which can be used to identify the chemical by its technical name), and hazard and precautionary statements. Active ingredients are identified by chemical name; these are the names which can be used if a responder is consulting references about health, fire, and reactivity hazards of the pesticide.

Shipping Papers

An emergency response brigade at an industrial facility may be involved in an incident which occurs during loading or unloading of hazardous materials for transport. In such cases, the responders should draw on hazard information contained in the shipping papers. Shipping papers consist of a Bill of Lading for highway transport, Waybill Consist or Wheel Report for rail movement, Dangerous Cargo Manifest for water transport, or Air Bill with Shipper's Certification for Restricted Articles for air shipment. If the material being shipped is regulated by EPA as a hazardous waste, shipment of the material must be accompanied by a Uniform Hazardous Waste Manifest. All of these would show, at a minimum, the proper DOT shipping name for the material, the DOT hazardous material identification number, and the material's DOT hazard class. Shipping papers can be found in the cab of a truck, most likely in the driver's side door pocket; on a train, they are carried in the engine car.

Additional information for the benefit of emergency responders is required to be printed on the shipping papers, or carried with them in the form of Material Safety Data Sheets or other written documents.

Emergency information documents must contain, at a minimum, the following:

- A description of the material (technical name)
- Immediate hazards to health
- Risk of fire or explosion
- Immediate precautions to be taken in the event of an accident
- Immediate methods for handling fires
- Initial methods for handling spills or leaks
- Preliminary first aid for exposure victims

A 24-hour emergency response contact telephone number must be entered on shipping papers. The telephone must be monitored at all times by someone who is knowledgeable about the hazards and characteristics of the material being shipped. The shipper may list the number of an agency, such as the Chemical Emergency Transportation Center (CHEMTREC), if the agency has been given all the required information on the material.

Placards

Placards are 10¾-inch signs which are affixed to both ends and both sides of vehicles carrying hazardous materials. These placards have the same writing, color, symbol, and IMO hazard class number as do labels. For purposes of placarding, hazardous materials are divided into two types: those for which placards are required when they are being transported in any quantity (Table 4-5), and those which must be placarded only when

TABLE 4-5. Materials Which Must Be Placarded When Shipped in Any Quantity

Category of Material	Placard Name
1.1	Explosives 1.1
1.2	Explosives 1.2
1.3	Explosives 1.3
2.3	Poison Gas
4.3	Dangerous When Wet
6.1 (PG I, inhalation hazard only)	Poison
7 (Radioactive Yellow III label only)	Radioactive

TABLE 4-6. Materials Which Must Be Placarded Only When Shipped in a Total Quantity of a Hazard Class Greater Than 1,000 Pounds

Category of Material	Placard Name
1.4	Explosives 1.4
1.5	Explosives 1.5
1.6	Explosives 1.6
2.1	Flammable Gas
2.2	Nonflammable Gas
3	Flammable
4.1	Flammable Solid
4.2	Spontaneously Combustible
5.1	Oxidizer
5.2	Organic Peroxide
6.1 (PG I or II, other than PG I inhalation hazard)	Poison
6.1 (PG III)	Keep Away from Food
8	Corrosive
9	Class 9

the sum of the materials in any hazard class on the vehicle is 1,000 pounds or greater (Table 4-6).

Placards provide information in four ways about the contents of the carrier (Figure 4-4):

- The color of the placard is correlated with the hazard class.
- The title of the hazard class is printed across the placard.
- A symbol indicating the hazard class is shown.
- The IMDG hazard class number is usually shown. After the revised DOT regulations take effect, it will always be shown.

In one situation, placards are different from labels: When bulk shipments are made in large containers such as tanker trucks, the placard includes the four-digit identification number printed across its center (Figure 4-5). This allows the initial responder on the scene to identify the material in the tanker truck or rail car from a distance. Since the potential quantity spilled may be large, maintaining a safe distance can be very important. For the same reason, bulk tanker cars must be placarded on both sides, but not on the ends, when they carry less than 1,000 pounds of a material listed in Table 4-6.

FIGURE 4-4. A DOT hazard class placard displays hazard information four ways (color, symbol, word, number).

FIGURE 4-5. A bulk container placard showing the four-digit identification number.

Special Placarding Situations

A placard must remain on the bulk container or bulk transport vehicle even when it is empty. Some volatile, flammable materials leave behind vapors which are quite dangerous, and many toxic chemicals are toxic upon exposure to small residues. When the tanker has been cleaned and purged, or cleaned and filled with a nonhazardous material, the placard may be removed.

If two or more hazard classes, each totaling 1,000 pounds or more, are being transported on the same truck, the "Dangerous" placard may be used in lieu of the individual placards. If 5,000 pounds of one of the materials are loaded at one facility, the placard for that hazard class must also be on the truck. The "Dangerous" placard is also used to indicate individual shipments of Explosives C or Irritants. A responder approaching a vehicle placarded this way would be at a loss to determine what was inside the truck without further information from the shipper or the shipping papers.

No placard is required on shipments of radioactive materials, combustible liquids in containers of less than 110 gallons, etiologic agents, Otherwise Regulated Materials (ORM), and small hazardous loads which total less than 1,000 pounds, with the following exceptions: the small load must be placarded when it contains Explosives A or B, Poison A, water-reactive flammable solids, gases which are poisonous by inhalation, or Radioactive III materials.

The placarding exception for loads in which the hazardous materials do not total 1,000 pounds may cause problems for emergency responders. Firefighters and hazmat teams sometimes encounter mixed loads carried by trucks belonging to companies such as Wal-Mart, United Parcel Service, and others which contain hazardous materials and are legally unplacarded.

DOT regulations are complicated, and there are currently two legal standards in place. For emergency responders whose work may include responding to transportation incidents, more study of these regulations is recommended.

SUMMARY

Two kinds of pre-incident preparation will help an emergency responder recognize hazardous materials terminology quickly in an emergency: training and practice with the colors, words, and symbols used on labels, placards and shipping papers; and providing ready access to a good terminology reference document at the scene of the incident.

REFERENCES

Currie, J.V. 1988. *Driver's Guide to Hazardous Materials*. Alexandria, VA: American Trucking Association.

National Fire Protection Association. 1986. *Fire Protection Guide to Hazardous Materials*, 9th ed. Quincy, MA: National Fire Protection Association.

U.S. Department of Transportation. Title 49 Parts 171–180. Washington, DC: U.S. Government Printing Office.

U.S. Environmental Protection Agency. Title 40 Part 302. Washington, DC: U.S. Government Printing Office.

5

Hazard and Risk Assessment

Barbara M. Hilyer, M.S., M.S.P.H., and
D. Alan Veasey, M.A. Ed.

Hazardous materials incidents include some of the same dangers encountered in other emergencies. There may be a potential for fire, traumatic injury, and physical and mental stress. However, hazmat emergencies are distinguished from other kinds of emergencies by the presence of chemicals or other substances that, by the nature of their quantity, concentration, or properties, can harm humans who encounter them.

A fire triggered by sparks from a metal saw in a magnesium processing plant in Los Angeles generated a large, noxious cloud, forcing evacuation of businesses in the surrounding square-mile area. Firefighters initially aimed water hoses at the fire without first assessing the hazards of the magnesium, which is water reactive. The water touched off a series of explosions. The fire eventually was left to burn itself out.

Hazardous materials responders at all levels make a series of important decisions based on their perception of the risks posed by released materials to themselves, other people, property, and the environment. From the first detection of a release to closure of the incident, risks are continually being evaluated. Risk evaluation should begin well before an incident occurs, and a good risk analysis must be done before response decisions can be made and acted upon.

OBJECTIVES

- Be able to detect the presence of hazardous materials in an incident
- Know the methods of scene survey and the means used to protect responders during the survey

- Know how to use visual clues and the information in available documents to identify the name or chemical group of a hazardous material and assess its flammability, reactivity, explosivity, toxicity, corrosivity, and radioactivity
- Recognize words, abbreviations, and acronyms which can be utilized to gain hazard information from relevant documents
- Understand how weather, topography, and the chemical and physical properties of the material itself influence its dispersion if it is released
- Know the role of direct-reading monitoring equipment in determining the concentrations of hazardous materials in air

DECISION MAKING IN EMERGENCIES

The National Fire Academy suggests Ludwin Benner's guide to decision making, called the "DECIDE process," to provide help in structuring the steps taken in response to an emergency. The DECIDE process includes six steps, the first two of which are undertaken during hazard and risk assessment:

- **D**etect the presence of hazardous materials
- **E**stimate likely harm without intervention
- **C**hoose response objectives
- **I**dentify options
- **D**o best option
- **E**valuate progress

Detect the Presence of Hazardous Materials

If the materials involved or likely to become involved in the emergency incident are hazardous, the response to the incident will be subject to applicable OSHA, DOT, and EPA regulations. If they are not hazardous, the incident can be treated as a normal fire or spill, and the regulations for response and reporting do not apply. The term "hazardous materials" in the DECIDE process refers to specific chemicals and other substances which are listed by name or by characterization criteria, which usually are physical or chemical properties. The term as used here may include hazardous materials, hazardous chemicals, hazardous substances, and hazardous wastes, as defined in Chapter 4.

Hazardous materials in a transportation incident are those listed by DOT in the Hazardous Materials Table in Section 172.101 of Title 49 of the CFR. If the materials are listed by EPA as hazardous substances, cleanup and reporting are regulated by that agency. If hazardous materials, including hazardous substances, are labeled and placarded correctly, the fact that they are hazardous is readily apparent. If materials are not labeled or placarded, it cannot be assumed that they are not hazardous without further investigation.

At fixed facilities, a working knowledge of the processes and chemicals involved is a good starting point for hazard assessment. Responders can identify, and thereby learn the hazards of, materials at a particular location by referring to data that should be developed prior to an incident. In the section on pre-incident assessment, this subject will be addressed further.

This step is the most important one to responder safety, and initiates the proper flow of decisions which will follow.

Estimate Likely Harm Without Intervention

In this step, responders try to determine what will happen if they do not intervene to prevent harm. Before this can be done, the material must be identified by name so that its hazardous properties can be determined. Failing this, at least a hazard class categorization must be accomplished. It is impossible to estimate possible harm without knowing the hazard class or name of the chemical.

To estimate likely harm, one must think of the possible undesirable outcomes. Ask yourself:

- What has been or may be released?
- How much has been or may be released?
- What are the hazards to people, property, and the environment?
- What are possible migration routes for the released product?

This step will require the use of reference materials which have been gathered in preparation for an incident.

Choose Response Objectives

Having figured out what the harm will be if nothing is done, responders can decide if that harm is acceptable. If it is not, what outcome is acceptable? This acceptable outcome becomes the goal, or objective, of whatever actions will be taken. Accomplishment of the objective(s) will prevent or minimize potential bad outcomes. Objectives might include stopping a leak, preventing fire if the chemical involved is flammable, preventing exposure to a nearby group of people, or keeping a spilled liquid out of a storm drain.

TABLE 5-1. Using the DECIDE Process in an Incident at a Pulp Mill

Process Step	What We Do	What We Decide
Detect the presence of hazardous materials	Look at the material being released Look at the source of the release Use a reference to identify the chemical	The release is a green gas cloud It is coming from a rail tank car placarded "UN 1017" and "Nonflammable Gas" It is chlorine gas
Estimate likely harm without intervention	Read a Material Safety Data Sheet or another source of information	Chlorine forms hydrochloric acid on contact with water, or with moisture on skin or inside lungs
Choose response objectives	Decide what we want to be the result of our response actions	No inhalation by responders, workers, or community members No contact damage to responders' skin Terminate release at source Minimize migration of released gas
Identify options	Think. What will work?	Respiratory and skin protection for responders; evacuation of workers and community members; protection in place; containment of release at source; confinement and/or enhancement of dispersal of released gas
Do best option	Think. What are our resources? Compare these with what we think will work	Combination: evacuate most plant workers; signal needed workers to don SCBA; respond to apply C kit for containment; use fog pattern to absorb gas; alert law enforcement personnel to stand by to evacuate downwind community
Evaluate progress	Observe response and results	Determine success of actions; act to increase success; determine need to evacuate further

In choosing response objectives, it is wise to consider what can reasonably be accomplished with available personnel, equipment, training, and personal protective equipment.

Identify Options for Achieving Response Objectives

Looking at all aspects of the incident, responders consider viable options for achieving the objectives of the response. In cases in which sufficient emergency response resources are available, offensive action to end the release at the source may be feasible. In situations in which an offensive response is not feasible, a defensive response with personnel kept away from the source of the leak may be possible. A defensive response might include calling for help from off-site personnel and evacuating potentially affected persons, performing remote confinement operations (like diking) to limit the spread of the contaminant, and gathering additional information about the incident while waiting for help to arrive.

Do Best Option

In this phase of the DECIDE process, the best option of all those that have been considered is chosen. It is only at this point that the team begins to put the selected response into effect. Responder safety is the critical concern. Remember the advice so often heard in the fire service: Don't become part of the problem.

Evaluate Progress

As the response unfolds, new information will be gained. Progress is evaluated as the team works, and the chosen plan is reevaluated. Is the plan working to accomplish the stated objectives? Are responders at risk? Other options can always be considered; for example, calling for additional assistance from off-site personnel may become necessary.

Table 5-1 illustrates the DECIDE process, using the example of a chlorine release in the chemical storage area of a pulp mill.

PRE-INCIDENT HAZARD ASSESSMENT

Hazard and risk assessment begins well before an emergency incident occurs. Study of several factors during the preemergency planning stage will enable an emergency response team to act more quickly and efficiently if a release occurs. The in-house hazardous materials response team at a fixed facility has an advantage over the mobile hazmat team in knowing their relatively small territory and the materials on the site.

Materials Present On Site

Two types of information regarding the chemicals on site should be available. The first is the Material Safety Data Sheet; the second is a storage and use area map of all chemicals in the area.

Material Safety Data Sheets

Material Safety Data Sheets are inconsistent sources of information, but are generally available and will be described later in this chapter. In lieu of these, some industrial facilities use computer systems to generate their own forms, which include the same information. A formatted system is usually better for in-plant emergency use, since workers become familiar with the way the documents are set up and can read and understand them more quickly. Hazmat team members should become thoroughly familiar with the documents used at each location for which they are responsible.

Storage and Use of Materials

When a release occurs, a response can be initiated much more quickly if documentation has been prepared showing the locations where hazardous materials are stored and used. Since the product must be identified before response decisions can be made, initial identification based on the location of the release will speed up the response. The documentation can be in the form of lists, maps, or computer records.

Contingency Plans

Contingency plans must be developed prior to the occurrence of an emergency. These plans can be consulted to take a lot of the guesswork out of hazard assessment. Contingency plans show step-by-step response actions, and are based on the storage and use locations of hazardous materials and the resources present at the facility and available through nearby assisting organizations. Specific steps which might be included in these plans are discussed in Chapter 7. In order to be effective, contingency plans must be thorough, updated as chemicals or conditions change at the site, and written well in advance of an incident.

Confinement Systems

Any location at which liquid hazardous materials are stored is required by law to have confinement systems in place which can confine either 10 percent of the total volume stored or the entire volume of the largest single storage container, whichever is greater. These systems should be checked frequently to be sure that they are in good order. If new bulk containers are added, the confinement system may need to be enlarged.

One of the largest drinking water pollution events in recent years happened because a breached diking system allowed almost a million gallons of diesel fuel from a collapsed above-ground tank to flow into the Monongahela River in Pennsylvania. The contaminant slug moved down the river, entered the Ohio River, and fouled drinking water supplies of over a million people in cities downstream.

Maps

Procurement of several types of maps should be a part of pre-incident hazard assessment. A topographic map is invaluable in predicting the land migration routes of flowing liquids. A transparent overlay can be drawn indicating the locations of large quantities of liquids on the site. A road map showing major highways and local roads can be used to determine routes of evacuation or additional help. A good medium-scale map can be used to draw predicted hazard zones for gas or vapor plume releases.

A site map is a useful adjunct to the storage and use documents. This map indicates the locations of named chemicals and can also show the locations of gates, communication systems, fire hydrants, and hazard document access. Figure 5-1 shows the map developed at one industrial facility.

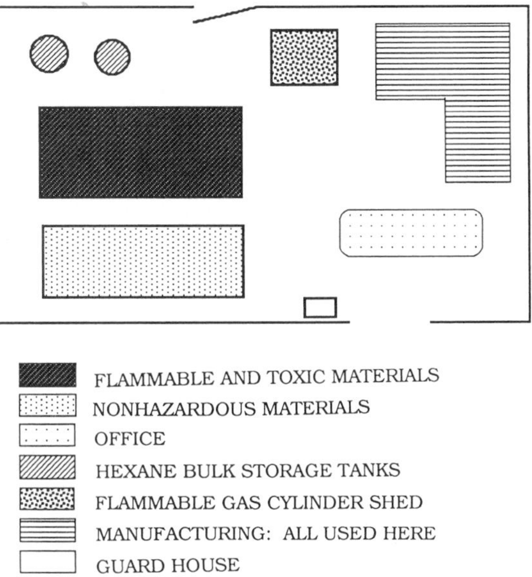

FIGURE 5-1. Storage and use map of a manufacturing facility. A detailed chemical list would be attached to the map.

INCIDENT PERIMETER SURVEY

The first responder on the scene will want to determine where, what, and how much material has been released, but avoid rushing in to become part of the problem rather than part of the solution. His initial survey around the safe perimeter of the release will attempt to answer the following questions.

Where is the Spill?

Exactly where is the released material coming from or, at least, where is the spilled material? The exact location of the leaking container may provide a valuable clue regarding its contents if a pre-incident map is available. Also, someone who works in that area can be contacted to find out what materials are stored or used there. In some cases, information can be gained from visible labels or markings.

Where is the Spill Headed?

Topography will determine the direction of liquid flow and may affect vapor movement by influencing air movement. For example, valleys may "funnel" wind currents, and ridges may block or divert the wind. If the release is a gas, vapor, or dust, wind speed and direction are important.

Temperature affects the volatility of a liquid and the rate at which the liquid will evaporate. High temperatures will also hasten loss of integrity of a container by expanding the chemical inside it. Rain will, of course, enhance runoff and disperse spilled materials more widely. Rain can also wash a gas plume from the air.

The ultimate destination of the spilled chemical will be determined by its chemical and physical properties. Predicting the dispersal of materials will be addressed more fully later in this chapter.

What is the Identity of the Leaking Material?

This important question may be answered by the initial perimeter survey if the storage and use map, or labeling on the breached container, names the contents of the container. In order to assess the risks inherent in the spill accurately and completely, the name of the leaking product must be known. A major portion of the risk assessment involves predicting the behavior of the product based on its chemical and physical properties. These are best obtained by looking up the product by name in one or several of the reference sources described in this chapter. Answers to these questions will help the responder accomplish Step 2 of the DECIDE process: Estimate likely harm without intervention.

What are the Hazards to Responders?

The final step to be taken before anyone enters the potential hazard zone is an evaluation of the risks the product presents to responders. Based on this evaluation, responders on the entry teams will protect themselves in a safe manner.

ENTRY AND ON-SITE SURVEY

If the perimeter survey is inconclusive or indicates the need for further evaluation of the hazard, a closer on-site survey may be deemed necessary. If the hazard has not been completely characterized, the maximum PPE should be worn and only experienced members of the team should enter.

The on-site survey is used to verify and confirm the findings of the perimeter survey; an entry plan setting out specific goals of the survey should be formulated. Key objectives of the survey are to monitor the air, observe the containers, and observe the released materials.

Monitor the Air

Prior to the selection of response action options and the choice of personal protection procedures and gear, intrinsically safe, portable instruments should be used to determine the presence of

- Toxic vapors or gases
- Oxygen-deficient or oxygen-enriched atmosphere
- Combustible or flammable gases or vapors
- Radioactive materials

Observe Containers

Often the types and sizes of containers offer clues about what may be in them (Table 5-2). For example, cylinders may contain liquids or gases under pressure, drums with bung tops are designed to hold liquids, and drums with removable tops are designed to hold solid materials.

The material from which the container is constructed can also provide information about the contents of the container. Since corrosive materials corrode carbon steel containers, they are shipped and stored in plastic

TABLE 5-2. Nonbulk Containers

Containers	Description	Possible Contents
Bags	Multiwall paper, plastic, paper lined with plastic	Solids (corrosive, flammable, poison, blasting agents, explosives)
Bottles (jars)	Glass or plastic with stopper or lid; hold up to several gallons	Any types of hazardous materials; solids or liquids
Carboys	Glass, plastic; usually inside cushioned boxes	Liquids; primarily used for corrosives
Cylinders	Metal; may be color coded, but this is not required by law, and therefore is not a reliable indicator	Pressurized gases
Pails	Metal, plastic; hold 1–5 gallons	Any types of hazardous materials; solids or liquids
Drums	Metal, fiberboard, plastic; hold 5–85 gallons;	Any types of hazardous materials; solids or liquids

containers or containers with plastic liners. Metals other than carbon steel, sometimes referred to as "exotic metals" because they are not commonly seen in a storage area, may indicate contents that are unusually hazardous. These containers may appear to be unusually shiny or different in some other way from ordinary carbon steel containers.

Bulk containers are found at most hazardous materials facilities and are built to specifications based on the products they are designed to hold. Safety features are included in the specifications. Figure 5-2 shows bulk containers which hold liquids. Materials in these tanks are stored at low or atmospheric pressure.

Some large-capacity storage containers maintain high pressure or very cold temperature to reduce the volume of the product by keeping it in the liquid state. If damaged, these containers may release a liquid or gas, depending on whether the point of release is above or below the liquid level; either may be hazardous due to the pressure or temperature of the released materials. These storage tanks are shown in Figure 5-3.

Hazards associated with materials transported in bulk containers can be roughly characterized by the configuration of the truck or rail car involved in an incident. DOT issues specifications for vehicles that are written to ensure safe transport of different kinds of hazardous materials.

Bulk transportation containers have an internal volume greater than 119 gallons for liquids, 882 pounds for solids, or a water capacity greater than 1,000 pounds for gases. Vehicles that function as bulk containers can be classified using factors such as container shape and fitting arrange-

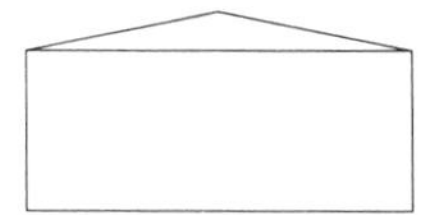

CONE ROOF TANK

Round with pitched or conical roof welded to the tank shell, with a weak seam or attachment designed to allow roof to separate in the event of internal explosion. Used primarily to store crude oil stocks. May contain flammable, combustible, or corrosive liquids.

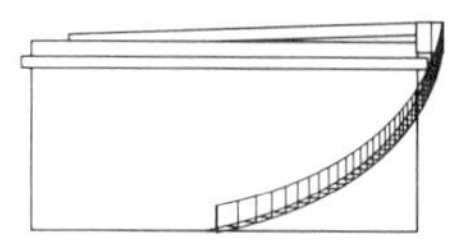

OPEN FLOATING ROOF TANK

Round with roof floating on surface of product, supported by pontoons or a double deck. Seal between roof rim and tank shell. Used to store low flash point liquids and crude oil. Weight of excess water or foam will sink the roof.

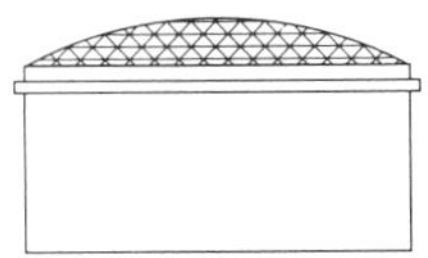

FLOATING ROOF TANK WITH DOME

Same as regular open floating roof tank, and stores same contents. Geodesic dome cover is added for protection from weather or to prevent vapor emissions.

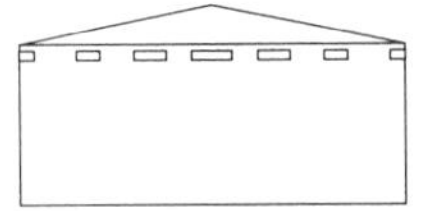

COVERED INTERNAL FLOATING ROOF TANK

Round with roof floating on surface of product. In addition, there is a pitched or conical roof. Used to store low flash point/high vapor pressure liquids. Vents at roof to shell joint allow for "breathing" during loading and unloading.

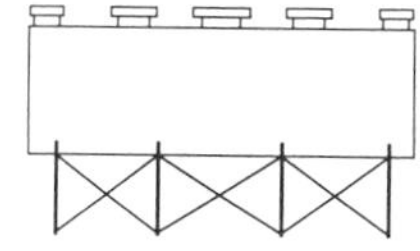

HORIZONTAL TANKS

Horizontal, cylindrical tank sitting on legs or blocks. Used to store flammable and combustible liquids, corrosives, poisons, and other hazardous materials.

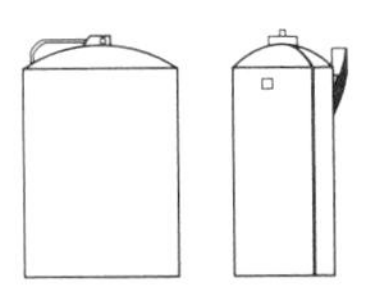

DOME ROOF TANKS

Vertical cylindrical tank with a dome shaped roof which is designed to fail in case of excess pressure. Operating pressure of 2.5 - 15 psi. Used to store flammable and combustible liquids, fertilizers, solvents, and other hazardous materials.

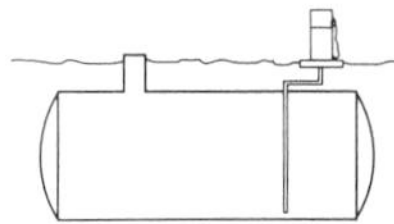

UNDERGROUND STORAGE TANK

Horizontal tank of steel, fiberglass, or coated steel. Must be protected against corrosion. Visible clues are vents, fill pipes. Most are found in retail service stations and contain petroleum.

FIGURE 5-2. Bulk liquid containers which store materials at low or ambient pressure.

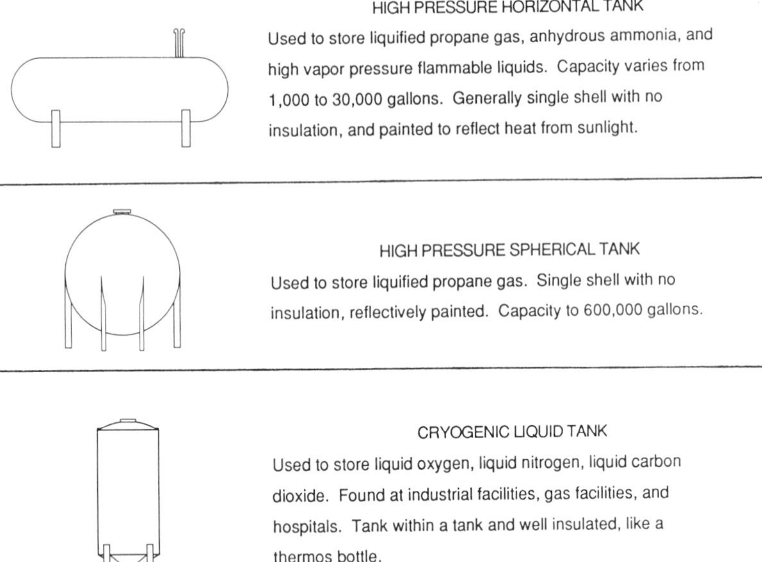

HIGH PRESSURE HORIZONTAL TANK
Used to store liquified propane gas, anhydrous ammonia, and high vapor pressure flammable liquids. Capacity varies from 1,000 to 30,000 gallons. Generally single shell with no insulation, and painted to reflect heat from sunlight.
HIGH PRESSURE SPHERICAL TANK
Used to store liquified propane gas. Single shell with no insulation, reflectively painted. Capacity to 600,000 gallons.
CRYOGENIC LIQUID TANK
Used to store liquid oxygen, liquid nitrogen, liquid carbon dioxide. Found at industrial facilities, gas facilities, and hospitals. Tank within a tank and well insulated, like a thermos bottle.

FIGURE 5-3. Bulk liquid containers which utilize high pressure or low temperature to liquefy contents.

ment. The classification of vehicles involved in an incident may provide initial clues to the nature of the materials involved (Figure 5-4). Further training will allow responders who expect to handle transportation incidents to become familiar with valves, piping, and emergency shutoff systems on bulk transport vehicles.

Containers should have labels and markings which indicate their contents. The DOT and NFPA labeling systems were presented in Chapter 4. In addition to hazard labels, the chemical name may be stenciled on the container, or there may be a label indicating appropriate emergency response action. On rail and motor carriers, the name and telephone number of the transporter may be visible.

All sizes and types of containers may be subjected to stress during an emergency. Look for containers that may be stressed by heat, if fire is a real or potential part of the incident, by mechanical factors such as metal damage, or by chemical stress like corrosion or chemical reaction. Stress can strain or deform a container beyond the limit of its ability to adapt to the stress, causing it to breach. It is vital to assess the integrity of a container before approaching it for any reason.

Observe Released Material

Observations of the released material may be helpful, especially if its identity is questionable. The physical state, color, visible vapor emission, bubbling, and other properties or behaviors may help to confirm the identity of a product.

Bubbling, smoking, and other visible behaviors may indicate that a chemical reaction between the spilled chemical and water, air, or another spilled chemical is occurring. If uncontrolled reactions are evident, responders should be wary and remain at a safe distance. Always be on the lookout for fire or other chemical reactions, as these will be important considerations for assessment of the hazard.

Sample unknown materials only if no other method exists to identify them. Do this only if personnel have been trained in safe sampling and rapid analysis can be made. A sample which has to be sent to a laboratory for analysis may take weeks to provide results, and obviously that is too long to wait in an emergency.

Kits are available including flow charts and materials which can be used to identify, or at least determine the chemical group of, spilled materials. The results are perhaps less accurate than those of a laboratory analysis, but they can be obtained in about 30 minutes. Users must be trained in the use of these kits. It is worth noting that opinions on the usefulness of these kits vary widely among people who have used them.

RESEARCHING IDENTIFIED MATERIALS

Once a material has been identified by name, the next step is to use a reference source to assess completely the risks it presents. References include books and other printed material, telephone hotlines, and computer databases.

Printed Materials

Material Safety Data Sheets

An employer is required by law to retain and make available to workers a Material Safety Data Sheet (MSDS) for each hazardous chemical to which workers may be exposed, and to send a copy of each, or a list of materials on site, to the Local Emergency Planning Committee or its designated emergency response agency and the state Emergency Response Commission. At a fixed facility, for example an electroplating shop, an MSDS for each chemical can be obtained when the chemical is ordered and filed for workers' use.

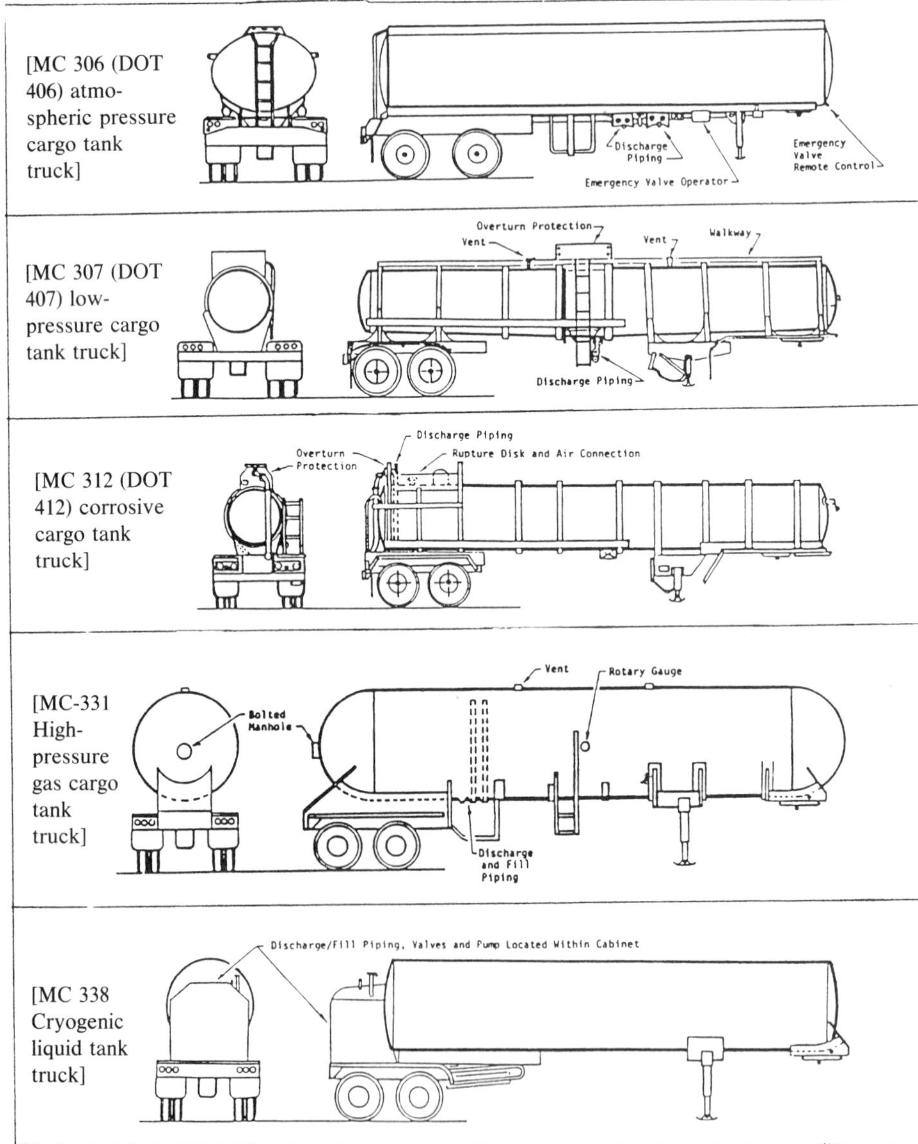

FIGURE 5-4. Bulk transportation containers. (Source: EPA)

Flammable and combustible liquids such as gasoline and diesel fuel, and class B poisons Capacity: 2,000 - 9,500 gallons	Oval or round rear cross section, pressure less than 3 psi. Usually single shell aluminum.
Products with vapor pressures not more than 40 psi at 70°F Flammable liquids and mild corrosives Capacity: 2,000 - 8,000 gallons	Circular cross section, pressures up to 25 psi. Double shell construction, usually steel
High density liquids and corrosives	Cylindrical cross section, narrow diameter, external ribs. Steel, stainless steel, or aluminum; often lined to resist degradation or reaction to contents.
Gases which are liquefied by pressure. Compressed gases and some very hazardous liquids	Cylindrical cross section, usually larger diameter than MC 312. Hemispherical heads; smooth surface without external ribs. Pressure 100 - 500 psi.
Cryogenic gases liquefied by refrigeration, such as liquid helium at -425°F	Cylindrical cross section, hemispherical heads, smooth surface. Container within a container, as a thermos bottle. Pressure 23.5 - 500 psi.

Although the format may differ, Material Safety Data Sheets are required to provide certain information. Examination will show that the sheet provides physical, chemical, and toxicological (health hazard) data which are useful to workers engaged in sampling, materials handling, spill control, and firefighting, and to persons responsible for choosing PPE. Table 5-3 shows an MSDS for 1,1,1-trichloroethane, a commonly used industrial solvent.

Name and Address
Material Safety Data Sheets will have the name and address of the manufacturer, importer, or other responsible party who can, if necessary, provide additional information about the chemical and appropriate emergency procedures.

Date Prepared
The date the information on the MSDS was finalized or approved should be shown in this section.

Material Identification
This section of the sheet indicates the name and synonyms of the material, as well as the trade names under which it may be provided. The Chemical Abstract Service number, a number that designates this chemical only, is noted.

Hazardous Ingredients/Identity Information
This section lists the various components of the material and, if established, the allowable exposure limits. The section is included for manufactured products which are mixtures. If a reaction-inhibiting chemical is present in the mixture in a concentration of less than 10 percent by volume, the manufacturer is not required to identify the inhibitor by name, although some sheets do list inhibitors. Some inhibitors are toxic.

Chemical Family/Formula
Many manufacturers include this information, although it is not provided on the MSDS in Table 5-3. The chemical family is the general class of the chemical, such as acid, solvent, halogenated hydrocarbon, organic amine or other class. For simple substances, the manufacturer may provide the chemical formula.

Physical/Chemical Characteristics
This section lists the chemical and physical properties of the substance, as determined by laboratory testing. Only those tests applicable to the product will be shown, and these can vary from substance to substance.

TABLE 5-3. Material Safety Data Sheet for 1,1,1-trichloroethane

Section I. Material Identification

MATERIAL NAME: 1,1,1-TRICHLOROETHANE
OTHER DESIGNATIONS: Methyl Chloroform, CCl_3CH_3, GE Material D5B79, CAS# 000 071 556
TRADE NAMES AND MANUFACTURER: BLACO-THANE (Baron-Blakeslee), CHLOROTHENE NU & NG (Dow), DOWLCENE WR (Dow), INHIBISOL (Penetone Corp.), TRI-ETHANE (PPG Ind. Inc.), TRITHENE (SRS, Inc.)

Section II. Ingredients and Hazards	%	Hazard Data
1,1,1-Trichloroethane[a]	>90	TLV 350 ppm[b]
Inhibitor	<10	Unknown
[a] High purity material is commercially available (DOWCLENE WR). Other commercial materials (Trade names, Sect. I) can contain up to 10% inhibitor and are designed for cold cleaning or vapor degreasing use or both (TRI-ETHANE).		*Human inhalation* LCLo 27 g/m³ for 10 minutes
[b] NIOSH has proposed a 10-hour TWA of 200 ppm with a 350-ppm ceiling concentration (15 minutes sampling time) and recently has recommended caution in use.		TCLo 920 ppm for 70 minutes (central nervous system effects)

Section III. Physical Data

Boiling point at 1 atm, deg F	ca 165[a]	Specific gravity, 25/25C	1.30–1.3
Vapor pressure at 20 C, mm Hg	100	Volatiles, %	ca. 100
Vapor density (Air = 1)	4.55	Evaporation rate (CCl_4 = 1)	1
Water solubility, g/100g H_2O	0.07 g	Molecular weight	133.41

Appearance and odor: Colorless liquid with a mild, ether-like odor which may be just perceptible (unfatigued) at about 100 ppm in air.

[a] Properties depend on the inhibitor and inhibitor level.

Section IV. Fire and Explosion Data			Lower	Upper
Flash Point and Method	Autoignition Temp.	Flammability Limits in Air		
None	—	(High energy ignition source at 25C) Vol. %.	8.0%	10.0%

This material is nearly nonflammable. High energy, such as electric arc, is needed for ignition, and the flame tends to go out when the ignition source is removed. Water for carbon dioxide, dry chemical, or foam may be used to fight fires.

Use self-contained or air-supplied breathing apparatus for protection against suffocating vapors and toxic and corrosive decomposition products.

TABLE 5-3. (Continued)

Section V. Reactivity Data

This material can be hydrolyzed by water to form hydrochloric acid and acetic acid. It will react with strong caustic, such as caustic soda or caustic potash to form flammable or explosive material.

It requires inhibitor content to prevent corrosion of metals; and when inhibitor is depleted, it can decompose rapidly by reaction with finely divided white metals, such as aluminum, magnesium, zinc, etc. (Do not use these metals for fabrications of storage containers for 1,1,1-trichloroethane.)

It will decompose at high temperature or under ultraviolet radiation to produce toxic and corrosive materials (phosgene and hydrogen chloride).

Section VI. Health Hazard Information	TLV 350 ppm or 1900 mg/m^3

Brief exposure at 800–1000 ppm causes mild eye irritation and a little loss of coordination due to the anesthetic properties of 1,1,1-trichloroethane. Skin contact can cause defatting and, when prolonged or repeated, can produce irritation and dermatitis. It can absorb through the skin. Eye contact can result in pain and irritation. This material is considered low in toxicity among the chlorinated hydrocarbons.

FIRST AID:

Eye contact: Flush eyes well with plenty of running water for 15 minutes.

Skin contact: Remove solvent-wet clothing promptly. Wash contact area with warm water and soap. Get medical attention for irritation.

Inhalation: Remove to fresh air. If needed, apply artificial respiration. Get medical assistance immediately. (*Note:* Advise physician not to use adrenalin.)

Ingestion: Get medical assistance! (If a physician is not immediately available and the amount swallowed was appreciable, give milk or water to drink and induce vomiting. Repeat several times. Estimated lethal dose for 150-lb man is 0.5 to 1 pint.)

PHYSICIAN: Avoid using sympathomimetic amines in treatment.

Section VII. Spill, Leak, and Disposal, Procedures

For small spills, mop, wipe, or soak up with absorbent material using rubber gloves. Evaporate outdoors or in an exhaust hood.

For large spills, inform safety personnel and evacuate area. Use protective equipment during cleanup (see Sect. VIII). Ventilate area. Contain liquid: pick up and place in closed metal containers. Do not allow to enter water supply sources.

DISPOSAL: Dispose of via a licensed waste solvent disposal company, or reclaim by filtration and distillation procedures.

Section VIII. Special Protection Information

Provide general and exhaust ventilation to meet TLV requirements. Gloves and apron (of neoprene, polyethylene, or polyvinyl alcohol) should be worn when needed to avoid skin contact. Remove solvent-wet clothing promptly. A safety shower should be available to use area.

Chemical goggles or a faceshield should be worn if splashing is possible. An eye wash station should be readily available if splashing is probable.

TABLE 5-3. (Continued)

In emergencies or nonroutine work use self-contained or air-supplied breathing apparatus for high or unknown vapor concentrations in air. NIOSH recommends use of a full facepiece respirator with an organic vapor cartridge or canister for limited time exposure below 1,000 ppm. (Full facepiece protection is not required below 500 ppm.)

Section IX. Special Precautions and Comments

Store in closed containers in a cool, well-ventilated area. Keep water-free. Monitor inhibitor level for vapor degreasing use. Use caution in cleaning operations involving white metal fines (see Sect. V). Trichloroethylene contamination may cause decomposition when aluminum is degreased.

Provide regular medical monitoring of those exposed to this material in the workplace. Preclude those with CNS, liver, or heart disease from exposure. Personnel using this solvent should avoid drinking alcoholic beverages shortly before, during, or soon after exposure.

Exposure of pregnant female rats to high levels of 1,1,1-trichloroethane may have caused birth defects in offspring.

DATA SOURCE(S) CODE: 1-8, 12, 19

Judgments as to the suitability of information herein for purchaser's purposes are necessarily purchaser's responsibility. Therefore, although reasonable care has been taken in the preparation of such information. General Electric Company extends no warranties, makes no representations and assumes no responsibility as to the accuracy or suitability of such information for application to purchaser's intended purposes or for consequences of its use.

Boiling Point: The temperature at which a liquid boils and rapidly changes to a vapor, generally at a pressure of 1 atmosphere (ambient pressure at sea level). In general, the lower the boiling point of a flammable liquid, the greater the fire hazard.

Melting Point: The temperature at which a solid substance changes to a liquid state. For mixtures, the melting range may be given.

Vapor Pressure: The pressure exerted on the inside of a container by the vapor in the space above the liquid. Most chemicals exert vapor pressure inside a container, even if the container is not artificially pressurized. Vapor pressure, described by a number, has several applications in assessing hazards.

Vapor pressure increases as temperature increases, as can be seen in Table 5-4. In case of fire, the vapor pressure inside intact containers may reach a point where the container ruptures violently, spewing the contents and releasing dangerous projectiles. Each liquid chemical has its own unique vapor pressure; this property is generally listed in reference sources.

Vapor pressure is an indicator of the volatility of the chemical. The higher the vapor pressure, the higher will be the rate of evaporation from

TABLE 5-4. Vapor Pressures of Some Liquids at Different Temperatures

	32°F	77°F	122°F	167°F
Water	4.6	23.8	92.5	300
Benzene	27.1	94.4	271	644
Methyl alcohol	29.7	122	404	1,126
Diethyl ether	185	470	1,325	2,680

a liquid release. The vapor pressure of 1,1,1-trichloroethane is 100 millimeters of mercury (mm Hg). This pressure indicates a relatively volatile chemical.

Vapor Density: The relative density or weight of a vapor or gas (with no air present) compared to an equal volume of air at ambient temperature. With air rated at 1.0, a density greater than 1.0 indicates a vapor or gas heavier than air; a density of less than 1.0 indicates that it is lighter than air. 1,1,1-trichloroethane, for example, produces vapors 4.55 times as heavy as air, and we would predict that these vapors would hover around spilled liquid, displacing the air and posing the threat of asphyxiation to unprotected responders.

Unfortunately, assessing the hazard presented by heavier-than-air vapors is not this simple, since it describes the density of the pure vapor which cannot exist in open air. A pure vapor would contain 1 million ppm of this product but this cannot happen in open air, even in a low-lying area, because of natural dilution by the air. For easy estimation of vapor migration, remember that the closer the density of the pure vapor is to 1.0, the more likely it is to become dilute and float away.

Solubility in Water: The percentage of a material that will dissolve in water at a given temperature. Solubility information can be useful in determining spill procedures and fire extinguishing agents and methods. An insoluble chemical like 1,1,1-trichloroethane can be expected to form a separate layer in water, its location dependent on its specific gravity.

Specific Gravity: The ratio of the mass of the product compared to the mass of an equal volume of water. This is an expression of the density of the product relative to the density of water. Insoluble materials with a specific gravity of less than 1.0 tend to float in or on water, while those with a specific gravity of more than 1.0 tend to sink in water. Most flammable liquids will float on water, which is an important consideration in firefighting. The specific gravity of 1,1,1-trichloroethane is 1.3. Since it is heavier than water (water = 1.0) it will sink in a container of water.

Appearance and Odor: A brief description of the material under normal room temperature and atmospheric pressure. Noting whether the appearance of an identified chemical matches its description provides an additional indication of the accuracy of the identification.

pH: The degree of acidity or alkalinity of a solution, with neutrality indicated as 7. The lower the number below 7, the more acidic the solution; the higher the number above 7, the more alkaline or basic the solution. Only aqueous (water) solutions have a pH. Since 1,1,1-trichloroethane is a hydrocarbon solvent, not a water solution, no pH is listed for it.

Fire and Explosion Hazard Data

This section describes factors that should be considered when assessing an incident involving fire or the potential for ignition of the chemical.

Flash Point: The flash point of a material is the lowest temperature at which vapor will be given off in sufficient quantity to form a flammable vapor-air mixture which will ignite in the presence of an ignition source. Since flash points vary with the test method, the method is shown. The Pensky-Martens Closed Cup Tester (PMCC) and the Setaflash Closed Cup Tester (SETA) are two of the more commonly used test methods.

Flammable or Explosive Limits: When flammable vapors are mixed with air in the proper proportions, the mixture can be ignited. The range of concentrations over which the ignition will occur is designated by the lower explosive limit (LEL) and the upper explosive limit (UEL). Below the LEL, the mixture is "too lean" to burn (without enough flammable vapors); above the UEL it is "too rich" to burn (without enough oxygen). Flammable limits are expressed as the percentage volume of vapor in air.

Extinguishing Media: The behavior of flammable or combustible chemicals when burning differs based on their physical and flammable characteristics. Therefore, the extinguishing medium must be selected for its ability to extinguish a fire without increasing the problems associated with the fire. Water, dry chemical, foam (AFFF, protein), CO_2, Halon 1211, and Halon 1301 are some commonly used extinguishing media.

Special Firefighting Procedures: General fire-fighting methods are not described, but special or "exception to the rule" procedures may be listed.

Unusual Fire and Explosion Hazards: Hazardous chemical reactions, changes in chemical composition, or by-products produced during fire or high heat conditions will be shown. Hazards associated with the application of extinguishing media will be shown if applicable.

Autoignition Temperature: The approximate lowest temperature at which a flammable vapor-air mixture will ignite spontaneously (i.e., without an ignition source such as spark or flame).

Reactivity Data

This section describes any tendency of the material to undergo a chemical change and release energy. Chemical reaction may produce undesirable effects such as temperature increase and formation of toxic or corrosive by-products. The Material Safety Data Sheet should describe these effects. Conditions which may cause a reaction, such as heating of the material or contact with other materials, should also be described.

Stability: An expression of the ability of the material to remain unchanged.

Incompatibility: An indication of the tendency of a material to react upon contact with other materials.

Hazardous Decomposition: An indication of the relative hazards associated with decomposition of the material. Notice that 1,1,1-trichloroethane may form phosgene gas in a fire. Phosgene is very toxic in small amounts; this explains the requirement for SCBA use in fighting a fire when the chemical is present.

Health Hazard Data

Routes of Entry: This section provides information on the ways the chemical may enter the body.

Health Hazards/Effects of Exposure: This section provides information on the health effects associated with overexposure. Both acute and chronic effects should be listed. In addition, toxicological information may be given.

Spill, Leak, and Disposal Procedures

Information is provided describing how to contain and handle the material properly in the event of spills or leaks that may damage the environment. This may include recommended cleanup materials, equipment, and PPE.

Waste Disposal: The manufacturer's recommended method for disposing of excess, spent, used, leaked, or spilled material.

Special Protection and Special Precautions

These sections will provide information regarding hazards unique to the material, as well as special measures for storage and/or handling which were not covered in other sections. Note that the MSDS is not always well organized: in the example, health hazard information is included in both of these sections.

Regulatory Status

Although this MSDS does not provide regulatory status information, newer ones have an addendum showing the SARA reporting requirements and indicating the SARA, CERCLA, and RCRA sections where regulations pertaining to the material are set forth.

NIOSH *Pocket Guide*

NIOSH publishes the *Pocket Guide to Chemical Hazards*. This document provides a great deal of information through which responder protection can be enhanced. In order to provide this information in a pocket-sized booklet, many specific symbols and abbreviations are used. In order to interpret these symbols and abbreviations, the glossary and codes located in the front of the *Guide* should be consulted by readers unfamiliar with this document.

Let us research the same chemical we reviewed in the previous section, 1,1,1-trichloroethane, using the *Pocket Guide to Chemical Hazards* (Table 5-5). Be warned that chemical names are highly specific and can be misleading if the reader is careless. For example, two chemical names which differ by only a single number of letter represent two different chemicals which may have widely differing hazardous properties.

Column I. Chemical, Formula, Etc.

Chemicals are listed alphabetically in column I. If numbers precede the name, as in the case of 1,1,1-trichloroethane, they are ignored in alphabetization. We find trichloroethane listed after tributyl phosphate and before trichloroethylene in column I. However, we see that the preceding numbers do not match: This chemical is 1,1,*2*-trichloroethane. How different can they be? The answer is, very different. The numbers describe the positions of the three ("tri") chlorine atoms on the ethane molecule, indicating whether they are all attached to the same carbon atom of the molecule (as they are in the 1,1,1- form) or whether two are attached to one carbon atom and one is attached to the other (as in the 1,1,2- form). This makes a significant difference in the chemical behavior of the molecule in this case, as one form is considerably more toxic than the other.

At this point, a careless reader may decide that the chemical is not listed in the guide, implying that it is not toxic. However, for safety's sake, a further search should be undertaken. From the MSDS we learned that 1,1,1-trichloroethane is called by another name, "methyl chloroform." The compound is listed by this name in the *Pocket Guide*. The new *Pocket Guide* (June 1990) contains a synonym index in the back, where 1,1,1-trichloroethane is listed and the reader is directed to methyl chloroform. Older editions do not have this index. We could also have learned the synonym by reading a label on a drum or consulting another reference, such as *Hawley's Condensed Chemical Dictionary*.

TABLE 5-5. ***NIOSH Pocket Guide*** **Information on 1,1,1-Trichloroethane, Listed by Its Synonym "Methyl Chloroform"**

Chemical Name Structure/Formula, CAS and RTECS Nos., and DOT ID and Guide Nos.	Synonyms, Trade Names, and Conversion Factors	Exposure Limits (TWA Unless Noted Otherwise)	IDLH	Physical Description	Chemical and Physical Properties: MW,BP,SOL FI.P,IP,Sp.Gr, Flammability	Chemical and Physical Properties: VP, FRZ UEL, LEL	Incompatibilities and Reactivities	Measurement Method
Methyl chloroform CH_3CCl_3 71-55-6 KJ2975000 2831 74	Chlorothene; 1,1,1-Trichloroethane; 1,1,1-Trichloroethane (stabilized) 1 ppm = 5.55 mg/m^3	NIOSH C 350 ppm (1900 mg/m^3) [15-min] OSHA 350 ppm (1900 mg/m^3) ST 450 ppm (2450 mg/m^3)	1000 ppm	Colorless liquid with a mild, chloroform-like odor.	MW: 133.4 BP: 165°F Sol: 0.4% FI.P: None IP: 11.00 eV Sp.Gr: 1.34 Noncombustible liquid; however the vapor will burn.	VP: 100 mm FRZ: −23°F UEL: 12.5% LEL: 7.5%	Strong caustics; strong oxidizers; chemically-active metals such as zinc, aluminum, magnesium powders, sodium & potassium; water [Note: Reacts slowly with water to form hydrochloric acid.]	Char; CS_2; GC/FID; III [#1003, Halogenated hydrocarbons]

Personal Protection and Sanitation	Recommendations For Respirator Selection—Maximum Concentration for use (MUC)	Health Hazards: Route	Symptoms	First Aid	Target Organs
Clothing: Repeat Goggles: Reason prob Wash: Prompt wet Change: N.R. Remove: Prompt non-imperv wet	NIOSH/OSHA 1000 ppm: SA*/SCBA* §: SCBAF:PD,PP/SAF:PD,PP:ASCBA Escape: GMFOV/SCBAE	Inh Ing Con	Head, lass, CNS depres, poor equi; irrit eyes; derm; card arrhy	Eye: Irr immed Skin: Soap wash prompt Breath: Resp support Swallow: Medical attention immed	Skin, CNS, CVS, eyes

Source: U.S. Department of Health and Human Services, 1990. *NIOSH Pocket Guide to Chemical Hazards*. Washington, DC: U.S. Government Printing Office, p. 148–149.

Under the name "methyl chloroform" we find the chemical formula CH_3CCl_3. The Cs represent carbon atoms. Note that there are two Cs, and therefore two carbon atoms: One has three hydrogen atoms attached (H_3), and the other attaches to three chlorine atoms (Cl_3). The two numbers listed below the chemical formula are the identification numbers given to 1,1,1-trichloroethane by the Chemical Abstract Service (CAS) and the *Registry of Toxic Effects of Chemical Substances* (*RTECS*) published by NIOSH. These numbers could be used to access information from reference sources in print or in a computer data base.

The two numbers at the bottom of column I are the DOT identification number (four digits) and the two-digit Emergency Guide number, which refers to a section in the *Emergency Response Guidebook*. These will be explained in a later section.

Column II. Synonyms

This column shows other names by which a chemical is known. Here we again see that methyl chloroform is also known as "1,1,1-trichloroethane," as well as some other synonyms.

Column III. Exposure Limits

In column III, the legal permissible exposure limit (PEL) is listed. This is the maximum airborne concentration of the chemical to which unprotected exposure is allowed. If higher atmospheric concentrations of the chemical are present in the work area, respiratory protection must be utilized to reduce the concentration in the inhaled air to, or below, the PEL. For methyl chloroform, the exposure limit recommended by NIOSH is also shown. Exposure limits established by other agencies may also be listed in this column. For example, several other chemicals listed on the page with methyl chloroform have exposure limits recommended by the American Council of Governmental Industrial Hygienists (ACGIH) in column III.

Column IV. IDLH Level

Column IV indicates concentrations of the chemical in air which are considered to be immediately dangerous to life and health (IDLH) and should never be inhaled. For some chemicals, the notation "Ca" is listed in this column. This designates a cancer-causing agent (carcinogen) and suggests that no exposure above the PEL should be permitted even though immediate death would not result. In fact, some scientists argue that there is no "permissible" safe exposure level to a carcinogen, so for carcinogens the IDLH levels are listed in brackets, indicating that they are thought to be hazardous at any level of exposure.

Column V. Physical Description

Column V provides a brief description of the appearance and odor of the substance. These descriptors can be used as clues for early identification or for confirmation after identification. Purposely sniffing a chemical to determine its odor should be avoided, as the detectable odor level may be higher than the safe breathing level.

Column VI. Chemical and Physical Properties

These are the same chemical and physical properties listed in the Material Safety Data Sheet. For a discussion of these properties, see the previous section of this chapter.

Column VII. Incompatibilities

Materials with which the chemical being researched may react are listed in column VII. The chemical should never be mixed with, or allowed to come into contact with, any material listed in this column. The resulting chemical reaction could lead to fire, explosion, or generation of a toxic gas or vapor. For example, methyl chloroform, when mixed with sodium hydroxide (a strong caustic), forms three reactants, one of which is hydrogen, a flammable and potentially explosive gas. This column is of special importance to emergency responders, who are likely to encounter the product outside of its container.

Column VIII. Measurement Method

Column VIII provides information on suggested sampling and analysis methods used to determine the atmospheric concentration of the chemical in the work area. The abbreviations used are listed in tables near the front of the *Guide*.

Column IX. Personal Protection and Sanitation

Column IX provides recommendations for preventing or minimizing exposure to the chemical being researched. Translations of the terms and abbreviations used in this column for methyl chloroform tell us that we should

- Wear protective clothing if we anticipate repeated or prolonged skin contact with the chemical
- Wear protective goggles if there is a reasonable probability that the chemical may contact the eyes

- Wash the skin promptly if the chemical gets on it
- Remove promptly any nonimpervious clothing which becomes wet with the chemical

Column X. Respirator Selection

Column X provides information on respiratory protection. The agency recommending respiratory protection is identified at the top of the column. The abbreviations used are explained in Table 3 near the front of the *Guide*. For methyl chloroform we see that NIOSH recommends the following:

Between 350 ppm (below which no protection is needed: see column III) and 1,000 ppm, wear supplied air (SA) or self contained breathing apparatus (SCBA), with eye protection (*).

In emergencies, or planned entry into unknown concentrations or IDLH conditions (§), use SCBA equipped with a full facepiece (F) and designed to operate in pressure demand (PD) or other positive pressure (PP) modes, or use a supplied-air respirator with a full facepiece and operated in pressure demand or another positive pressure mode and equipped with an auxiliary escape SCBA operated in a pressure-demand or other positive pressure mode.

For protection during escape from an area in which work has been taking place without a respirator, following a sudden increase in contaminant concentration, suitable respirators are a gas mask (GM) with full facepiece (F) equipped with organic vapor (OV) cartridges, or any appropriate escape-type SCBA.

Columns XI–XIV. Health Hazards

Columns XI through XIV provide information on potential adverse health effects resulting from chemical exposure.

Route of Entry

Column XI lists the routes of entry by which chemicals may enter the body. Abbreviations used in this column refer to:

- Inh: inhalation (breathing in)
- Ing: ingestion (swallowing)
- Con: contact (with skin or eyes)
- Abs: absorption (through the skin into the blood vessels and into internal body tissues and organs)

Symptoms

Symptoms that may result from chemical exposure are listed in column XII. Table 5 in the front of the NIOSH *Guide* explains the abbreviations used in this column. Methyl chloroform exposure may cause headache, lassitude (slowing down), poorly functioning central nervous system, poor equilibrium and balance, irritated eyes, dermatitis (skin irritation), and cardiac arrhythmia (irregular heartbeat).

First Aid
Actions which should be taken immediately following accidental exposure to chemicals are described in column XIII. The abbreviations used in this column are explained in Table 6 of the *Guide*. The following first aid procedures are recommended for exposure to methyl chloroform: Irrigate (wash) the eyes or skin to remove the chemical; give artificial respiration to someone who has inhaled it and is not breathing; and seek medical attention immediately for someone who has swallowed any of the chemical.

Target Organs
Column XIV lists the organs of the body most likely to be affected by chemical exposure. The abbreviations used are explained in Table 5 of the *Guide*. Target organs for methyl chloroform are the skin or eyes if splashed, and the central nervous system (primarily the brain) or the cardiovascular system (heart and blood vessels) if the chemical is inhaled, absorbed, or swallowed.

Emergency Response Guidebook

This *Guidebook* is published by DOT for use by firefighters, police, and emergency services personnel during the initial stages of response to hazardous materials incidents. It is a guide for immediate action and lists sources of information for decisions on further action and cleanup.

In the front of the book is a depiction of placards. If the name or ID number of the material cannot be determined, the placard on the truck can be matched with a guide number, using these pages, and the proper guide consulted for appropriate action.

The introduction to the *Guidebook* explains words and terms, discusses fire control agents, and makes suggestions for approaching a hazardous materials incident. The remaining pages are color-coded for quick use.

Yellow Pages: ID Number Listing
Containers such as tank trucks should be placarded with a four-digit DOT identification number intended to allow identification of its contents. These numbers are listed in the yellow pages in numerical order, along with the name of each material (Table 5-6). The guide numbers in this

TABLE 5-6. 1,1,1-Trichloroethane Listed by Its Four-Digit DOT Identification Number, 2831, in the Yellow Pages of the *Emergency Response Guidebook*

ID No.	Guide No.	Name of Material
2796	39	BATTERY FLUID, acid
2796	39	ELECTROLYTE, BATTERY FLUID, acid
2797	60	BATTERY FLUID, alkali
2797	60	BATTERY FLUID, alkali, with battery
2797	60	BATTERY FLUID, alkali, with electronic equipment of actuating device
2798	39	BENZENE PHOSPHORUS DICHLORIDE
2798	39	PHENYL PHOSPHORUS DICHLORIDE
2799	39	BENZENE PHOSPHORUS THIODICHLORIDE
2799	39	PHENYL PHOSPHORUS THIODICHLORIDE
2800	60	BATTERY, electric, storage wet, nonspillable
2800	60	BATTERY, wet, nonspillable (electric storage)
2801	60	DYE, n.o.s. (corrosive)
2801	60	DYE INTERMEDIATE, n.o.s. (corrosive)
2802	60	COPPER CHLORIDE
2803	60	GALLIUM, metal
2805	40	LITHIUM HYDRIDE, fused, solid
2806	37	LITHIUM NITRIDE
2809	60	MERCURY
2809	60	MERCURY METAL
2810	55	POISON B LIQUID, n.o.s.
2810	55	POISONOUS LIQUID, n.o.s. (Poison B)
2811	53	FLUE DUST, poisonous
2811	53	LEAD FLUORIDE
2811	53	POISONOUS SOLID, n.o.s.
2811	53	SELENIUM OXIDE
2812	60	SODIUM ALUMINATE, solid
2813	40	LITHIUM ACETYLIDE ETHYLENEDIAMINE COMPLEX
2813	40	SUBSTANCES, which, when in contact with water, omit flammable gases, n.o.s.
2813	40	WATER REACTIVE SOLID, n.o.s.
2814	24	ETIOLOGIC AGENT, n.o.s.
2814	24	INFECTIOUS SUBSTANCE, affecting humans
2815	60	AMINOETHYLPIPERAZINE
2817	60	AMMONIUM BIFLUORIDE, solution
2817	60	AMMONIUM HYDROGEN FLUORIDE SOLUTION
2818	60	AMMONIUM POLY-SULFIDE SOLUTION
2819	60	AMYL ACID PHOSPHATE
2820	60	BUTYRIC ACID
2821	55	PHENOL SOLUTION
2822	54	CHLOROPYRIDINE
2823	60	CROTONIC ACID
2825	68	DIISOPROPYL-ETHANOLAMINE
2826	59	ETHYL CHLOROTHIO-FORMATE
2829	60	CAPROIC ACID
2830	41	LITHIUM FERROSILICON
2831	74	METHYL CHLOROFORM
2831	74	TRICHLOROETHANE
2834	60	PHOSPHOROUS ACID (ortho)
2835	40	SODIUM ALUMINUM HYDRIDE
2837	60	SODIUM BISULFATE SOLUTION
2837	60	SODIUM HYDROGEN SULFATE SOLUTION
2838	26	VINYL BUTYRATE
2838	26	VINYL BUTYRATE, inhibited

Source: U.S. Department of Transportation. 1990. *Emergency Response Guidebook*. Washington, DC: U.S. Government Printing Office.

section refer the reader to a hazard and action guide in the orange pages. Materials that are highlighted in yellow in this section are also listed in the Table of Initial Isolation and Protective Action Distances.

Blue Pages: Material Name Listing

If the name of a chemical is known, this section can be used to look it up. The names are arranged in alphabetical order. Following the name of each chemical are the action guide number and chemical identification number (Table 5-7). Materials highlighted in blue in this section are listed in the Table of Initial Isolation and Protective Action Distances.

Orange Pages

Materials listed in the yellow and blue pages are grouped according to their expected chemical behavior during a spill or fire into hazard groups, with a generalized hazard and action guide included for each group. The guides can be read quickly, and give important facts regarding fire and explosion hazards, health hazards, and emergency actions recommended for small or large fires, small or large spills, and first aid treatment of exposure victims (Table 5-8).

Each of the guides recommends calling the Chemical Transportation Emergency Center (CHEMTREC) for information to aid in dealing with the problem. Help should also be available from the Local Emergency Planning Committee in each region, the local Civil Defense Office, or the National Response Center of the U.S. Coast Guard.

Green Pages (White in Older Editions)

The green pages of the *Guidebook* contain the Table of Initial Isolation and Protective Action Distances (Table 5-9). In this table are listed materials which are seriously hazardous by inhalation. 1,1,1-Trichloroethane is not listed here. The table is useful for the first 30 minutes of an accident involving volatile hazardous liquids or gases which are not on fire. After 30 minutes, or sooner on a hot day, the vapors or gases will move beyond these distances.

The first responder should begin at the source of the release and protect or remove unprotected persons from the dangerous area, remaining alert for wind shift or other changes in the situation.

Isolation distances are delineated for a circular area and are listed for both small and large spills or potential spills. Protective action proceeds following isolation, with direction and measurements dependent on wind direction. Protective action may include evacuation, or protection of people inside buildings (occasionally even in vehicles) by instructing them to close windows, shut off air systems, and stay away from windows.

TABLE 5-7. 1,1,1-Trichloroethane Listed by Name in the Blue Pages of the *Emergency Response Guidebook*

Name of Material	Guide No.	ID No.	Name of Material	Guide No.	ID No.
TRIAZINE PESTICIDE, liquid, flammable, poisonous, n.o.s.	28	2764	TRIFLUOROCHLORO-ETHYLENE	17	1082
TRIAZINE PESTICIDE, liquid, poisonous, flammable, n.o.s.	28	2997	TRIFLUOROCHLORO-ETHYLENE, inhibited	17	1082
TRIAZINE PESTICIDE, liquid, poisonous, n.o.s.	55	2998	TRIFLUOROCHLORO-METHANE	12	1022
TRIAZINE PESTICIDE, solid, poisonous, n.o.s.	55	2763	TRIFLUOROETHANE, compressed	22	2035
TRI(1-AZIRIDINYL) PHOSPHINE OXIDE	55	2501	TRIFLUOROMETHANE	12	1984
TRIBUTYL ALUMINUM	40	1930	TRIFLUOROMETHANE, refrigerated liquid (cryogenic liquid)	21	3136
TRIBUTYLAMINE	68	2542	TRIFLUOROMETHANE and CHLOROTRIFLUORO-METHANE MIXTURE	12	1078
TRICHLORFON	55	2783	TRIFLUOROMETHANE and CHLOROTRIFLUORO-METHANE MIXTURE	12	2599
TRICHLOROACETIC ACID	59	1839	2-TRIFLUOROMETHYL-ANILINE	55	2942
TRICHLOROACETIC ACID SOLUTION	59	2564	3-TRIFLUOROMETHYL-ANILINE	55	2948
TRICHOROACETYL CHLO-RIDE	59	2442	3-TRIFLUORO-METHYLPHENYL-ISOCYANATE	55	9268
TRICHLOROBENZENE, liquid	54	2321	TRIISOBUTYL ALUMINUM	40	1930
TRICHLOROBUTENE	54	2322	TRIISOBUTYLENE	27	2324
TRICHLOROETHANE	74	2831	TRIISOCYANATOISO-CYANURATE of ISO-PHORONEDIISO-CYANATE, 70% solution	26	2906
TRICHLOROETHYLENE	74	1710	TRIISOPROPYL BORATE	26	2616
TRICHLOROISOCYANURIC ACID, dry	45	2468	TRIMETHOXYSILANE	57	9269
TRICHLOROPHENOL	53	2020	TRIMETHYLACETYL CHLORIDE	29	2438
2,4,5-TRICHLORO-PHENOXYACETIC ACID	55	2765	TRIMETHYL ALUMINUM	40	1103
2,4,5-TRICHLORO-PHENOXYPROPIONIC ACID	55	2765	TRIMETHYLAMINE, anhy-drous	19	1083
TRICHLOROSILANE	38	1295	TRIMETHYLAMINE, aque-ous solution	29	1297
TRICHLORO-S-TRIAZINE-TRIONE, dry	45	2468	TRIMETHYLBENZENE	26	2325
TRICRESYLPHOSPHATE	55	2574	TRIMETHYLBORATE	26	2416
TRIETHYLAMINE	68	1296	TRIMETHYLCHLORO-SILANE	29	1298
TRIETHYLENE TETRAMINE	60	2259	TRIMETHYLCYLCO-HEXYLAMINE	29	2326
TRIETHYL PHOSPHITE	26	2323			
TRIFLUOROACETIC ACID	60	2699			
TRIFLUOROACETYL CHLORIDE	16	3057			
TRIFLUOROAMINE OXIDE	20	9271			

Source: U.S. Department of Transportation. 1990. *Emergency Response Guidebook*. Washington, DC: U.S. Government Printing Office.

TABLE 5-8. Action Guide 74 for a Group of Chemicals, Including 1,1,1-Trichloroethane, From the *Emergency Response Guidebook*

Guide 74 ERG90

POTENTIAL HAZARDS

HEALTH HAZARDS

Vapors may cause dizziness or suffocation.
Exposure in an enclosed area may be very harmful.
Contact may irritate or burn skin and eyes.
Fire may produce irritating or poisonous gases.
Runoff from fire control or dilution water may cause pollution.

FIRE OR EXPLOSION

Some of these materials may burn, but none of them ignites readily.
Most vapors heavier than air.
Air/vapor mixtures **may explode** when ignited.
Container may explode in heat of fire.

EMERGENCY ACTION

Keep unnecessary people away; isolate hazard area and deny entry.
Stay upwind, out of low areas, and ventilate closed spaces before entering.
Positive pressure self-contained breathing apparatus (SCBA) and structural firefighters' protective clothing will provide limited protection.
Isolate for ½ mile in all directions if tank, rail car or tank truck is involved in fire.
Remove and isolate contaminated clothing at the site.
CALL CHEMTREC AT 1-800-424-9300 FOR EMERGENCY ASSISTANCE.
If water pollution occurs, notify the appropriate authorities.

FIRE

Small Fires: Dry chemical or CO2.
Large Fires: Water spray, fog or regular foam.
Apply cooling water to sides of containers that are exposed to flames until well after fire is out. Stay away from ends of tanks.

SPILL OR LEAK

Shut off ignition sources; no flares, smoking or flames in hazard area.
Stop leak if you can do it without risk.
Small Liquid Spills: Take up with sand, earth or other noncombustible absorbent material.
Large Spills: Dike far ahead of liquid spill for later disposal.

FIRST AID

Move victim to fresh air and call emergency medical care; if not breathing, give artificial respiration; if breathing is difficult, give oxygen.
In case of contact with material, immediately flush eyes with running water for at least 15 minutes. Wash skin with soap and water.
Remove and isolate contaminated clothing and shoes at the site.
Use first aid treatment according to the nature of the injury.

Source: U.S. Department of Transportation. 1990. *Emergency Response Guidebook*. Washington, DC: U.S. Government Printing Office.

TABLE 5-9. A Portion of the Table of Initial Isolation and Protective Action Distances from the *Emergency Response Guidebook*. 1,1,1-Trichloroethane is Not Listed in This Table Since it is Not Considered an Extremely Dangerous Inhalation Hazard.

USE THIS TABLE WHEN THE MATERIAL IS *NOT* ON FIRE.		SMALL SPILLS (Look or spill from a small package or small leak from a large package.)		LARGE SPILLS (Look or spill from a large package or spill from many small packages.)	
ID No.	NAME OF MATERIAL	First ISOLATE in all directions- (Feet)	Then, PROTECT those persons in the DOWNWIND direction (Miles)	First, ISOLATE in all directions- (Feet)	Then, PROTECT those persons in the DOWNWIND direction- (Miles)
2810	POISON B LIQUID, n.o.s. (When "Inhalation Hazard" is on a package or shipping paper.)	1200	4	1500	5
2810	POISONOUS LIQUID, n.o.s. (Poison B) (When "Inhalation Hazard" is on a package or shipping paper.)	1200	4	1500	5
2845	ETHYL PHOSPHONOUS DICHLORIDE, anhydrous	150	0.2	150	0.2
2845	METHYL PHOSPHONOUS DICHLORIDE	150	0.2	150	0.2
2901	BROMINE CHLORIDE	1500	5	1500	5
2922	CORROSIVE LIQUID, poisonous, n.o.s. (When "Inhalation Hazard" is on a package or shipping paper.)	1500	5	1500	5
2927	POISONOUS LIQUID, corrosive, n.o.s. (When "Inhalation Hazard" is on a package or shipping paper.)	1500	5	1500	5
2929	POISONOUS LIQUID, flammable, n.o.s. (When "Inhalation Hazard" is on a package or shipping paper.)	1500	5	1500	5
3023	tert-OCTYL MERCAPTAN	150	0.2	150	0.2
3070	DICHLORODIFLUOROMETHANE and ETHYLENE OXIDE MIXTURE	150	0.2	150	0.8
3070	ETHYLENE OXIDE and DICHLORODIFLUOROMETHANE MIXTURE, with not more than 12% ETHYLENE OXIDE	150	0.2	150	0.8

Source: U.S. Department of Transportation. 1990. *Emergency Response Guidebook*. Washington, DC: U.S. Government Printing Office.

Emergency Action Guides, American Association of Railroads

This document presents fold-out pages in a looseleaf binder for commonly shipped hazardous materials. The information given is a combination of all the information discussed previously. The book is available with paper pages, or with pages made of Tyvek, which is more durable than paper. Purchase of the manual entitles the buyer to buy update pages as knowledge or regulations change.

CHRIS Manual, U.S. Coast Guard

The *Chemical Hazard Response Information System (CHRIS) Manual* is another large document containing information regarding hazards and emergency response procedures for hazardous materials. Materials are listed on separate sheets which are inserted into a binder. Manual 2, Hazardous Chemical Data, is the most relevant of four volumes to emergency response.

Telephone Hotlines

A number of 24-hour manned telephone services will provide hazard information in an emergency.

CHEMTREC is supported by the Chemical Manufacturers Association. CHEMTREC can also provide conference calls with manufacturers of the chemical in question. The number is 1-800-424-9300.

The Agency for Toxic Substance and Disease Registry (ATSDR), an agency of the U.S. Public Health Service, staffs a telephone line which provides information about toxicological effects of chemicals. On-site assistance can also be requested. The number is 1-404-639-0615.

The U.S. Coast Guard National Response Center (NRC), along with DOT, supplies assistance in identification, technical information, and initial response action. Spills which are above the RQ should also be reported to them. The number is 1-800-424-8802.

The Centers for Disease Control (CDC) of the U.S. Public Health Service is available for assistance in handling infectious disease related incidents. The number is 1-404-633-5313.

The Association of American Railroads Hazmats System provides assistance in handling railroad emergencies. The number is 1-202-639-2222.

Computer Networks and Data Bases

Many hazmat units equip their mobile or stationary facilities with computers which can be used to access data bases or interactive agencies for help during an incident. If the computer has a printer attached, some of these resources can provide a hard copy of information. See Chapter 17 for more information about computer resources.

PREDICTING DISPERSAL OF HAZARDOUS MATERIALS

Using reference documents to identify the health, fire, reactivity, corrosivity, and radioactivity hazards of chemicals involved in an emergency response incident is a basic step in assessing the risks posed by the spilled chemicals. The chemical and physical properties of hazardous materials determine these risks, as well as how the material will move (disperse) in land, water, and air following a release.

Predicting the dispersal of released hazardous materials is an important part of risk assessment and must be done in order to "estimate likely harm without intervention." Dispersal is highly dependent on the physical state of the material during and immediately after the release.

Solids

Solids will not disperse very far unless they change state, as a refrigerated solid will do if it reaches its melting point, a subliming solid will do if it reaches its sublimation temperature, or a water-soluble solid will do if it spills into water or encounters rain. Subliming solids will release vapors which disperse as a gas, and dissolved solids will behave as liquids. It is possible for a finely divided solid to be dispersed as a dust by wind, but this is not usually a serious concern in a hazardous materials incident.

The only events likely to generate dangerous dispersal of hazardous solids are those in which a large quantity of fine powdered solid is released into air in a chemical reaction, especially a fire. Flammable solids such as white or red phosphorus, which burn readily in air, have been known to produce large clouds of hazardous dusts which float on air and collect on surfaces when they settle to earth. The pure metal changes form in the fire, becoming an oxide or other compound, but may retain hazardous properties and can damage skin or respiratory tract linings.

Liquids

The majority of industrial chemicals are liquids, and liquids are the most commonly used and transported hazardous materials. Materials in the liquid state may be released on land or into water; both types of spills may allow volatile liquids to evaporate into air.

Liquids Released on Land

Liquids may percolate through soil, and will flow downhill or down sewers and storm drains. Because they flow, liquids may contaminate groundwater, surface water, and soil. Liquids may also vaporize, thereby contaminating the atmosphere (Figure 5-5).

Whether released liquid chemicals move readily through soil depends upon the properties of the chemical and the properties of the soil onto which it spills. An emergency responder is not expected to determine the transport of liquids through soil, but should instead attempt to prevent chemicals from remaining on soil long enough to penetrate.

Liquids which are stored or shipped at cold temperatures may change to the vapor state when released. If the boiling point of the liquid is below ambient (outside the container) temperature, the liquid will begin to boil upon release. Even if the boiling point is above ambient temperature, the rate of evaporation will increase as the liquid warms.

Materials which are liquefied due to pressure in the containers in which they are stored and shipped will behave differently upon release, depending on whether the point of release is below or above the surface of the liquid. Liquids under pressure will jet from a hole, and large amounts will probably vaporize. If the container is punctured in the space above the

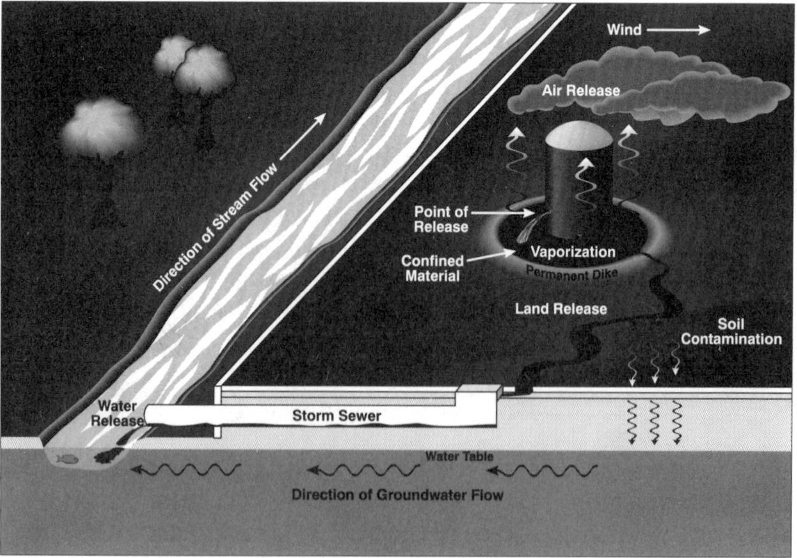

FIGURE 5-5. Liquids released on land may contaminate air, soil, groundwater, and surface water.

liquid, the gas will vent at high velocity, with the velocity slowing as the pressure drops. Examples of materials shipped and stored in pressurized containers are liquid anhydrous ammonia, ethylene, chlorine, vinyl chloride, and liquid petroleum gas (LPG) or propane.

Liquids Released into Water

In predicting the dispersion of liquids which are spilled or flow into water, several characteristics of the chemical must be known. Boiling point, vapor pressure, solubility, and specific gravity will determine where the liquids will go.

The boiling point and vapor pressure of the material will determine whether part of the material will boil off or vaporize from water. A container dropped into water will release its contents if it ruptures. If the water temperature is above the chemical's boiling temperature, the chemical will vaporize rapidly as it enters the water from the breached container, and the vapors will bubble up through the water into air. Volatilization may occur well below the boiling point, but at a much slower rate.

The solubility of the material will determine whether or not it will dissolve and the rate of dissolution. Solubility is quantified by number in some references, where the portion which will dissolve in water is given as a decimal fraction or percentage. Some references use qualitative terms such as "insoluble" or "partially soluble." Designation of insolubility should not be taken as absolute, as many so-called insoluble materials will partially dissolve after enough time has elapsed. Both liquids and vapors can dissolve in water. In bodies of water which are turbulent, the mixing action may increase the rate of dissolution. Also, insoluble materials may become physically mixed with water if sufficient turbulence or wave action is present.

The insoluble portion of a liquid will sink or float in still water, depending on the density or specific gravity of the material. Specific gravity and density do not describe the same measurement but are sometimes reported interchangeably, since the numerical values are the same at certain temperatures. These terms can be defined as follows. Density is the mass of a substance divided by the volume it fills; specific gravity is the density of a substance divided by the density of water.

Water is assigned a specific gravity of 1.0 at normal temperature, so a liquid with specific gravity greater than 1.0 will tend to sink and a liquid with specific gravity less than 1.0 will tend to float. Materials with specific gravity close to 1.0 may be dispersed throughout the water column. In turbulent water, mixing will occur and decrease or break up the floating layer. The "slick" will be less visible, but it is still present and is still dispersing.

When a flammable liquid enters a sewer or storm drain, it can pose a fire or explosion hazard. A water-reactive liquid may produce the same hazards. Toxic liquid contaminants pose a threat to all life forms exposed to them on land or in water.

The effects of boiling point, vapor pressure, specific gravity, and solubility on the dispersion of chemicals in water can be seen in Table 5-10.

Gases and Vapors

Gases and vapors may be released into the air by direct venting or from volatile solids or liquids. Factors an emergency responder must consider when attempting to predict the pathway of gas and vapor releases are

- Travel distance and direction
- The duration of the discharge
- Mixing of the gas or vapor with air (dilution)

Travel distance can be predicted by multiplying current wind speed by time. Direction is assessed by observation of a wind sock. In an emergency, anything which will blow from a pole may be used but will not give a totally accurate wind direction.

"Duration" refers to the length of time the release continues. The two basic types of discharges are considered to be instantaneous (a puff or cloud) and continuous (a plume).

A puff or cloud will move in the direction of the wind at a similar speed. It will begin to mix with air at its edges. The cloud will become larger as it mixes with air, but the concentration of the contaminant at the edges will go down as a result of the mixing. As the cloud mixes and grows, the concentration of the contaminant toward the center will decrease, and at some point will drop below the level of concern (LOC), the concentration considered to be safe for exposure of the general public. LOC and the formula for its calculation are discussed in Chapter 14.

Ground-level contaminant concentration in an instantaneous discharge decreases as the cloud moves away from the point of release. Figure 5-6 illustrates the decrease. At the same time, the cloud expands and the hazard zone grows larger, as shown in Figure 5-7.

In a continuously released plume, the concentration downwind will be relatively constant for a period of time approximately equal to the duration of the release. A period of time will elapse before the leading edge of the plume reaches a certain location, as is true for an instantaneous emission, and a similar length of time will pass after the release is ended before the trailing edge passes the same location.

TABLE 5-10. Predicting Dispersal of Chemicals in Water

Boiling Point	Vapor Press.	Spec. Grav.	Solubility	Expected Behavior in Water
Below ambient	Very high	Any	Insoluble	All liquid will rapidly boil from surface of water.
Below ambient	Very high	Below that of water	Low or partial	Most liquid will rapidly boil off but some will dissolve. Some of the dissolved liquid will evaporate.
Below ambient	Very high	Any	High	At least 50% will rapidly boil off; the rest will dissolve. Some of the dissolved liquid will evaporate later.
Above ambient	Any	Below that of water	Insoluble	Liquid will float, forming a slick. Those with significant vapor pressure will evaporate over time.
Above ambient	Any	Below that of water	Low or partial	Liquid will float but will dissolve over time. Those with significant vapor pressure may simultaneously evaporate over time.
Above ambient	Any	Below that of water	High	Liquids will rapidly dissolve in water up to the limit (if any) of their solubility. Some evaporation may take place over time if vapor pressure is significant.
Above ambient	Any	Near that of water	Insoluble	Difficult to assess. May float on or beneath surface or disperse through the water column. Some evaporation may occur from surface over time if vapor pressure is significant.
Above ambient	Any	Near that of water	Low or partial	Will behave as above at first and eventually dissolve. Some evaporation may take place over time.
Above ambient	Any	Any	High	Will rapidly dissolve up to the limit (if any) of their solubility. Some evaporation may take place over time.
Above ambient	Any	Above that of water	Insoluble	Will sink to the bottom and stay there. May collect in deep water pockets.
Above ambient	Any	Above that of water	Low or partial	Will sink to the bottom and then dissolve over time.
Above ambient	Any	Above that of water	High	Will rapidly dissolve up to the limit (if any) of their solubility. Some evaporation may take place from the surface over time if vapor pressure is significant.

Source: Federal Emergency Management Agency, the U.S. Department of Transportation, and the U.S. Environmental Protection Agency 19. *Handbook of Chemical Analysis Procedures*. Washington, DC: U.S. Government Printing Office, p. 3-8, 3-9.

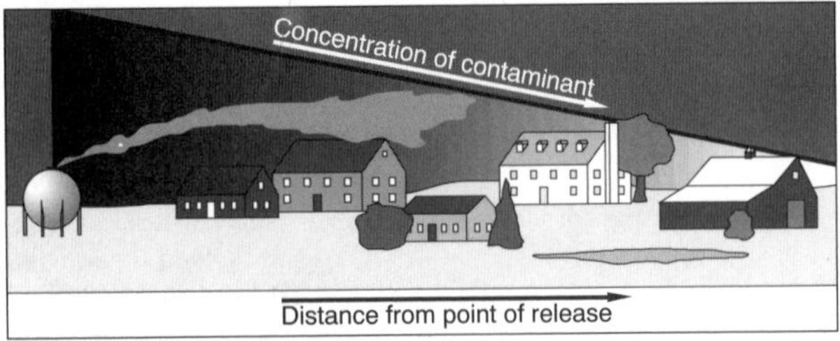

FIGURE 5-6. Plume concentration decreases as the plume mixes with air.

Someone within the ICS will have to predict the hazard zone, the area in which the concentration of contaminant is too high to allow people to remain there unprotected. Selecting the hazard zone depends on the three factors listed above (wind distance and direction, release duration, and plume dilution), the predicted chemical concentration and the size of the cloud, and the following four factors:

- Amount of the discharge. Generally speaking, the larger the release, the longer and wider the zone.
- Prevailing atmospheric conditions. These conditions include temperature, strength of sunlight, and wind speed and direction.
- Gas or vapor density relative to air.
- Height of the discharge.

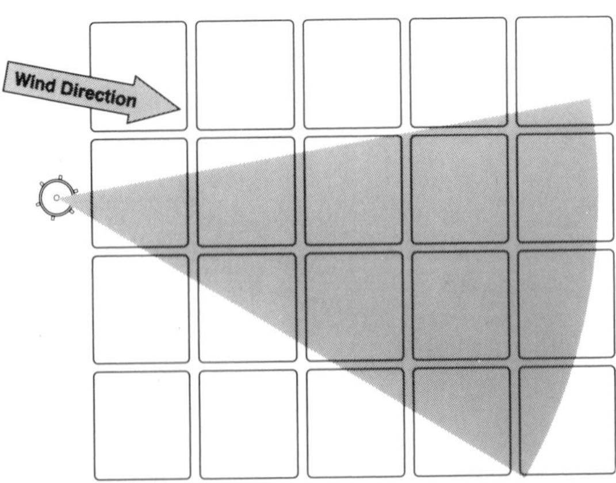

FIGURE 5-7. Plume hazard area enlarges as the plume moves downwind.

It must be remembered that terrain or buildings affect wind direction and atmospheric stability, and that wind speed and direction often change. Changing or "meandering" winds can greatly enlarge the hazard zone. A "vulnerable zone" should be selected and kept in mind in case winds change. In a location subject to changing winds from all directions, the vulnerable zone may be a circle around the point of release.

Lighter-than-air plumes will float up if atmospheric conditions do not prevent this, and soon will dilute or disperse to a nonhazardous concentration. Heavy plumes will hug the ground as they move and provide higher concentrations, possible breathing-air displacement, and the potential for ignition. Typical ignition sources are vehicles, spark-generating friction, pilot lights, and cigarettes.

If the point of release is on an elevated portion of a tank or from a pressure relief pipe of a stack, the ground-level concentration of the contaminant may be affected. In this case, it will make a great deal of difference whether the relative vapor density is higher, lower, or the same as the density of air.

Hazard zone size and direction can be calculated based on the factors listed. Computer programs, including LOTUS and ARCHIE, will also do this.

SUMMARY

The dangers of actually or potentially released hazardous materials must be assessed before personnel enter the area and before response action decisions are made. The DECIDE process will help organize decision making in an emergency. Planning for a possible spill will greatly enhance the ability of a hazmat team to respond to the incident quickly, effectively, and, most important, safely. There is a wide variety of sources of help that can be utilized in planning for and responding to an emergency involving hazardous materials. The release and route of migration of these materials can be predicted accurately if emergency responders have good training, equipment, and resources.

REFERENCES

Andrews, L.P., editor. 1990. *Worker Protection during Hazardous Waste Remediation*. New York: Van Nostrand Reinhold.

Bureau of Explosives, Association of American Railroads. 1990. *Emergency Action Guide*. Washington, DC: Association of American Railroads.

Cashman, J.R. April 1991. *Hazardous Materials Newsletter*, p. 8.

Federal Emergency Management Agency, U.S. Department of Transportation,

U.S. Environmental Protection Agency. *Handbook of Chemical Analysis Procedures*. Washington, DC: U.S. Government Printing Office.

——— 1985. *GATX Tank Car Manual*, 5th ed. Chicago: General American Transportation Corporation.

Germann, R. 1988. Oil spill on the Monongahela: as the story unfolded. *EPA Journal*, April 1988.

Henry, M.E., editor. 1989. *Hazardous Materials Response Handbook*. Quincy, MA: National Fire Protection Association.

Klaussen, C.D., M.O. Amdur, and J. Doull, editors. 1986. *Casarett and Doull's Toxicology*, 3rd ed. New York: Macmillan Publishing Co.

Noll, G.G., M.S. Hildebrand, and J.G. Yvorra. 1988. *Hazardous Materials—Managing the Incident*. Stillwater, OK: Fire Protection Publications.

Sax, N.I., and R.J. Lewis, Sr. 1987. *Hawley's Condensed Chemical Dictionary*, 11th ed. New York: Van Nostrand Reinhold.

Tinsley, I.J. 1979. *Chemical Concepts in Pollutant Behavior*. New York: John Wiley & Sons.

U.S. Department of Health and Human Services. 1990. *NIOSH Pocket Guide to Chemical Hazards*. Washington, DC: U.S. Government Printing Office.

U.S. Department of Transportation. 1990. *Emergency Response Guidebook*. Washington, DC: U.S. Government Printing Office.

U.S. Department of Transportation and U.S. Coast Guard. 1984. *Chemical Hazard Response Information Systems*, Vol II. Washington, DC: U.S. Government Printing Office.

Whitten, K.W., and K.D. Gailey. 1984. *General Chemistry with Qualitative Analysis*, 2nd ed. Philadelphia: WB Saunders Publishing Co.

Windholz, M., S. Budavari, L.Y. Stroumtsos, and M.N. Fertig, editors. 1976. *The Merck Index*. Rahway, NJ: Merck & Co.

6

Site Control

D. Alan Veasey, M.A. Ed., and
Kenneth W. Oldfield, M.S.P.H.

A chemical spill occurred when a delivery hose ruptured while a 10,000-gallon railcar of chlorosulfonic acid was being off-loaded at an industrial facility in Birmingham, Alabama. Approximately 200 gallons of the acid spilled onto the ground before the release was terminated by closing the railcar's off-loading valve. The spilled acid generated a dense acidic cloud which threatened plant personnel and the off-site population. Injuries occurred due to chemical exposure and an automobile accident resulting from the release. Company employees made the required notifications and initiated the emergency response sequence. Site control was established by creating an isolation perimeter and evacuating areas of potential exposure. Evacuation of 250 off-site residents was required. Access to the scene was restricted to personnel directly involved in the response operation, and control zones were established on site. Facility emergency response personnel began confining and neutralizing the spilled chlorosulfonic acid using sand, soda ash, and lime. Although access to the scene was adequately controlled around the site's isolation perimeter, site security was threatened by a helicopter operated by a local television news team. In attempting to film the incident, the helicopter made a direct approach to the scene. This created the potential for exposure of media personnel in the helicopter. Also, turbulence from the helicopter's propeller could have pushed contaminants from the acidic cloud into the designated cold (support) zone at the scene, thereby contaminating response personnel and equipment located in that zone. Personnel responsible for coordinating the response operation contacted Federal Aviation Administration officials at the local airport in order to restrict air space over the area of the incident. The helicopter was ordered to remain a safe distance from the area of release, and control of the incident scene was restored.

By their nature, hazardous materials incidents, such as the chlorosulfonic acid release in Birmingham, tend to be out of control. Initial responders to an incident may find a great deal of disorder and confusion in and around the area affected by a release. For example, site personnel, some of whom require medical treatment, may be in the process of fleeing areas immediately affected by the release. Injured personnel remaining in the hazardous area may require rescue. If the incident involves an ongoing, large-quantity release, the migrating hazardous material may represent a significant threat to personnel at the facility, to the environment, and (in some cases) to the public.

Emergency situations can be worsened by uncontrolled actions taken in response to the incident. For example, facility personnel, in attempting impromptu rescues of injured co-workers, may become additional casualties. Also, curious members of the public who are attracted to the event may wander into hazardous areas or get in the way of responders. As the incident described above demonstrates, members of the press may be an even greater problem than the public. Uncontrolled actions of responders may serve to worsen the incident, for example, by tracking contaminants out of hazardous areas and into uncontaminated areas.

In order to minimize the harm resulting from a hazmat incident, responders must establish control of the incident scene as soon as possible and maintain that control throughout the response operation. Properly utilized site control procedures will isolate people from hazards related to the incident and allow an orderly, efficient, and safe response operation.

OBJECTIVES

- Know the objectives of incident site control
- Understand the importance of the ICS as a vehicle for site control enforcement
- Be familiar with concepts related to site control, including zoning, access control, evacuation, and protection in place
- Understand the vital role of communications in site control

OBJECTIVES OF SITE CONTROL

The very nature of the hazardous materials incident makes site control both difficult and necessary to establish and maintain. In order for site control measures to be effective, they must successfully address a variety of objectives.

To begin with, site control measures must be designed to minimize chaos and to provide direction and efficiency to the response operation. Site control procedures should allow for accountability so that the location and status of all personnel and equipment on site are known at all times during the response operation. Successful site control requires pre-emergency planning (see Chapters 7 and 8) and strict adherence to an ICS that coordinates and controls the efforts of all personnel and agencies on site (see Chapter 3).

Site control measures must effectively isolate the incident scene to prevent harm to response personnel, facility employees, and members of the public such as well-intentioned volunteers, media crews, and curious bystanders. Only personnel directly involved in the response should be present at the scene of the incident. The presence of unnecessary people and equipment will only add to the confusion of the incident and increase the likelihood of needless injury. Response personnel not directly involved in activities in the hazard area should be kept at a safe distance to minimize the effects of unexpected events, such as explosions or sudden large-quantity releases that may enlarge the hazard area.

Site control activities must also prevent or minimize contamination of response personnel and the tracking of hazardous materials. The uncontrolled movement of personnel and equipment through contaminated areas will result in the tracking of contaminants into clean areas, thus endangering the health of others and increasing the property damage and mitigation costs. Site control procedures must prevent or minimize contact between contaminated and uncontaminated personnel and equipment. As seen in the incident described at the beginning of the chapter, this aspect of site control may be complicated by unexpected problems such as helicopters and changing wind direction.

ENFORCEMENT OF SITE CONTROL

The importance of enforcing site control procedures cannot be overemphasized. In an emergency response operation, all boundaries and security measures must be strictly adhered to, without regard to an individual's status or position. Persons not involved in the emergency response must be kept outside the site perimeter. These include onlookers, media representatives, and plant workers not involved in the response operation. Law enforcement or plant security personnel are generally better trained and equipped for crowd control and site security than hazmat team members. Law enforcement officers are also recognized off site as having authority to control access to an incident scene.

It is an unfortunate reality that in addition to the initial victims of a hazardous materials incident, others may subsequently become victims through their own actions. The latter group may include hazmat team members and civilians alike. For example, would-be rescuers, acting on impulse rather than in accordance with standard operating procedures (SOPs) may become additional casualties of an incident. While rescue of injured personnel from hazardous areas should be the first priority considered in initiating a response operation, rescues should be undertaken only after it has been determined that they can be performed without undue risk to responders.

The appropriate mechanism for controlling the activities of response personnel on site is the ICS. It is critical that all activities be coordinated and controlled through this system to prevent response personnel or other involved parties from acting independently and endangering the health and safety of themselves and others. Utilization of a preestablished ICS in managing the incident should promote adherence to SOPs and a highly organized response (see Chapters 3 and 7).

ISOLATION

The best way to protect people and property from the potential harm of a hazardous material release is to separate them from the released material. This is the purpose of isolation techniques, which may range from using flagging tape to mark off a small area around a leaking drum to complete evacuation of an entire industrial facility and the surrounding residential area. Isolation allows response personnel to plan and conduct their activities without having to perform unnecessary rescue operations.

Access Control

The first responder's initial actions should be to control access to the hazard area and to establish an isolation perimeter. The area located within the perimeter can then be secured in preparation for hazard mitigation operations. Securing occupied areas may involve evacuation or protection in place, as described in the following section. Isolation of the incident scene allows responders an unimpeded field of operations and provides protection to people in case the incident suddenly worsens.

The isolation perimeter can be considered the outer boundary of the site or incident scene (see Figures 6-1, 6-2, and 6-3). The distance from the point of release to the incident perimeter may be measured in feet or in miles, depending on the specifics of the incident. Two opposing considerations will affect the decision on how much ground to include within the initial perimeter.

On the one hand, it is desirable to make the isolation area as large as possible. It is easier to reduce the isolation area than to expand it as the response operation progresses. As crowds and traffic increase around a site, expansion may involve much more than simply moving barriers. Also, the unpredictable nature of hazmat incidents makes a large buffer zone desirable.

On the other hand, an isolation perimeter requires valuable manpower to patrol. Inadequate perimeter control that allows unauthorized entry to the site may result in a breakdown of site control. A smaller perimeter that is secure may therefore be preferable to a larger perimeter that is not well controlled.

In establishing the isolation area, it may be possible to utilize applicable geographic and physical barriers such as walls, fences, and bodies of water that will reduce the need for perimeter patrol (see Figures 6-2 and 6-3). It is important to gain control of access points such as doors, gates, bridges, and intersections as soon as possible. If physical barriers are not available, use barrier tape, rope, sawhorses, traffic cones, or whatever is available to mark the perimeter.

In planning and initiating site control procedures, response personnel should be aware that ensuring unimpeded access by response personnel or equipment from off site is just as important as preventing unauthorized entry. Provisions must be made to control traffic and keep vital roads, intersections, and access ways open. Coordination of activities with law enforcement agencies during preemergency planning should address this concern.

The concept of isolation also applies to incidents inside buildings or structures. Response personnel should initially gain control of entry points (such as doors) to the building and deny unnecessary access; then work can begin to secure the building as necessary. As in outdoor incidents, trying to isolate a larger area than can be adequately controlled may lead to a breakdown of site control.

Evacuation and Protection in Place

For spills or releases of solids or liquids with low evaporation rates, outlining and controlling the area of hazard may be a relatively simple process. However, for releases of gases or highly volatile liquids, the contaminant may travel in gas or vapor form over a much greater distance (see Chapter 5). For this reason, the potential hazard area may include inhabited areas within the plant or off the facility's property. In this case, a decision must be made about how to protect local inhabitants. If the individuals are outdoors and the likelihood that the air concentrations will reach a dangerous level is high, complete evacuation of the area is certainly appropriate. Evacuation should be performed in a rapid and orderly fashion by personnel responsible for site control.

Evacuation can be a difficult and time-consuming activity. Preparations must be made to move people out of the affected area to a safe location. If the evacuation area extends beyond the plant and into nearby residential or commercial areas, evacuation must be coordinated through local law enforcement agencies and provisions must be made for temporary shelter and food. Evacuation plans should be carefully considered during emergency response planning.

Under certain circumstances, it may not be practical or even safe to evacuate all people from a structure within the evacuation area. In the workplace setting, industrial processes cannot always be quickly shut down and abandoned without causing serious consequences in other areas of the plant. In the community, institutions such as hospitals, detention facilities, and nursing homes present serious problems for evacuation. In some instances, a vapor cloud may be migrating too rapidly to allow complete evacuation of a building. Once a vapor cloud has surrounded a building, safe evacuation may not be possible.

In these cases, it may be appropriate to use protection-in-place procedures. These procedures involve sealing off a structure so as to minimize the movement of air contaminants into it. Simply closing doors and windows and turning off heating, air conditioning, or ventilation systems can greatly reduce the level of contamination within a building. Protection-in-place procedures within the industrial facility may involve placement of appropriate protective equipment in vital process control areas to allow time for proper shutdown.

The feasibility of using protection-in-place procedures must be assessed during preemergency planning. The decision to use this approach during an incident should be made only after careful consideration of the type and concentration of contaminant involved (see Chapter 5).

ZONING

Once an isolation perimeter has been established, the area within it can be subdivided into control zones with distinct lines of demarcation (as shown in Figures 6-1, 6-2, 6-3, and 6-4). Zoning is a very useful concept for establishing and maintaining site control for a response operation. Site zones should be plotted on site maps based on information gathered during assessment of an incident and used in planning and conducting response operations. Zones are established based on factors such as the following:

- Type and degree of hazard (such as contaminant's identity and concentration, presence of flammable vapors, etc.) in a given area
- Type and level of PPE required for safe entry into a given area
- Type of response or response-related activities carried out in a given area

Though various names are used to describe them, three zones are almost universally recognized at hazardous materials incidents—the hot zone, the warm zone, and the cold zone.

Figure 6-1 shows a typical site zoning layout. This figure illustrates the zones and related features discussed below.

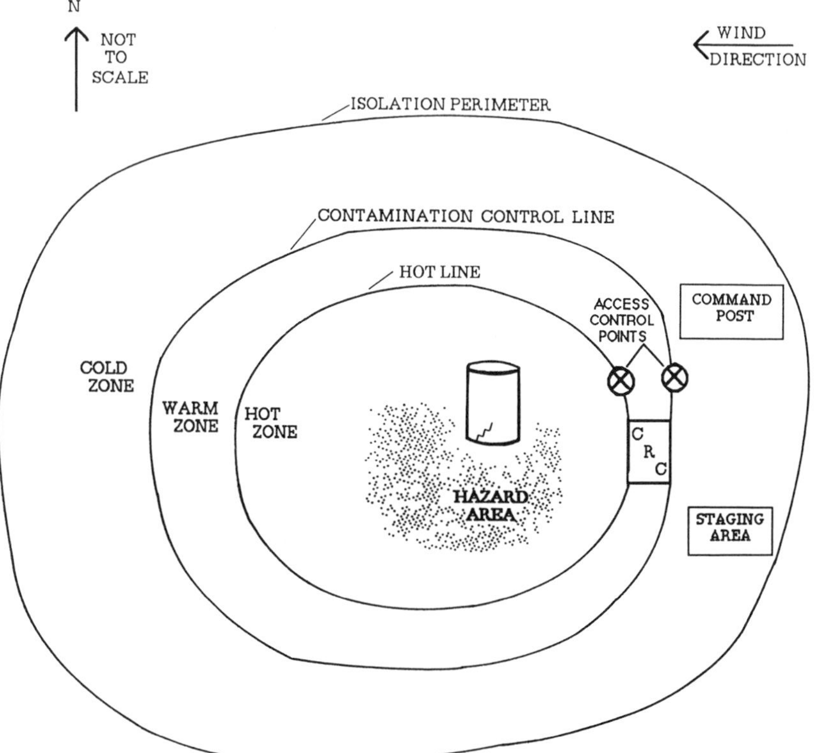

FIGURE 6-1. Control zones and related features.

Hot Zone

The hot zone (or exclusion zone) contains the actual hazard area (see Figure 6-1). This includes the location of the release and any areas to which hazardous substances have migrated or are likely to migrate in hazardous concentrations. The hot zone is the area where primary response operations are carried out in order to mitigate the incident. This is the most hazardous location on site, and entry requires the use of appropriate protective gear and close adherence to emergency response SOPs. Subareas may be specified within the hot zone if different levels of protective gear are required for different tasks and/or work areas within the zone (Figure 6-2).

The hot line marks the outer boundary of the hot zone (see Figures 6-1, 6-2, and 6-3). At any given location, the hot line must be far enough from the point of release to prevent exposure of personnel outside the hot zone to unsafe concentrations of hazardous materials released within the zone.

Location of the hot line should be determined based on hazard and risk assessment (see Chapter 5) of the current site conditions. Any predictable events which may expand the boundaries during response activities should also be considered. Considerations for establishing the hotline include:

- The identity and characteristics of the released material
- The location of the point of release
- The status of the release (i.e., whether the leak is continuing or has stopped)
- The total quantity of material expected to be released as leakage progresses (e.g., based on the size of the container, the location of the breach on the container, etc.)
- The current location of released material and the expected migration pattern as the incident progresses
- The presence and extent of airborne hazards, including oxygen-deficient or enriched atmospheres; combustible or flammable atmospheres; toxic gases, vapors, or particulates at harmful concentrations; and ionizing radiation

Obviously, direct-reading hazard detection equipment can be invaluable in establishing the hot line's location (see Chapter 15).

Geographic and man-made features must be considered in setting the hot line. Site topography may control the route of migration that a released product will follow. Also, there may be existing structures or natural features that can be utilized in confinement operations to limit contam-

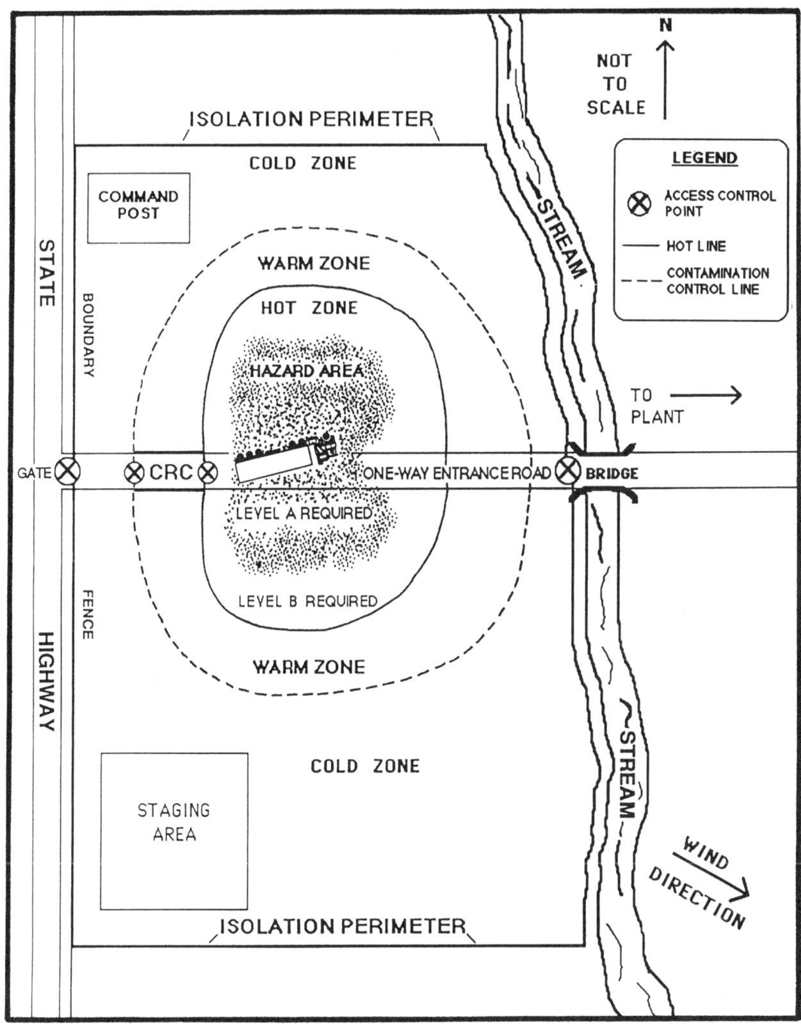

FIGURE 6-2. Use of existing geographic and man-made features in site control.

inant movement. If so, these locations should be included within the hot zone. If fire fighting with water may be involved during incident mitigation, the hot zone must include provisions for containing contaminated runoff. In some instances, it may be possible to adopt existing fence lines, walls, or other active barriers as the hot line in order to promote adherence to site control procedures (Figure 6-3).

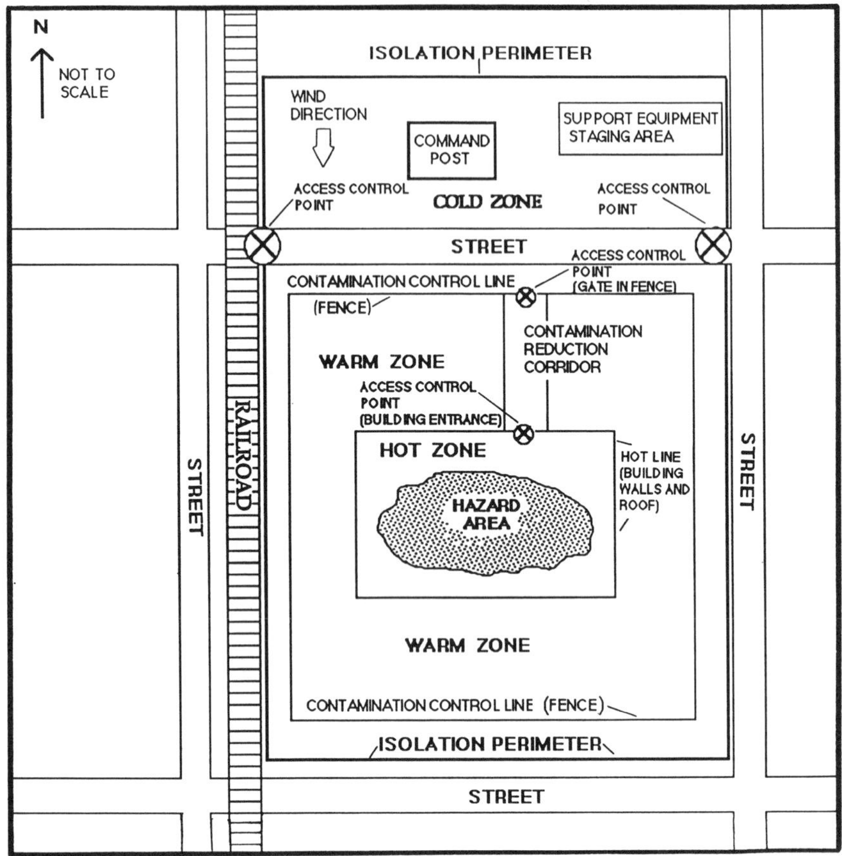

FIGURE 6-3. Site control for an incident involving a release inside a building.

The hot zone must include enough room for mitigation activities to take place and must be large enough to provide protection for on-scene personnel outside the zone in the event of an explosion, a fire, or an unexpected release during response activities. However, the distance responders must travel from the access control point to the hazard area must also be considered.

The potential effects of changing meteorological conditions such as wind, rain, or high temperatures on contaminant behavior and migration are important considerations in establishing the hot zone. Changes in conditions such as temperature and wind speed and direction may directly affect the generation and dispersion of airborne contaminants (see Chapter 5). A wind sock should be used for accurate determination of wind direction during an incident.

Warm Zone

The warm zone, or contamination reduction zone (CRZ), is located beyond the hot line and serves as a buffer zone between the hot zone and the cold zone or uncontaminated area of the site (see Figures 6-1, 6-2, and 6-3). The warm zone provides an extra margin of safety from the primary hazards of the incident for support and CP personnel located in the cold zone.

Decontamination activities are performed within the contamination reduction corridor (CRC), which is a subpart of the warm zone (see Figures 6-1, 6-2, and 6-3). Equipment needed to support the primary response operation (such as spare air cylinders, tools, sorbents, fire-fighting equipment, first-aid supplies, etc.) may be staged within the warm zone. As with the hot zone, traffic to and from the warm zone, must be controlled to prevent the spread of contamination. The contamination control line marks the outer boundary of the warm zone. Contaminated materials should never be transported beyond this line.

The CRC is located upwind from the hot zone within the warm zone, as shown in Figures 6-1, 6-2, and 6-3. This is the area where decontamination operations are performed on personnel, tools, and equipment exiting the hot zone prior to entry into clean areas of the site (see Chapter 10). For heavy equipment (such as loaders or backhoes) used in the hot zone, a separate decontamination corridor should be designated and equipped. Personnel performing decontamination operations in the CRC must use prescribed protective gear, depending on the type and degree of hazard present.

If the site is properly zoned and site control is maintained, no contaminants should be present in hazardous concentrations outside the hot zone, except for contaminants which are transported into the CRC by personnel undergoing decontamination. However, this may not always hold true. If an incident suddenly worsens or wind direction changes abruptly, parts of the warm zone may become dangerously contaminated, necessitating enlargement of the hot zone to incorporate these areas.

Cold Zone

The area of an incident scene located beyond the contamination control line is the cold zone or support zone (see Figures 6-1, 6-2, and 6-3). Command functions and supporting operations are carried out here. The cold zone should remain free of contamination so that no chemical protective equipment is required for response personnel working in this area. The CP and staging areas for equipment and personnel are located in this zone, and first aid/medical procedures, administrative operations, and various other supporting operations are performed here.

The placement of support facilities and functions within the cold zone should take these factors into consideration:

- Wind direction—support activities should always be located upwind of the hazard area
- Topography—support activities should be located uphill of the hazard area, if possible
- Space required—while support functions are interrelated and should be in good communication, crowding will cause confusion
- Accessibility—necessary vehicles and personnel must be able to reach the cold zone readily
- Distances—the cold zone should be located far enough away to provide a sufficient buffer from hazardous activities, but no further than necessary, so that excessive travel will not be required for entry into the hot zone

Access Control Points

Access control points are the only locations at which it is permissible to cross zone boundaries (see Figures 6-1, 6-2, and 6-3). This allows inbound and outbound responders to be checked in and out of specific zones. It is recommended that separate control points be used for inbound and outbound personnel. Access control points allow the use of required PPE and decontamination procedures to be verified. Also, the locations of responders can be closely monitored in the event that rescue is required.

Other Zoning Considerations

While the idealized zoning layouts shown in Figures 6-1, 6-2, and 6-3 may be appropriate, with some modification, for most hazardous materials incidents, there will be incidents where the geography or site conditions will require the use of judgment and creativity in establishing zones. For instance, a release of a low-volatility chemical in the loading dock area of a building may not require a warm zone around the entire area (see Figure 6-4). Rather, the access to the dock from the building area could be controlled and doors and air vents sealed to create a hot zone. A CRC

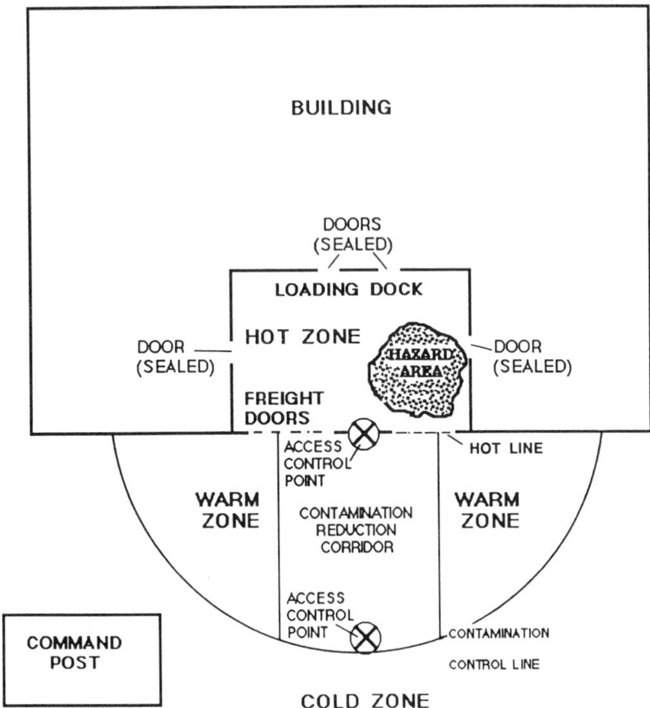

FIGURE 6-4. Zoning for a loading dock incident.

could be set up outside the dock. As long as the objectives of each zone are met, an appropriate zoning plan may take any form.

COMMUNICATION

Communication is the key to maintaining control during a response operation. Few things will throw an incident into chaos faster than a breakdown in communication. Ideally, all site activities should be visually monitored by command staff personnel. However, this is not always possible. Communication must be maintained between response personnel on site throughout the incident. Additionally, communication must be available between the CP and off-site agencies and resources.

Large industrial facilities usually have internal radio communication in place. During emergency response activities, radio channels should be dedicated solely to the use of the IC and team or sector officers. When entry teams are in the hot zone, a dedicated channel should be used solely for communication between them and their team leader. In all cases, unnecessary radio communication should be eliminated.

Communication procedures should be established during preparation of the facility's ERP. Backup communication procedures such as horn blast, bell, or hand signals should be developed and learned by all employees for use in the event of primary communication failure. The signals should be brief, limited in number, distinct from ordinary signals, and rehearsed regularly to be effective. Table 6-1 gives examples of emergency communication procedures.

Establishing SOPs before an incident, and clearly stating entry objectives in preentry briefings, can reduce the need for communication during an entry. However, if the entry team will be required to work out of sight of the ICS officer directing the team, radio check or other procedures should be set up to confirm communication at regular intervals. If communication cannot be confirmed during an entry, it may be appropriate to consider that the team members are in trouble and the backup team may need to enter. SOPs should address procedures to be followed in the event that communication equipment fails during an entry.

If outside agencies or multiple departments are involved in the response, the IC must ensure that a unified communication system exists among all parties. This may include telephones with conference calling

TABLE 6-1. Emergency Communication Procedures

Devices[a] and Signals	Example
Two-way radio	Established code words
Noisemakers, including:	
Bell	One long blast: Evacuate area by nearest emergency exit
Compressed air horn	Two short blasts: Localized problem (not dangerous to workers)
Megaphone	
Siren	Two long blasts: all clear
Whistle	
Visual signal, including:	
Hand signals	Hand clutching throat: Out of air/can't breathe
Whole body movements	Hands on top of head: Need assistance
Sign board	Thumbs up: OK/I'm all right/I understand
Flag	Thumbs down: No/negative
Lights	Grip partner's wrist or both hands around waist: Leave area immediately

[a] All devices and equipment used in the hot zone must be intrinsically safe and not capable of sparking if flammable atmospheres are possible.

Source: NIOSH, 1985.

capabilities, dedicated radio channels, or even dedicated runners/messengers. An ICS similar to that described in Chapter 3 should facilitate communication among officers and with the IC.

SUMMARY

Site control procedures range from very simple to very complex based on the severity and complexity of the incident involved. In many cases, isolation procedures may be the most effective action needed to prevent or limit casualties resulting from an incident. The use of effective site control procedures serves not only to prevent the tracking of contaminants and cross-contamination, but also to provide for accountability of personnel and equipment and to promote an orderly, well-organized response operation. The concept of site control is therefore very important for use in responding to hazardous material emergencies.

REFERENCES

Henry, M.F., ed. 1989. *Hazardous Materials Response Handbook*. Quincy, MA: National Fire Protection Association.

International Fire Service Training Association. 1984. *HazMat Response Team Leak and Spill Guide*. Stillwater, OK: Fire Protection Publications.

International Fire Service Training Association. 1988. *Hazardous Materials for First Responders*. Stillwater, OK: Fire Protection Publications.

NIOSH/OSHA/USCG/EPA 1985. *Occupational Safety and Health Guidance Manual for Hazardous Waste Site Activities*. NIOSH Publication No. 85-115. Washington, DC. U.S. Government Printing Office.

Noll, G.G., M.S. Hildebrand, and J.G. Yvorra. 1988. *Hazardous Materials: Managing the Incident*. Stillwater, OK: Fire Protection Publications.

7

SOPs and Termination

John E. Backensto and Lori P. Andrews, P.E.

There are many guides or procedures for performing the wide variety of tasks associated with responding to emergencies involving hazardous material releases. These may be administrative, technical, or managerial. All of these procedures are intended to provide uniform instructions for accomplishing a specific task. In addition to other types of procedures, standard operating safety procedures (SOPs) and termination procedures are needed. The purpose of this chapter is to provide an awareness of these plans and directives.

An example of a incident that illustrates the importance of developing and using effective SOPs occurred several years ago in a semirural area of Kentucky. The site was a natural gas distribution facility. At 11 o'clock one weekday morning, an explosion occurred at the facility, killing the entire crew. The crew included the maintenance personnel, an engineer, a computer specialist, and the plant manager. The fire department arrived to find the site devastated, with no apparent survivors. Since the SOPs for the facility were all tied to the computer system, which did not survive the explosion, no one knew exactly what to do to shut down the system. After several hours of searching for an individual with knowledge of the facility, a night watchman was brought in to assist in locating the shutoff valves. If any possibility that this type of accident could occur, then an SOP should have been developed to notify the other shift personnel for assistance. The several hours of delay could have been crucial in preventing further mitigation of the gas.

OBJECTIVES

- Have an understanding of relevant SOPs
- Understand the importance of following uniform SOPs
- Know where to find the generic SOPs
- Know the differences between termination procedures
- Know the job descriptions of hazmat responders
- Know what signs constitutes an emergency
- Know when to initiate and terminate a response

STANDARD OPERATING PROCEDURES

Emergency response personnel generally rely on verbal safety instructions and use existing SOPs until, time permitting, the plan can be modified to fit the site-specific application. OSHA Standard 29 CFR 1910.120 states that the first responder at the operations level, at a minimum, shall have an understanding of relevant SOPs. Federal regulations require facilities that utilize hazardous materials to provide Local Emergency Planning Committees (LEPCs) with the information needed to prepare and maintain emergency plans. Those facilities that store quantities of extremely hazardous substances (EHSs) in excess of designated threshold planning quantities (TPQs) are required to appoint a facility emergency coordinator to assist the LEPC in its planning efforts.

Other facilities may also be required by State Emergency Response Committees (SERCs) to participate in the planning process under Title III of SARA. In addition, hazardous material technicians and specialists are required to meet all requirements of the preceding levels; therefore, each responder must have a working knowledge of all relevant SOPs applicable to an emergency response.

Considerations

A major consideration in responding to an accidental release of hazardous materials/waste is the health and safety of the response personnel. Planning involves having an in-place protocol or SOP for handling various types of incidents. This should include the roles and responsibilities of responders, available equipment, evacuation procedures, and decontamination capabilities, among others. A good plan will include accurate information such as the identity, location, and characteristics of hazardous materials, related processes, and all response capabilities available. Maintaining an up-to-date plan is of prime importance. When corrections, additions, or changes are made, they should be recorded in a simple bookkeeping style so that all plan users will be aware that they are using a current plan. See Figure 7-1 for a detailed view of the SOP process. Not only must a variety of technical tasks be conducted efficiently to mitigate an incident, but they must be accomplished in a manner that protects the worker. Appropriate equipment and trained personnel, combined with SOPs, help reduce the possibility of harm to response workers.

For procedures to be effective:

- They must be written in advance. Developing and writing safe, practical procedures is difficult, and is virtually impossible when done under stress while responding to an incident.
- They must be based on the best available information, operational principles, and technical guidance.
- They must be field-tested, reviewed, and revised when appropriate by competent safety professionals.
- They must be concise, understandable, feasible, and appropriate.

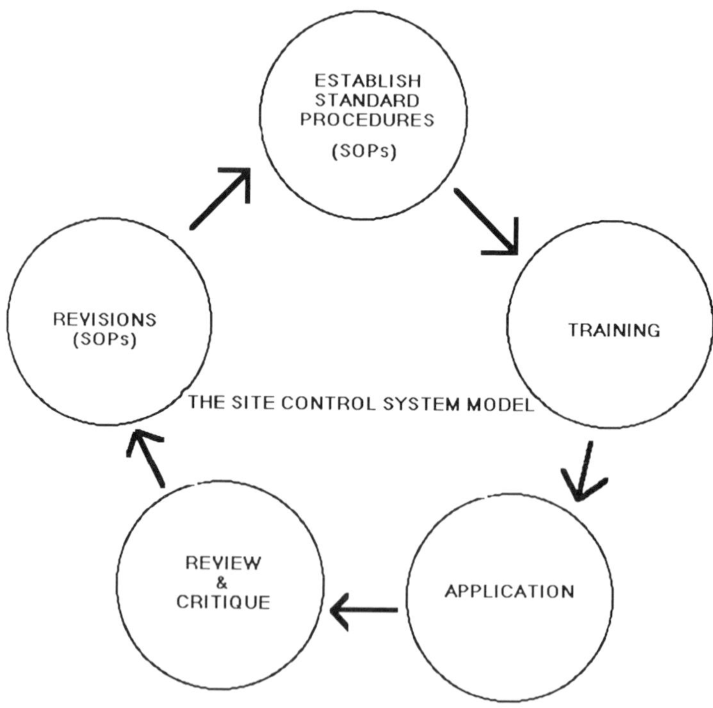

FIGURE 7-1. The site control system model contains the activities required to manage SOPs.

- All personnel involved in response activities must have copies of the safety procedures and be briefed on their use.
- Response personnel must be trained and periodically retrained in personal protection and safety procedures based on the applicable EPA standards and OSHA Standard 29 CFR 1910.120.

Response Activities

Many of the procedures involved in response activities are primarily concerned with health and safety. In concept and principle, these are generic and independent of the type of incident. They are adapted or modified to meet site-specific requirements.

Each potential hazardous materials incident at the fixed site must be evaluated to determine its hazards and risks. A hazard analysis is used to describe the overall procedure for evaluating the hazards, consequences, vulnerabilities, probabilities, and risks associated with the presence of hazardous materials within any given locality or jurisdiction. There are four basic steps required to conduct of a hazard analysis; see Chapter 8 for the details.

A related fifth step of the hazard analysis, which takes advantage of the knowledge gained during the writing of the plan or in an actual response, can be used to modify the comprehensive emergency plan for hazardous materials. This step focuses attention on the known threats to a community or facility while maintaining sufficient flexibility to deal effectively and efficiently with unforeseen events. Most of the time, effective SOPs should be developed around this fifth step. The location, identification, and characterization of potential spill sources and accident sites in the jurisdiction or locality of concern is a process referred to as "hazard identification." These compiled data are relevant to specific SOPs for applicable industries with EHSs. These planning efforts permit examination and/or prioritization of potential scenarios in terms of overall risk for inclusion in the SOPs. SOPs must be developed for personnel to approach the site of the emergency for containment and other control activities. Each of these activities requires that safety procedures emphasized in the risk analysis and existing SOPs be adapted so that response personnel are adequately protected.

Operating Guides—Generic Plans

Generic SOPs available for use as a guide in developing site-specific plans are included in Appendices 1–3. These guides highlight the major components required in SOPs and illustrate the technical considerations necessary in developing hazardous material release incident responses. For a given incident, the procedures recommended should be adapted to conditions imposed by that specific situation. A checklist can be useful as a summary SOP to aid the hazmat members in following the prescribed process for each chemical emergency response. Appendix 1 is a suggested SOP in the checklist format. This list is a simple SOP that can be effective as long as there is a technical support document to substantiate the response actions. See Appendix 3 for an example of a simple plan with backup data. It should be mentioned that training in SOPs at the site should extend to each technical support document—not just the checklist.

Another way to simplify the process of establishing and implementing a *uniform* response is to develop the SOP utilizing the decision-making process called DECIDE. In brief, the DECIDE process involves the following steps:

- Detect the presence of a hazardous material
- Estimate the likely harm without intervention
- Choose response objectives
- Identify action options
- Do the best option
- Evaluate progress

Another simple process in use is the following:

- Establish command
- Survey the scene
- Get help
- Identify the product
- Seal off the area
- Make a safe entry

TYPICAL STANDARD OPERATING PROCEDURES

Every incident or environment will have specific procedures to follow for safe responses. The following procedures are examples of typical SOPs that can be incorporated into specific emergency response plans.

- Walk cautiously to avoid tripping.
- Never walk on drums.
- Take special care when working near stacked drums.

- Always test an object before attempting to lift and carry it.
- If practical, drums and containers should be inspected to ensure their integrity prior to being moved.
- Unlabeled containers will be assumed to contain hazardous substances until proven otherwise.
- Portable dock boards must be secured in position with devices which will prevent their slipping during loading and unloading.
- Any attempt to open a vehicle involved in a response operation should be carried out in such a way that the person(s) making the attempt are not in the path of materials or containers which might spill out of the part of the vehicle being opened.
- Any driver-operated equipment used on a site with uneven terrain must have some form of rollover protection.
- Use equipment to detect an explosive or flammable atmosphere.
- Use explosion-proof instruments and nonsparking tools.
- In areas where flammable liquids are present, smoking and carrying lighters, matches, and other spark-producing devices (including warning flares) is prohibited.
- All containers involved in flammable liquids transfer should be bonded and grounded.
- If a fire starts, workers should leave the area unless they have been assigned fire-fighting responsibilities.
- Allow only medically and physically fit responders to be exposed to heat or other stress.
- Drink liquids to replace body water lost during sweating.
- Rest frequently.
- Allow a 10-foot clearance area between raised equipment and electric power lines.
- Before entering a confined space, test the atmosphere of oxygen and toxic and combustible levels of gases or vapors.
- A second worker (buddy) should be on hand at the entrance to the confined space throughout the entry time.
- The buddy should wear all PPE required for the worker inside the confined space.
- The buddy should be equipped with some means of communication to be used in summoning help if needed.
- Under no condition should the buddy enter the confined space unless other workers are standing by.
- Before response operations begin, provide for prompt medical attention in case of serious injury.
- At least one person at the site should be trained in first aid.
- First aid supplies should be readily available.

- Proper equipment should be available for prompt transportation of injured persons to medical facilities.

TERMINATION PROCEDURES

Termination procedures provide the comprehensive data from the incident which may be required to comply with local, state, and federal laws. Termination activities are divided into three phases: incident debriefing, post-incident analysis of actual activities, and formal critique designed to emphasize successful as well as unsuccessful operations.

Debriefing should begin as soon as the emergency phase of the operation is completed. Debriefing is the process of reviewing a hazmat incident, focusing on the following factors:

- Informing responders of exactly what hazardous materials they were (possibly) exposed to, and their signs and symptoms.
- Identifying equipment damage and unsafe conditions.
- Assigning information-gathering responsibilities for a critique.
- Summarizing the activities performed by each sector.

A good debriefing should reinforce the positive aspects of the response. In a large incident, representatives must convey information from this debriefing back to involved responders. The debriefing should be chaired by someone who has an overall view of the incident. This person is not necessarily the IC. Ideally, the debriefing should be limited to 30 minutes, with a 45-minute maximum. Post-incident analysis is the reconstruction of a hazardous materials incident to establish a clear picture of the events that took place during the incident. The post-incident analysis begins with the assignment of one person to collect all information during the debriefing. The purpose of the post-incident analysis is to determine who pays and to establish a clear picture of the emergency response for further study.

The critique is designed to share the lessons learned and to improve ERP performances and safety. The critique should identify both the good and bad things that happened, *taking care not to assign blame or to discipline personnel.* The critique leader must be well suited and comfortable working in front of a group. This person does not have to be the IC and can be a neutral third party. The critique should start and end with a positive statement.

Depending on manpower, the debriefing, post-incident analysis, and critique may have to be done in one session. This combination of sessions does not limit the information gained during the critique. This portion of the response is primarily established to document the safety procedures, site operations, hazards faced, and lessons learned. Termination procedures are initiated when the actual emergency response is complete.

Usually the IC deems the response action completed when the chemicals released have been safely confined, contained, and pose no threat to health and the environment. Normal operations may begin after the following steps are completed:

- All personnel and reusable equipment are properly decontaminated.
- All equipment used in the emergency response has been replenished and restocked.
- The IC has approved all actions.
- The EPA/state have approved the cleanup procedures and other response activities.

Note: Detailed procedures for confinement, containment, and decontamination procedures are discussed later in this text. The IC is the safety manager at the fixed facility and is usually the fire chief in a community hazmat team.

Other Types of Termination Procedures

As discussed earlier in this chapter, "termination procedures," as referred to by most emergency responders, refer to completion of the actual incident. Termination procedures terminology can be confusing due to the previous practices by OSHA and EPA in their standards and guidance documents. Most of the other uses for the termination procedures in emergency response refer to fixed-facility shutdown procedures in conjunction with beginning the actual response activities.

Included in Appendix 2 is a generic SOP for emergency response initiation and hazardous material emergency release notification procedures. This plan is site specific and can be used to establish safe working conditions based on the contamination concentrations levels that indicate

- Emergency operations codes
- Cessation of the type of work
- Notification procedures

This guideline can be used in conjunction with the specific task SOPs to develop an effective emergency response SOP. Termination activities in the OSHA standard refer to the steps taken to stop normal work and to initiate an emergency response. The previously mentioned SOP can be used to determine when it is necessary to discontinue or terminate normal work and begin response procedures.

Specific Procedures

Termination procedures include but are not limited to the following actions:

- Shut off equipment if possible.
- Shut off the electrical supply to the affected area.
- Shut off gas-fired ovens if possible.
- Locate gas mains and the shutoff valve if it is safe to do so.
- Shut off other pipelines containing chemicals upstream.
- Close manual fire doors.

Job Descriptions

It is typical to find that hazmat teams at fixed facilities have designated electricians to cut off electrical power supplies and millwrights to cut off gas mains and other pipeline systems. Since the basic shutoff or termination procedures are straightforward, the remaining tasks can be easily identified at each facility and delegated to specific job categories, such as the following:

- Electrical equipment shutoff procedures should be accomplished by their designated operator.
- Oven operators should shut off ovens.
- Warehousemen or forklift truck operators should close manual fire doors.
- The millwright should shut off the gas main.

REFERENCES

NFPA. 1989. *Hazardous Materials Response Handbook*. Quincy, MA: National Fire Protection Association.

Noll, G.C., M.S. Hiderbrand, and J.G. Yvorra: 1988. *Hazardous Materials: Managing the Incident*. Annapolis, MD: Peake Production, Incorporated.

OSHA. 1989. Title 29 CFR 1910.120. Hazardous Waste Operations and Emergency Response: Final Rule. US Occupational Safety and Health Administration. Washington, DC: Office of the Federal Register.

Standard Operating Safety Guides. 1988. Washington, DC: U.S. Environmental Protection Agency, Office of Emergency and Remedial Response, Emergency Response Division, Environmental Response Team.

Appendix 1

Chemical Spill Response Procedures

This checklist, from *Chemical Engineering*[1] magazine, is intended mainly for use by company personnel. It can help you evaluate your company's overall handling of chemical emergencies.

- Know the company emergency response plan
- Notify the chemical emergency team
- Evacuate the building if necessary
- Verify head counts of evacuated personnel
- Notify plant and other appropriate authorities
- Determine the chemicals or materials involved
- Determine the source
- Decide how to stop the spill
- Determine how to handle the spilled chemical
- Contain and clean up the spilled substance
- Assess the health and environmental problems caused by the spill. This includes health effects experienced by employees and the community, contamination of water supplies, and structural damage to plant
- Evaluate the cause of the accident
- Formulate and initiate recommendations to prevent recurrence
- Test or validate revised systems or procedures

---

[1] Chemical Engineering Magazine, Vol 95 Jan 18, 1988 pp 87–91. Martins, Keith,

Appendix 2

Emergency Operation Codes[2]

Real-Time Monitor
(suggested minimum action plan)
(Site Name)

Code Designations

1. *Green*
   A. Normal operations
2. *Yellow A*
   A. Cessation of specific work activity on site because of:
      (1) Continuous organic readings on direct-reading instrument of __*__ppm above background (measured 20–30 ft from the point of the suspected release), and
      (2) Current or projected meteorological conditions indicate a probable impact on work activity.
   B. If background readings above __*__ppm are obtained during cessation of activity, redesign the activity to lower releases and/or delay that on-site activity until off-site air monitoring indicates an accepted off-site concentration.
   C. Site personnel will immediately notify the EPA/state of site condition.

[2] Environmental Protection Agency, 1984 Standard Operating Safety Guides, Annex 7 Emergency Operation Codes.

 Richard R. Paul

3. *Yellow B*
   A. Termination of all work on site because of:
      (1) Continuous organic readings on direct-reading instruments of __*__ppm above background (measured approximately 1,000 ft from work area of site property limits), and
      (2) Current or projected meteorological conditions indicate a potential impact on inhabited areas.
   B. Site personnel will immediately notify the EPA/state of site conditions.
   C. EPA/state will modify off-site air monitoring to meet the needs of the contingency plan.
4. *Red*
   A. Termination of all work on site because of:
      (1) Continuous organic readings on direct-reading instruments of __*__ppm above background (measured downwind at the nearest occupied area off site, and
      (2) Current or projected meteorological conditions indicate a potential impact on inhabited areas.
   B. Site personnel will immediately notify the EPA/state of site conditions.
   C. Local officials making evacuation/public health decisions will be advised by the EPA/state to:
      (1) Release a public health advisory to potentially affected areas since on-site control methods will not rescue the source of contamination; and/or
      (2) Implement a temporary relocation plan because on-site activities indicate a potential for continuous above background/acceptable readings at the nearest inhabited area(s).

---

* Concentration should be determined by appropriate response personnel.

Appendix 3

SOPs for HAZ-1

HAZARDOUS MATERIALS ACCIDENT/INCIDENT EMERGENCY NOTIFICATION AND INITIAL EMERGENCY RESPONSE

I. Application

This SOP applies to all persons who receive notification, witness, or initially respond to a spill or release of a hazardous material. It applies more specifically to fire and law personnel, who are the persons most likely to witness, observe, receive initial notification of, or initially respond to a spill or release of hazardous materials.

II. General

A. DEFINITION

1. Hazardous materials: Any substance or material which poses an unreasonable risk to safety, health, property, or the environment.
2. Spill or release: Any leaking, dumping, pouring, emptying, emitting, spraying, discarding, injecting, escaping, leaching, pumping, or unauthorized manner of discharge or as the result of an equipment failure or accident.
3. Notification: The dispatch of response personnel and equipment. Also included is the alerting and dispatching of support personnel as appropriate.
4. Response: Agencies or individuals involved in emergency operations.

5. Level of response: A response level classification system that allows decision makers to determine quickly what agencies will be involved; the three levels are described below. See Attachment B for definitions.
   Level 1 Potential emergency
   Level 2 Limited emergency
   Level 3 Full emergency

B. BACKGROUND
Hazardous materials are stored, used, and transported throughout Madison County. The city and county have experienced numerous hazardous material incidents and accidents in the past and can expect them in the future.

C. SITUATION
1. Hazardous materials are transported by railcars, vehicles, watercraft, and aircraft.
2. Hazardous materials are present in homes, educational facilities, retail facilities, industrial facilities, and warehouse facilities.
3. Hazardous materials incidents or accidents can occur anyplace, at any time, without notice.
4. Hazardous materials incidents or accidents may vary in scope from extremely minor to catastrophic.
5. Government and industry have established a capability to handle hazardous material incidents or accidents.
6. Appropriate response and support agencies must be notified and respond to minimize the effect of a hazardous material accident/incident.

D. ASSUMPTION
1. Hazardous material incidents or accidents will continue to occur.
2. The local level of response capability and training will continue to improve.
3. Police and fire personnel are adequately trained to make an initial assessment of the relative severity of the situation and recommend timely actions to minimize loss of life or property.
4. Adequately trained personnel and communications equipment are available to complete these notification procedures.

III. Procedures

All hazardous material incidents will be considered to be a Level 1 response, potential emergency, until confirmed to be a higher level by the IC.

A. An Initial Emergency Notification Checklist is shown in Attachment A.
B. An Initial Emergency Response Checklist is shown in Attachment B.
C. A copy of the Hazardous Materials Incident Data Sheet is shown in Attachment C.
D. An Emergency Agency Telephone Listing is shown in Attachment D.

IV. Approval and Adoption

This SOP has been approved and adopted in writing by the following: County Commission, County Sheriff, County Association of Volunteer Fire Departments, and corresponding mayors of affected cities.

V. Changes

Recommended changes to this SOP should be submitted in writing to the LEPC Focal Point. Changes approved by the LEPC will be incorporated in future revisions.

ATTACHMENTS

Attachment A	Initial Emergency Notification Checklist
Attachment B	Initial Emergency Response Checklist
Attachment C	Hazardous Materials Incident Data Sheet (not included)
Attachment D	Emergency Agency Telephone Listing (not included)

Attachment A: Initial Emergency Notification Checklist

These procedures are required for all hazardous material accidents or incidents, regardless of the level of response classification.

Agency Receiving Report	Actions
Any agency (other than 911)	• Complete "Hazardous Materials Incident Data Sheet" Contact Information and items 1–6 of Incident Specific Data. • Dispatch agency units as required to scene. • Notify 911.

911 Emergency	• Complete "Hazardous Materials Incident Data Sheet" Contact Information and items 1–6 of Incident Specific Data. • Dispatch appropriate fire department. • Request dispatch of appropriate law enforcement agency if deemed necessary. • Request dispatch of hazardous material units if deemed necessary. • Notify Emergency Management Agency (EMA). Request coordinator to scene and Emergency Operations Center (EOC) activation if deemed necessary. • Determine the fire department incident commander and the command post (CP) Location as soon as possible. • Notify all response agencies of the IC's name, CP location, and level of response classification as soon as possible.

Attachment B: Initial Emergency Response Checklist

Agency Receiving Report	Actions
Initial response unit	• Establish incident (CP). • Senior fire department response officer will assume the position of IC. • Identify the type and quantity of hazardous materials involved. • Determine the level of response classification, using the following descriptions. • Level 1: Potential Emergency Description—An incident that can be controlled by the first response agencies and does not require evacuation of places other than the involved structure or the immediate outdoor area. The incident is confined to a small area and does not pose an immediate threat to life or property. • Level 2: Limited Emergency Description—An incident involving a greater hazard or larger area which poses a potential threat to life or property and which may require a limited evacuation of the surrounding area. The EOC may or may not be activated. • Level 3: Full Emergency Description—An incident involving a severe hazard or a large area which poses an extreme threat to life and property and will probably require a large-scale evacuation; or an incident requiring the expertise or resources of county, state, federal, or private agencies/organizations. The EOC will be activated.

Incident commander

- Establish Hot Zone Definition—The zone where the most significant contamination exists and where entry shall be limited to those authorized by the IC. Unprotected exposure to this zone may be injurious to health. This zone is delineated by the IC. Entry to and exit from this zone shall be recorded, and all persons will wear proper protective clothing.
- Establish Warm Zone Definition—A zone where no contamination or low levels of contamination exist; only necessary personnel shall be permitted in this zone. The perimeter of this zone shall be determined in consultation with the IC. Access to this zone shall be permitted only with the permission of the CP.
- Establish Cold Zone Definition—A zone where no contamination exists and no personnel or equipment controls are required.
- Determine support required
  - Law enforcement
    - Site security
    - Traffic control
    - Evacuation
  - Medical support
    - Ambulance
    - Physicians on scene
    - Hospital alert
  - Heavy equipment for damming or diking
  - Absorbent or diking material
  - Neutralizing materials
  - Emergency management assistance
    - On-scene coordinator
    - EOC activation
  - Other support
    - Air pollution control
    - Water pollution control
    - County health dept.
    - Utilities
    - ____________________
    - ____________________
  - Ensure that a "Hazardous Materials Incident Data Sheet" (Contact Information and items 1–14 of Incident Specific Data) is completed.

City fire communications

- Coordinate all requests from the IC.
- Notify support agencies requested by the IC.
- Coordinate request from support agencies if the EOC is not activated.

Support agencies coordinators

- Dispatch an on-scene coordinator to the CP when requested.
- Provide assistance as requested by the IC.
- Coordinate response and support activities using internal

	procedures, or through fire department, or through the EOC when activated.
IC after action	• Perform post-incident cleanup inspection • Forward a completed copy of the "Hazardous Materials Incident Data Sheet" to emergency management.
EMA coordinator after action	• Complete a Hazardous Materials Accident/Incident Report • Coordinate and schedule an after-action incident critique if appropriate. • Report the incident to the state.

8

Sara Title III—State ERPs

John E. Backensto and Lori P. Andrews, P.E.

Emergency response personnel have a particularly dangerous job which present unique situations at every hazmat incident. The hazmat team member at a fixed facility and the public responder can benefit from a well-conceived and organized plan of action for safe execution. In recognition and support of this, the SARA Title III legislation of 1986 was established to address emergency first-response planning for hazardous materials. Understanding the key components of this legislation will help responders in implementing an effective hazmat emergency response plan. This chapter reviews those portions of the law that describe the Local Emergency Response Plans (LERPs) and list resources and methods for developing practical yet thorough plans. Additionally, communication requirements are discussed, with practical suggestions on compliance.

OBJECTIVES

- To understand the requirements under Title III for the planning, notification, and reporting of hazardous materials storage, emissions, and spills
- To understand the Extremely Hazardous Substance List and the importance of the Tier I, Tier II, and Form R reports from a practical viewpoint
- To know the criteria for creating and evaluating a local Emergency Response Plan (ERP)
- To be able to access pertinent reference sources

- To understand the role of effective communications in the execution of the ERP
- To understand the basic components of an ERP

HISTORY

Prior to the promulgation of SARA in 1986, no law existed to protect workers, the community, or the environment from accidental releases of hazardous materials, substances, or chemicals. After a series of industrial accidents that were highly publicized, like

- the near mishap at the Three Mile Island nuclear power plant in Pennsylvania in 1979, and
- the escape of a cloud of methyl isocyanate gas from a Union Carbide chemical plant in Bhopal, India, in 1984, when more than 2,500 people lost their lives and another 200,000 were injured, some suffering permanent disabilities,

the public became painfully aware of the need for legislation to prepare communities for such disasters.

The Bhopal tragedy started a chain of events in the United States. In June 1985, the EPA established the voluntary Chemical Emergency Preparedness Program (CEPP). At around the same time, an industrial representative organization, the Chemical Manufacturers Association (CMA), set up a prototype program called "Community Awareness and Emergency Response (CAER)." This program was established as a guidance recommendation for communities and their industries, and was completely voluntary. Its main purpose was to inform workers of the hazards in the workplace.

SARA was signed into law by President Reagan on October 17, 1986. The Emergency Planning and Community Right-to-Know Act is found in Title III of the act. This provision established requirements for reporting hazardous chemicals. The emergency planning requires all 50 states to have a State Emergency Response Commission (SERC) with Emergency Planning Districts. The established districts must each appoint a Local Emergency Planning Committee (LEPC).

The LEPC includes elected state and local officials, police, fire, civil defense, public health professionals, environmental, hospital, and transportation officials, as well as representatives of facilities, community groups, and the media. The LEPC's primary responsibility is to maintain an ERP.

The plan includes the identification of facilities and EHS transportation routes. A facility, sometimes called a "fixed facility," is a facility that manufactures, processes, or stores chemicals. A number of these chemicals are hazardous. The facility has to receive chemicals and ship them from the plant. The transportation route is the route the truck driver takes. To determine what hazardous chemicals and how much of each chemical are being routed through a community, a survey is usually carried out. A statistical sample of the actual traffic hauling hazardous materials or hazardous waste is taken by a group of individuals placed along strategic transportation routes. The vehicle's cargo is identified by the DOT placards and by looking up the ID numbers in DOT Table 49 CFR 172.101. Once the survey is completed, all the data are typically collected, sorted, and logged into a data management computer program. From this survey one can establish the types and quantities of hazardous materials and the busiest transportation routes and time periods.

Another element of the plan is the emergency response procedures for both on-site and off-site facilities. On-site facilities are fixed facilities or industrial sites, as described previously. The off-site setting is the transportation corridors affected.

A coordinator has to be established for the community, as well as at each facility. This person is the one to contact in case of a chemical release. Communication is very important during a chemical release, and timing is critical. The coordinator is the individual responsible for initiating and implementing the ERP.

Proper notification procedures must be executed. Depending on the specific chemical and the quantity released, the EPA and state agencies have to be notified by law based on reportable quantities (RQ).

Methods to determine how the release occurred and the probable harm the particular chemical might create, as well as the potential area affected, must be developed. Population areas of high density are a major concern during a chemical release. There are numerous computer programs that can project and hypothesize the probable harm; see Chapter 16 for details.

Emergency equipment should be itemized and prioritized for budget purposes. Planning should address the urgent purposes and long-term purchases. Also, the equipment list should address any available equipment owned by private companies. These pieces of equipment should be defined in term of quantity and by specific location at each participating company. Thus, if an industrial site operates on a one-shift-a-day basis, the equipment to be used can be located quickly. During an emergency, it is not unusual for the equipment required to handle the release, to be provided by an outside vendor. A list of these resources should be available to all emergency coordinators. A contingency (secondary) list should also be on hand in case availability problems occur with the primary list.

Evacuation plans must be well thought out in advance and alternate routes documented. These plans should be available to all individuals involved. They should be rehearsed on a fairly regular basis (annually is the typically interval).

A training program for responding to chemical emergencies should be described in detail, with training course schedules. Training and exercising the ERP will provide insight into the changes necessary to mount effective response.

The methods and schedules for exercising ERPs involve many agencies. See Chapter 20 for details.

To assist the LEPCs, the National Response Team (NRT) was established. It is composed of 14 federal agencies with emergency preparedness responsibilities. The NRT publishes guidance documents on emergency planning.

RELEVANT PROVISIONS OF SARA TITLE III

SARA Title III is a free-standing law that is divided into three subtitles: Subtitle A—Emergency Planning and Notification of Releases; Subtitle B—Reporting Requirements; and Subtitle C—General Provisions. Although a part of the Superfund Amendment Reauthorization Act of 1986, this title addresses a broad spectrum of activities involving community/environmental protection and has established separate agencies other than EPA for compliance (i.e., SERC, and LEPC)

Subtitle A (Including Sections 301–305)

The SERC, Planning Districts, and LEPC are established based on Section 301.

Section 302 includes hazardous substances and RQs, facilities covered, and notification procedures. In order to understand this section better, we must first define some terms. A "facility" is any building, structure, installation, equipment, pipe or pipeline (including any pipe into a sewer or publicly owned treatment works), well, pit, pond, lagoon, impoundment, ditch, landfill, storage container, motor vehicle, rolling stock, or aircraft, or any site or area where a hazardous substance has been deposited, stored, disposed of, placed, or otherwise come to be located. A "hazardous substance" is any substance designated pursuant to 40 CFR Part 302. Approximately 360 chemicals are included on this list. A "release'' is any spilling, leaking, pumping, pouring, emitting, emptying, discharging, injecting, escaping, leaching, dumping, or disposing into the environment. The ''Reportable quantity (RQ)'' is that quantity, as set forth in 40 CFR Part 302, the release of which requires notification. ''Notification requirements'' state that any person in charge of a facility shall, as soon as he has knowledge of any release (other than a federally permitted release or application of a pesticide) of a hazardous substance from the facility in a quantity equal to or exceeding the RQ determined by this section in any 24-hour period, immediately notify the NRC.

Comprehensive ERPs are also included in Section 303. All covered facilities must designate a facility representative to participate in the local emergency planning effort. Facilities must provide the LEPC with information for the response plan, such as a list of the chemicals used, their locations, response equipment, and an inspection of these items.

These sections are designed to develop government response and preparedness capabilities. LEPCs are designated and emergency response or contingency plans must be developed. The plan must include all items specified in Sections 302 and 303.

Emergency notification (releases), the CERCLA list (Superfund), the EHS list, RQs, notification of applicable agencies, and the follow-up written report (for releases of EHSs) are included in Section 304. Figure 8-1 shows the interdependence of the chemical lists in SARA.

Any release of a hazardous substance exceeding its RQ must be reported to the LEPC, the SERC, and the NRC immediately. This emergency notification must include the chemical's name, whether or not it is extremely hazardous, how much was released, the time and duration of the release, what the chemical spilled into, possible health risks, precautions such as evacuation, and the name and telephone number of the contact person. A follow-up report must be written confirming all actions taken during the emergency and submitted within 15 days after the initial release. This report should include any information that would update the information originally reported. It should also include additional information with respect to actions taken to respond to and contain the release, any known or anticipated acute or chronic health risks associated with the release, and, where necessary, advice regarding medical attention necessary for exposed individuals.

How does one figure out the threshold planning quantity (TPQ) for a mixture or a solid? If a container or storage vessel holds a mixture or solution of an EHS, then the concentration of the EHS, in weight percent (greater than 1 percent), should be multiplied by the mass (in pounds) in the vessel to determine the actual quantity of EHS therein. The EHSs that are solids are subject to either of two TPQs (i.e., 500 or 10,000 pounds).

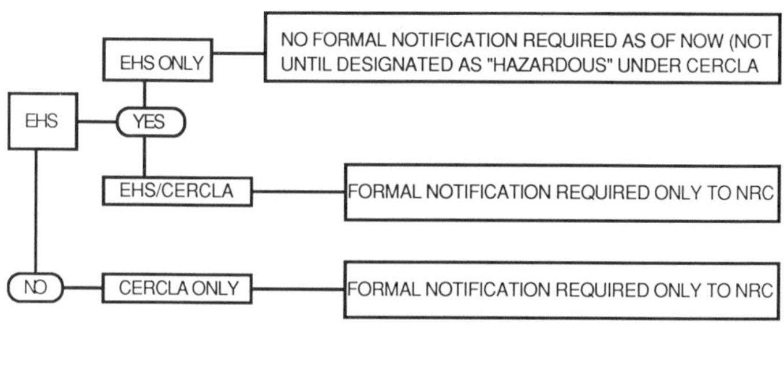

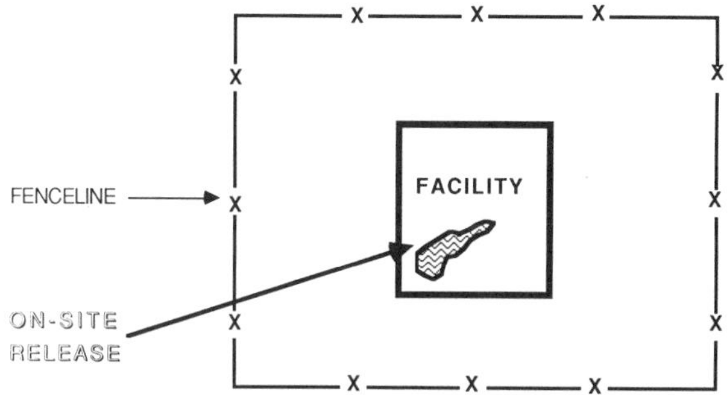

NOTE:
EHS = EXTREMELY HAZARDOUS SUBSTANCE
CERCLA = CERCLA HAZARDOUS SUBSTANCE
NRC = NATIONAL RESPONSE CENTER
(1-800-424-8802)

FIGURE 8-1. Interdependence of chemical lists in SARA Title III.

The lower quantity applies only if the solid exists in powdered form and has a particle size less than 100 microns; or is handled in solution or in molten form; or meets the criteria for an NFPA rating of 2, 3, or 4 for reactivity. If the solid does not meet any of these criteria, it is subject to the upper (10,000 pound) TPQ. See Figure 8-2 for a flow chart of an on-site release and Figure 8-3 for a flow chart of an off-site release.

The penalty for not reporting a release can be either civil or criminal, depending on the circumstances. Under civil penalties, a person can be fined up to $25,000 for each day the violation continues. In a second or subsequent violation, he can be fined up to $75,000 per day. In cases involving criminal penalties, any person who knowingly and willfully fails to provide notice can, upon conviction, be fined up to $25,000 or imprisoned for up to 2 years, or both. In cases of a second or subsequent

SARA Section 311 - 312
Hazardous (As per OSHSA - HCS)
Over 500,000

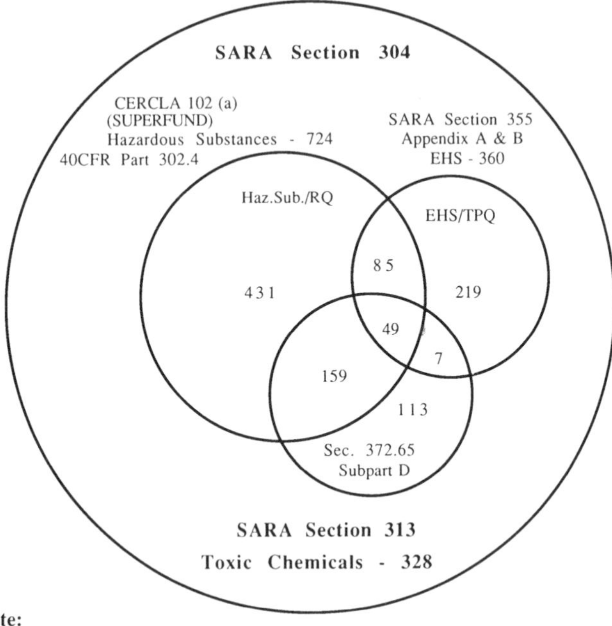

Note:

49 chemicals are the same in all three lists
134 are the same in the CERCLA/SARA(EHS) lists
208 are the same in the CERCLA/Toxic lists
56 chemicals are the same in the SARA(EHS)/SARA(TOXIC) lists

FIGURE 8-2. Emergency release notification flow chart (on site).

conviction, the person can be fined not up to $50,000 or imprisoned for up to 5 years, or both.

To obtain a chemical list for CERCLA hazardous substances and RQs, see a current copy of 40 CFR Part 302.4. To obtain a SARA chemical list of EHSs and TPQs, see a current copy of 40 CFR Part 355 Appendix A.

A training grant administered through the Federal Emergency Management Agency (FEMA) for training emergency responders is defined in Section 305. Also, as a part of this section, EPA is required to review emergency systems to determine their effectiveness.

Section 311 requires facilities to have Material Safety Data Sheets for on-site chemicals and to provide copies of these sheets or a list of the chemicals to the LEPC, the SERC, and the local fire department. The chemical inventory is preferred due to storage constraints and paper usage.

Section 312 requires submission of a chemical inventory form describing the amounts of hazardous chemicals stored, used, and their locations, in amounts exceeding 10,000 pounds. Upon request, further information must be submitted to a commission, department, or the public. For Extremely Hazardous substances (EHS) and hazardous chemical inventory forms, facilities subject to Sections 311 and 312 must submit a Tier I or Tier II inventory form for each chemical requiring a Material Safety Data Sheet (MSDS) to the SERC, LEPC, and local fire department annually (due March 1 of each year); inventory period January 1–December 31 of each year). Tier I and Tier II inventory forms, with instructions on how to fill them out, can be found in a current copy of 40 CFR Part 370.40.

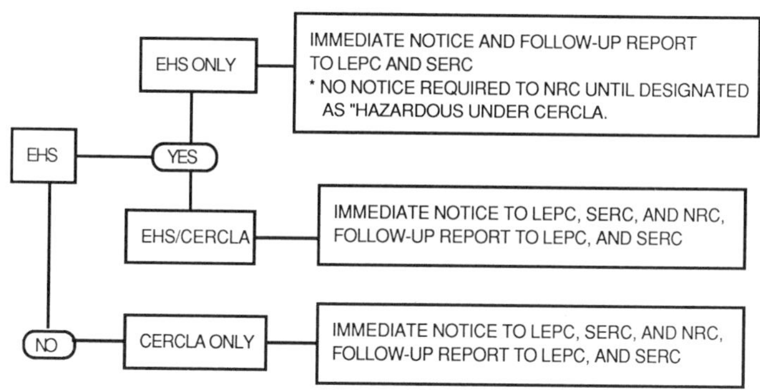

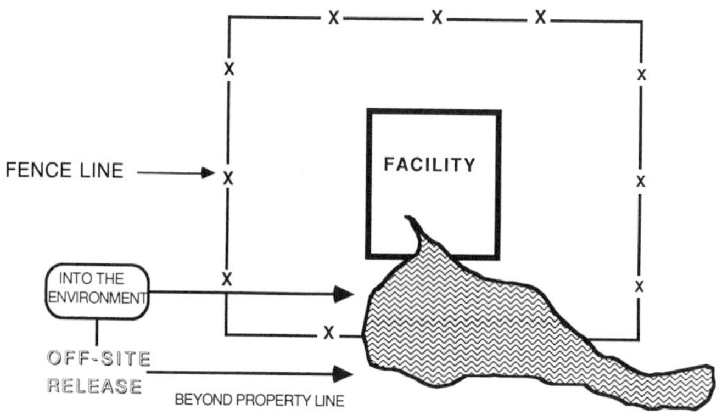

FIGURE 8-3. Emergency release notification flow chart (off site).

SARA Title III Section 313 establishes an EPA inventory of toxic chemicals emissions. These emissions are reported so that government officials and the public are aware of them and may evaluate the potential environmental damage that may result. Toxic chemical release form should be completed by facilities defined by Sec. 313 as follows: has 10 or more full time employees; is in Standard Industrial Classification (SIC) codes 20–39; and/or have releases that exceed the threshold limit for toxic chemicals from manufacture, process, or general use. Also, submitted to form R should be the SERC and EPA on an annual basis for each chemical that exceeds the threshold (due July 1 of each year). TPQs established for reporting with respect to a toxic chemical manufactured (including imported) or processed at a facility during the following calendar years are: 1987–75,000 pounds; 1988–50,000 pounds; 1989 and thereafter—25,000 pounds. The threshold for reporting with respect to a chemical otherwise used at a facility is 10,000 pounds. SARA's toxic chemical list can be located in a current copy of 40 CFR Part 372.65 Subpart D. EPA's Form R (No. 9350-1), with instructions on how to fill it out, can be found in a current copy of 40 CFR Part 372.85 Subpart E. See Table 8-1 for a summary of Title III chemical lists and their purposes.

General provisions are found in Subtitle C and include the following: Section 321, relationship to other laws; Section 322, trade secrets; and Section 323, provisions of information to health professionals, doctors, and nurses. In addition, Subtitle C includes enforcement, civil actions, and exemptions. Subtitle C comprises Sections 321–330, with the final sections covering regulations, definitions, and authorizations of appropriations.

Figure 8-4 is a flow chart which illustrates the intended communications from the facility level via the local level to the state level and ultimately to the federal level. For a better understanding of this chart, the following description follows the paths for planning and emergency reporting procedures. For planning information, the report is initiated at the facility. It includes the Material Safety Data Sheet or the chemical inventory. As mentioned earlier, the inventory list is preferred due to space constraints. Also, for planning purposes, the emergency equipment list must be submitted by the facility. All facility-generated reports are sent to the local fire department, the LEPC, and the SERC. The LEPC contact is typically the county emergency management agency, and the SERC contact is the state emergency management agency. Additional information for emergency planning purposes (i.e., identification of in-house emergency notification procedures and personnel responsible for emergency incidents) is sent to the LEPC and the SERC for inclusion in the state ERP. On an annual basis, the Toxic Chemical Release Form (R) must be

TABLE 8-1. Title III Chemical Lists and Their Purposes

List	Section	Purpose
Extremely hazardous substances (Federal Register, November 17, 1986; 402 chemicals listed in CEPP Interim Guidance)	Emergency Planning Section 302	• Facilities with more than established, initial planning quantities of these substances must notify the SERC. • Initial focus for preparation of emergency plans by LEPCs.
	Emergency Notification Section 304	• Certain releases of these chemicals trigger Section 304 notification to the SERC and the LEPC.
Substances requiring notification under Section 103 (a) of CERCLA (717 chemicals)	Emergency Notification Section 304	• Certain releases of these chemicals trigger Section 304 notification to the SERC and the LEPC, as well as CERCLA Section 103 (a) requirement to notify the NRC.
Hazardous chemicals considered physical or health hazards under OSHA 1910.1200 (This is a standard; there is no specific list of chemicals.)	Emergency Notification Section 304	• Identifies facilities subject to emergency notification requirements.
	Material Safety Data Sheets Section 311	• Material Safety Data Sheet or list of chemicals on the sheet provided by facilities to the SERC, LEPC and local fire department.
	Emergency and Hazardous Chemical Inventory Section 312	• Covered facilities must provide site-specific information on the quantity and location of chemicals to the SERC, LEPC, and local fire department to inform the community and assist in plan preparation.
Toxic chemicals identified as chemicals of concern by the states of New Jersey and Maryland (329 chemicals/chemical categories)	Toxic Chemical Release Section 313	• These chemicals are reported on an emissions inventory to inform government officials and the public about releases of toxic chemicals in the environment.

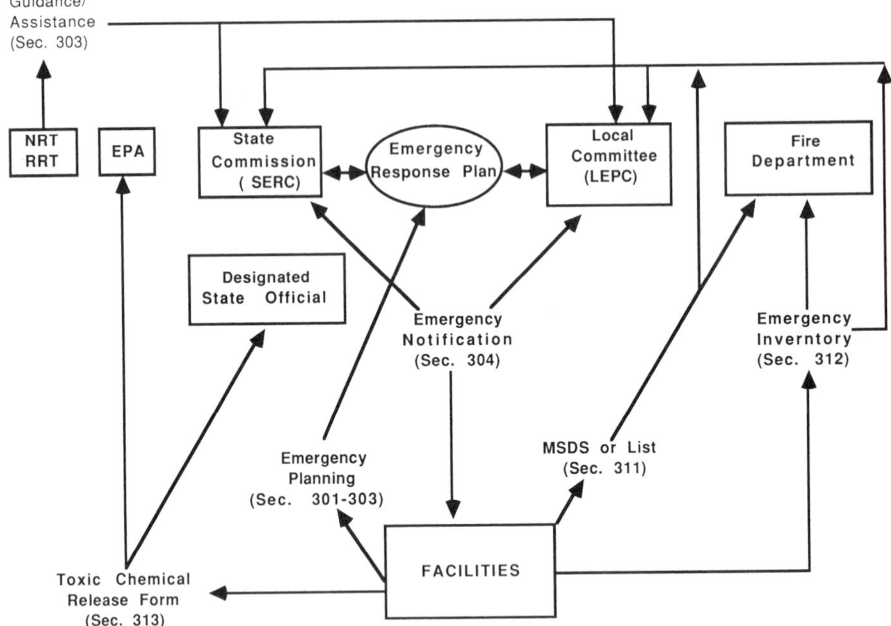

FIGURE 8-4. Title III major information flow/requirement.

submitted by qualifying facilities to EPA and the state-designated official (who typically is a member of the state environmental protection agency). An actual incident is reported from the facility to the LEPC and the SERC. The EPA is notified directly by the SERC office, and the National Response Team/Regional Response Team are notified by EPA as deemed appropriate.

STATE AND LOCAL ERPs

Major Components

The following are the necessary components of an ERP: Introduction; Incident Information Summary; Promulgation Document; Legal Authority and Responsibility for Responding; Table of Contents; Abbreviations and Definitions; Assumptions/Planning Factors; Concept of Operations (Governing Principles; Organizational Roles and Responsibilities; Relationship to Other Plans); Instructions on Plan Use (Purpose and Plan Distribution); and Record of Amendments.

An emergency assistance telephone roster should be included. It should list participating agencies, technical and response personnel, CHEMTREC, public and private sector support groups, and the NRC. This list should stand alone in the plan so that copies can be distributed. See Table 8-6 for a complete listing.

The third component of an ERP should consist of response functions. They are listed in Table 8-2.

Containment and cleanup involve techniques for handling spills and resources for cleanup and disposal. Areas with the highest risk should be mapped out in the planning stages.

An accurate data log should be kept by someone on the emergency response team. An investigative follow-up report should determine how the release occurred, what equipment and supplies were used, and what worked or did not work. Drills and various exercises should be planned carefully and practiced. These may be tabletop exercises or actual drills simulating the transport of contaminated victims to the hospital by ambulance. Once the exercise or drill is completed, a review or critique should be done. During the critique, things that went wrong should be specified. The plan can now be modified and updated.

Hazard analysis, references, and a technical library are all closely interrelated. Hazard analysis includes hazards identification, vulnerability analysis, and risk analysis. To assist in preparing a hazard analysis, see EPA's (CEPP) technical guidance document and DOT's Community Teamwork and Lessons Learned. Computer programs to assist you can

TABLE 8-2. Response Functions

- Initial notification of response agencies
- Direction and control
- Communications (among responders)
- Warning systems and emergency public notification
- Public information/community relations
- Resource management
- Health and medical services
- Response personnel safety
- Personal protection of citizens (indoor protection, evacuation procedures, and other public protection strategies)
- Fire and rescue
- Law enforcement
- Ongoing incident assessment
- Human services
- Public works
- Other forms of involvement that are site specific

be obtained by calling the nearest federal EPA office. Technical support and a list of specific people to contact should go in the reference section of the plan. A telephone directory of technical support services should include environmental and public health laboratories, private consultants, colleges or universities, and local chemical plants. The technical library should consist of a list of references, their locations, and their availability. A second reference list should be produced for backup. If one or more items on the primary list fail to come through, then the secondary list can be activated.

Selected Detailed Elements

The elements combined together make up an incident information summary sheet. This summary should be brief (no longer than one page in length). The initial information is critical. It is important to gather as much information as possible quickly in order to make decisions on public notification and evacuation. Identification numbers, shipping manifests, and placard information are essential to identify any hazardous materials involved in transportation incidents and to take initial precautionary and containment steps. See Table 8-3 for a complete list of these elements.

A promulgation document is a letter drafted up and signed by the community chief executive giving the authority to put the plan into action. There are different organizations involved in the community in both the private and public sectors. It may be appropriate to include letters of agreement signed by officials from these organizations. The community may choose to enact legislation in support of its plan. Make sure to identify any agencies required to respond to particular emergencies. Authorizing legislation and regulations include federal, state, regional, and local components.

Practical Implementation Procedures

Suggestion for an ERP include a Table of Contents (all sections of the plan should be listed here and clearly labeled with a tab for easy access), Abbreviations and Definitions (list frequently used abbreviations, acronyms, and definitions here for easy reference).

Assumptions and planning factors are a summary of the conditions that make an ERP necessary. The Hazards Identification and Analysis section is needed to complete this section. Appropriate maps that show water intake, environmentally sensitive areas, major chemical manufacturing or storage facilities, population centers, and the location of response resources should be provided. Assumptions are the advance judgments concerning what would happen in the case of an accidental spill or release. See Table 8-4 for a list of these planning factors.

TABLE 8-3. Selected Elements of An Incident Information Summary

- Date and time
- Name of person receiving call
- Name and telephone number of on-scene contact
- Location
- Nearby populations
- Nature (e.g., leak, explosion, spill, fire, derailment)
- Time of release
- Possible health effects/medical emergency information
- Number of dead or injured; where dead/injured have been taken
- Name of material(s) released, if known
- Manifest/shipping invoice/billing label
- Shipper/manufacturer identification
- Container type (e.g., truck, railcar, pipeline, drum)
- Railcar/truck four-digit identification numbers
- Placard/label information
- Characteristics of material (e.g., color, smell, physical effects) only if readily detectable
- Present physical state of the material (i.e., gas, liquid, solid)
- Total amount of material that may be released
- Other hazardous materials in the area
- Amount of material released so far/duration of release
- Whether significant amounts of the material appear to be entering the atmosphere, nearby water, storm drains, or soil
- Direction, height, color, odor of any vapor clouds or plumes
- Weather conditions (wind direction and speed)
- Local terrain conditions
- Personnel at the scene

TABLE 8-4. Assumptions/Planning Factors

- Geography
- Sensitive environmental areas
- Land use (actual and potential, in accordance with local development codes)
- Water supplies
- Public transportation networks (roads, trains, buses)
- Population density
- Particularly sensitive institutions (e.g., schools, hospitals, homes for the aged)
- Climate/weather statistics
- Time variables (e.g., rush hour, vacation season)
- Particular characteristics of each facility and the transportation routes for which the plan is intended
- On-site details
- Neighboring population
- Surrounding terrain
- Known impediments (e.g., tunnel, bridges)
- Other areas at risk

Concept of Operations (Governing Principles)

There should be a statement of precisely what is expected to be accomplished if an incident should occur. If, for instance, a railcar hauling anhydrous ammonia should derail at the intersection of Interstate 59 and Interstate 65 south bound, then a list of objectives would include confining the release, evacuating affected areas of the community, and providing traffic control.

Concept of Operations (Organizational Roles and Responsibilities)

This is a list of all the organizations and officials responsible for planning and/or executing the preresponse, response, and postresponse activities

TABLE 8-5. List of Involved Agencies And Personnel

- Municipal government
- Chief elected official
- Emergency management director
- Community emergency coordinator (Title III of SARA)
- Communications personnel
- Fire service
- Law enforcement agency
- Public health agency
- Public works
- County government
- Officials of fixed facilities and/or transportation companies
- Facility emergency coordinators (Title III of SARA)
- Nearby municipal and county governments
- Indian tribes within or near the affected jurisdiction
- State government
- Environmental protection agency
- Emergency management agency
- Public health agency
- Transportation organization
- Public safety organization
- Federal government
- Environmental Protection Agency
- Federal Emergency Management Agency
- Department of Transportation
- Department of Health and Human Services/Agency for Toxic Substances and Disease Registry
- U.S. Coast Guard
- Department of Labor/Occupational Safety and Health Administration
- Department of Defense
- Department of Energy
- Regional Response Team

TABLE 8-6. List of Basic Resources and Capabilities

- Predetermined arrangements
- How to use outside resources
- Response capabilities
- Procedures for using outside resources

in a hazardous materials incident. See Tables 8-5 and 8-6 for a list of involved agencies and personnel.

Concept of Operations (Relationship to Other Plans)

A major task of the planning group is to combine planning for hazardous materials incidents with already existing plans. In larger communities, it is likely that several ERPs have been prepared. It is essential to coordinate these plans. When two or more plans are put into action simultaneously, there is a real potential for confusion among response personnel unless the plans are carefully coordinated.

See the discussion of incident command in Chapter 3 for these issues. All ERPs (including facility and hospital plans) that might be employed in the event of an accidental spill or release should be listed. The community plan should include the methods and procedures to be followed by facility owners and operators and local emergency response personnel in responding to any releases of such substances. The National Contingency Plan (NCP), the Federal Regional Contingency Plan, any on-scene coordinator's (OSC) plans for the area, and any state plan should be referenced; of special importance are all local emergency plans. Even where formal plans do not exist, various jurisdictions often have preparedness capabilities. Planners should seek information about informal agreements involving private industries, cities, counties, states, and countries.

Instructions on Plan Use (Purpose)

Maintaining an up-to-date version of a plan is of prime importance. When corrections, additions, or changes are made, they should be recorded in a simple bookkeeping style so that all plan users will be aware that they are using a current plan.

An up-to-date emergency telephone roster is a vital piece of information for the emergency responder. The list should include names of contact persons and their alternates along with their telephone numbers. A brief description about each person or agency should be included. The emergency responder needs access to this list to make the appropriate phone calls when services, supplies, equipment, and other resources are

TABLE 8-7. Emergency Assistance Telephone Roster

- List of telephone numbers for:
  - Participating agencies
  - Technical and response personnel
  - CHEMTREC
  - Public and private sector support groups
  - National Response Center
- Community assistance
  - Police
  - Fire
  - Emergency management agency
  - Public health department
  - Environmental protection agency
  - Department of transportation
  - Public works
  - Water supply
  - Sanitation
  - Port authority
  - Transit authority
  - Rescue squad
  - Ambulance
  - Hospitals
  - Utilities
    - Gas
    - Phone
    - Electricity
  - Community officials
    - Mayor
    - City manager
  - County executive
  - Councils of government
- Volunteer groups
  - Red Cross
  - Salvation Army
  - Church groups
  - Ham radio operators
  - Off-road vehicle clubs
- State assistance
  - SERC (Title III of SARA)
  - State environmental protection agency
  - Emergency management agency
  - Department of transportation
  - Police
  - Public health department
  - Department of agriculture
- Response personnel
  - Incident Commander
  - Agency coordinators
  - Response team members

TABLE 8-7. (Continued)

- Bordering political regions
  Municipalities
  Counties
  States
  Countries
  River basin authorities
  Irrigation districts
  Interstate compacts
  Regional authorities
  Bordering international authorities
  Sanitation authorities/commissions
  Industry
  Transporters
  Chemical producers/consumers
  Spill cooperatives
  Spill response teams
- Media
  Television
  Newspaper
  Radio
- Federal assistance
  Federal on-scene coordinator (OSC)
  U.S. Department of Transportation (DOT)
  U.S. Coast Guard (USCG)
  U.S. Environmental Protection Agency (EPA)
  Federal Emergency Management Agency (FEMA) (24 hours: 202-646-2400)
  U.S. Department of Agriculture (USDA)
  Occupational Safety and Health Administration (OSHA)
  Agency for Toxic Substances and Disease Registry (ATSDR) (24 hours: 404-452-4100)
  National Response Center (NRC) (24 hours: 800-424-8802); 202-426-2675 in Washington, DC, area or 202-267-2675
  U.S. Army, Navy, Air Force
  Bomb Disposal and/or Explosive Ordnance Team, U.S. Army
  Nuclear Regulatory Commission (24 hours: 301-951-0550)
  U.S. Department of Energy (DOE)
  Radiological Assistance (24 hours: 202-586-8100)
  U.S. Department of the Treasury
  Bureau of Alcohol, Tobacco, and Firearms
- Other emergency assistance
  Chemical Transportation Emergency Center (CHEMTREC) (24 hours: 800-424-9300)
  Chemical Network (CHEMNET) (24 hours: 800-424-9300)
  Chlorine Produces Mutual Aid (CHLOREP) (24 hours: 800-424-9300)
  National Agracultural Chemical Association (NACA) Pesticide Safety Team (24 hours: 800-424-9300)
  Association of American Railroad/Bureau of Explosives (24 hours: 202-639-2222)
  Poison control center
  Cleanup contractor

needed in an emergency. A note stating what times of the day or night the number will be answered should be included. This list should be verified and updated at least every 6 months. An example of this telephone roster is found in Table 8-7.

REFERENCE SOURCES

Hotline Telephone Numbers

- Bureau of Explosives, Association of American Railroads 24-hour emergency number: 1-202-639-2222. (Assistance for hazardous materials incidents involving railroads, often contacted through CHEMTREC.)
- CHEMTREC: 1-800-424-9300. (24-hour emergency phone number for the Chemical Transportation Identification number of unknown chemicals; advice on proper initial response methods and procedures for specific chemicals and situations; and assistance in establishing contact with shippers, carriers, manufacturers, and special product response teams such as CHLOREP or the Pesticide Safety Team Network as necessary and appropriate.)
- DOT hotline: 1-202-366-4488. (Provides informational assistance pertaining to the federal regulations for transportation of hazardous materials under CFR 49.)
- EPA Chemical Emergency Preparedness Program hotline: 1-800-535-0202. (For communities to call to obtain 140-page interim guidelines regarding acutely toxic chemicals, which include "Organizing a Community," "Developing a Chemical Contingency Plan," and "Gathering Site-Specific Information"; also provided is a list of more than 400 acutely toxic chemicals. Guideline documents may be used to obtain Material Safety Data Sheets for such chemicals.)
- EPA RCRA hotline: 1-800-424-9346. (To respond to any citizen's request for specific information and to clear up confusion regarding RCRA and Superfund regulations; also, to respond to requests for certain documents printed in the Federal Register for which this telephone number is given as a contact point. In addition, in response to policy questions from the regulated communities and state/local governments, personnel will attempt to seek out the correct person to provide guidance.)
- EPA Small Business hotline: 1-800-368-5888. (To respond with advice and information to problems encountered by small-quantity generators of hazardous waste.)

- Hazardous Materials Newsletter information line: 1-802-479-2307. (To respond to any first responder—public safety agency, industrial firm, or commercial firm—that needs to obtain information or advice relative to hazardous materials tools/equipment/materials, planning, protocols, methods, strategies, tactics, training and research sources, and resources.)
- National Animal Poison Control Center of the University of Illinois: 1-217-333-3611. (Provides 24-hour consultation in diagnosis and treatment of suspected or actual animal poisonings or chemical contamination, staffs an emergency response team to rapidly investigate such incidents in North America, and performs laboratory analysis of feeds, animal specimens, and environmental materials for toxicants and chemical contaminants.)
- Texas Tech University Pesticide hotline: 1-800-858-7378. (The National Pesticide Telecommunications Network provides information on pesticide-related health, toxicity, and minor cleanup to physicians, veterinarians, fire departments, government agency personnel, and the general public.)
- U.S. Coast Guard/DOT NRC: 1-800-424-8802. (For required reporting of incidents in transportation where hazardous materials are responsible for death, serious injury, property damage in excess of $50,000 or continuing danger to life and property.)
- University of Alabama at Birmingham, Center for Labor Education and Research, Hazardous Waste Worker Training Program, for training compliance and information hotline (205-934-8016).

Manuals Widely Used By Response Personnel

- *Fire Protection Guide on Hazardous Materials*. Price: $45.00. Available from:

  National Fire Protection Association
  Batterymarch Park
  Quincy, MA 02269
  (800) 344-3555

  This guide provides fire hazard properties and hazardous chemical data.
- *Hazardous Materials Emergency Response Guidebook*. Price: $6.00. Developed by DOT and available from commercial suppliers. This manual gives sample emergency response guides for chemicals and includes an index by UN identification number; an index by alphabetic order; information on health hazards, fire or explosion, emer-

gency action procedures (fire, spill, or leak), and first aid; a table of isolation and evacuation distances; applicable response by placard information; UN class number; and general considerations.

- *Emergency Handling of Hazardous Materials in Surface Transportation*. Price: $25.00 nongovernmental; $20.00 government/all levels. Available from:

  Publication Services
  Bureau of Explosives
  Association of American Railroads
  50 F Street, N.W.
  Washington, DC 20001
  Tel. 202-639-2232

  Provides recommendations for response—general rules response by hazard class.

- *Occupational Health Guidelines for Chemical Hazards*, Supplement 1-$11.00 (DHHS/NIOSH, Publication No. 81-123 and 89-104-Supplement 2-$10.00). Developed by NIOSH and OSHA and available from:

  The Superintendent of Documents
  U.S. Government Printing Office
  Washington, DC 20402
  Tel. 202-783-3238

  This manual is compiled from various sample guides and includes chemical as well as physical properties.

- CHRIS (Chemical Hazards Response Information System). Price: $42.50. Developed by the U.S. Coast Guard and available from:

  U.S. Government Printing Office
  Washington, DC 20402
  Tel. 202-783-3238

  Although the CHRIS system is composed of four separate manuals, only Manual 2, "Hazardous Chemical Data," is of interest here. This Manual lists the specific chemical, physical, and biological data for 1,000 chemicals.

- *Handbook of Hazardous Materials: Fire/Safety/Health*. Price: $10.00. Available from:

  Alliance of American Insurers
  Publications/Order Department
  Suite 400 West
  1051 Woodfield Road
  Schaumburg, IL 60195-4980 (708) 330-8500

  Developed by the Industrial Hygiene Subcommittee of the Alliance of American Insurers, this manual provides information on physical properties, flammability characteristics, toxicity specifics, and threshold limit values for over 300 substances. Included as an outline of the principles or recognition, evaluation, and control of workplace contaminants as they relate to occupation disease.

- *Emergency Action Guides*. Price: $60.00 ($50.00 for emergency response personnel) Developed by and available from:

  Publication Services
  Bureau of Explosives Association of American Railroads
  50 F Street, N.W.
  Washington, DC 20001
  Tel. 202-639-2232

  Each of the 134 separate *Guides* covers a single commodity in great detail. (The 134 commodities account for 98 percent, by volume, of the hazardous materials transported by rail.)

- *A Guide to the Safe Handling of Hazardous Materials Accidents* (STP-825). Price: $15.00. Available from:

  American Society for Testing and Materials
  1916 Race Street
  Philadelphia, PA 19103
  Tel. 215-299-5400

  This *Guide* is intended for planning responses and training personnel to ensure the safest and most effective handling of hazardous materials incidents.

- *EnviroTIPS: Environmental and Technical Information for Problem Spills Manuals*. Available from:

  Technical Services Branch
  Environmental Protection Program Directorate
  Environmental Protection Service
  Ottawa, Ontario, Canada
  (613) 953-5921
  variable prices ($9.60–$22.00)

  Environment Canada has recently completed and is distributing the first chemical-specific manuals (50) in its EnviroTIPS (Technical Information for Problem Spills) series. EnviroTIPS manuals are available to persons with a strong interest in the environmental aspects of hazardous materials spills.

- *The Firefighter's Handbook of Hazardous Materials* by Charles J. Baker, 4th ed., 1984. Price: $19.95. Available from:

  Maltese Enterprises, Inc.
  P.O. Box 34048,
  Indianapolis, IN 46234.
  Tel. 317-243-2211.

  This work designed as a pocket reference manual to quickly determine relative toxicity, flammability, thermal stability, permissible extinguishing agents, and other pertinent data on a given substance. Listed in alphabetical order are over 5,600 substances, many of which are common or brand names.

- *Manual for Spill of Hazardous Materials* (Environment Canada, March 1984). Price: $18.95 (Canada), $22.75 (outside Canada). Catalog No. EN40-320-1984E (may use VISA or Master Card). Available from:

  Canadian Government Publishing Center
  Ottawa, Canada K1A 0S9
  Tel. 810-997-2560

  This manual is a comprehensive and practical guide to environmentally hazardous products prepared by the Technical Services Branch of Environment Canada. It includes quantitative data on chemical and physical properties, fire properties, human health and toxicity, and reactivity and environmental toxicity, as well as qualitative response information. Covered individually are 150 top-priority chemical substances, as well as fuels/oils and other frequently spilled substances.
- *NIOSH Pocket Guide to Chemical Hazards*, fifth printing, September 1985 (DHHS/NIOSH Publication No. 85-114). Developed by NIOSH and available from:

  The Superintendent of Documents
  U.S. Government Printing Office
  Washington, DC 20402.
  Tel. 202-783-3238

  This manual presents general industrial hygiene and medical surveillance information for workers, employers and occupational health professionals.

REFERENCES

EPA. 1988. *Chemicals in Your Community, A Guide to the Emergency Planning and Community Right-to-Know Act.* Publication No. OS-120 Washington, DC: U.S. Environmental Protection Agency.

EPA. 1988. "Memo EPA Title III Fact Sheet," U.S. Environmental Protection Agency. Washington, DC 20460.

EPA. 1990. Title 40 CFR Parts 300–399. U.S. Environmental Protection Agency. Washington, DC: Office of the Federal Register.

Handbook of Chemical Hazard Analysis Procedures. 1989. Federal Emergency Management Agency, U.S. Department of Transportation, U.S. Environment Protection Agency. Washington, DC: Federal Emergency Management Agency Publication Office.

Hazardous Materials Emergency Planning Guide (Publication No. NRT-1). 1987. Washington, DC: National Response Team. You can obtain a free copy by calling the Emergency Planning and Community Right-to-Know Hotline at 1-800-535-0202.

NIOSH/OSHA/CG/EPA. 1985. *Occupational Safety and Health Guidance Manual for Hazardous Waste Site Activities*, Chapters 3 and 12 (NIOSH Publication No. 85-115). Washington, DC: U.S. Government Printing Office.

NFPA. 1989. *Hazardous Materials Response Handbook*. Quincy, MA: National Fire Protection Association.

OSHA. 1989. Title 29 CFR 1910.120. Hazardous Waste Operations and Emergency Response: Final Rule. U.S. Occupational Safety and Health Administration. Washington, DC: Office of the Federal Register.

Standard Operating Safety Guides. 1988. Washington, DC: U.S. Environmental Protection Agency, Office of Emergency and Remedial Response, Emergency Response Division, Environmental Response Team.

Technical Guidance for Hazards Analysis, Emergency Planning for Extremely Hazardous Substances. 1987. Washington, DC: U.S. Environmental Protection Agency, Federal Emergency Management Agency, U.S. Department of Transportation. You can obtain a free copy by calling the Emergency Planning and Community Right-to-Know Hotline at 1-800-535-0202.

Worker Protection During Hazardous Waste Remediation. 1990. New York: Van Nostrand Reinhold.

9

Personal Protective Equipment

D. Alan Veasey, M.A. Ed., and
Kenneth W. Oldfield, M.S.P.H.

An uncontrolled release of chlorine gas originated from an open valve on a 1-ton chlorine cylinder at a water treatment facility located in East Aurora, New York. An employee of the facility used SCBA to enter the hazard area in an attempt to stop the release. However, before the worker could reach the point of release, he was injured by inhalation of chlorine gas due to a poorly fitting SCBA facepiece. As the release continued, the local fire service agency was notified and responders were dispatched to the incident scene. Hundreds of area residents located near the facility were evacuated. The release was finally terminated by two fire fighters who used appropriate personal protective equipment (PPE) to enter the hazard area and close the cylinder valve.

As the worker injured during the chlorine incident can attest, hazmat team members have a high potential for exposure to chemical contaminants and other hazards during emergency response operations. Due to the time-critical nature of emergency response activities, the use of engineering controls to create a safe work environment prior to entry is usually impractical. Thus, the effective use of PPE is typically vital to the safety and health of emergency responders. For this reason, it is critical that all hazmat team members have sufficient knowledge and hands-on experience to utilize PPE effectively and safely during emergency response operations. This chapter is intended as an introduction to PPE.

OBJECTIVES

- Know the distinguishing characteristics of the major types of respiratory protective equipment

- Know the selection considerations, advantages, and disadvantages of the different types of respirators
- Understand the importance of respirator facepiece-to-face fit and know the methods of fit testing
- Be aware of the available types of protective clothing and accessories
- Be aware of the selection considerations for chemical protective clothing (CPC)
- Appreciate the complexities and importance of CPC selection
- Be familiar with the EPA levels of protection for PPE ensembles
- Be aware of the NFPA certification standards for chemical protective garments
- Be aware of the various hazards related to PPE use and the requirements for safe use of PPE, with special emphasis on heat stress–related hazards
- Know and understand the topics pertaining to the use and care of PPE which OSHA requires to be incorporated into a written PPE program under 29 CFR 1910.120

RESPIRATORS

Respiratory Protection Requirements

Respirators are designed to protect personnel from the inhalation and ingestion of airborne contaminants. These contaminants may be present in a hazardous materials incident as dusts, fumes, mists, gases, and/or vapors. OSHA and EPA require that responders be protected from exposure to atmospheric contaminants in excess of applicable exposure limits (as described in Chapter 14). Since engineering controls and/or work practices are typically not a practical option for providing this protection, respirators must be used.

Respiratory protection is critical to hazmat team members because respiratory hazards are frequently encountered in the course of emergency response operations. For this reason, it is critical that all hazmat team members have a working knowledge of respirators.

Classification of Respiratory Protective Equipment

Respiratory protective equipment which may be used in chemically contaminated environments can be generally categorized based on the following factors.

Facepiece Type

Various styles of facepieces are used with respiratory protective equipment. Based on the amount of facial coverage provided, facepieces suitable for use in chemically contaminated work areas may be divided into the following major types:

- *Half-masks* cover the face from below the chin to the bridge of the nose (Figure 9-1).
- *Full facepieces* cover the entire face, thus offering eye protection as well as a fit which is not easily disturbed (Figures 9-2 and 9-3). For this reason, the full facepiece is the type most commonly used in emergency response.

Method of Protection

Respirators can be classified, based on the method of protection, as either air-purifying or atmosphere-supplying types.

Air-Purifying Respirators

Air-purifying respirators (APRs) use filters, neutralizing agents, and/or sorbent materials to purify the ambient atmosphere of the work area for

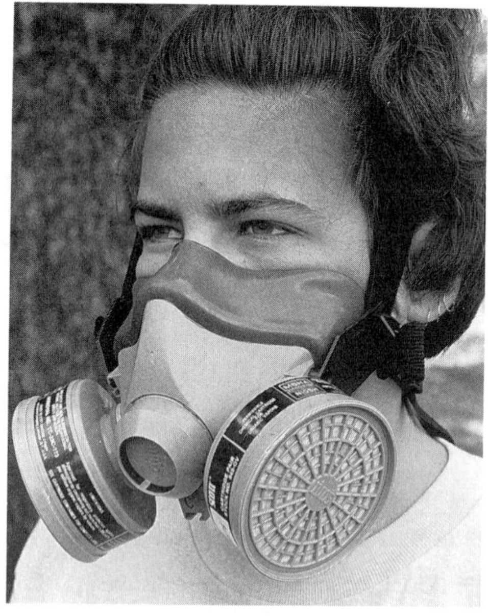

FIGURE 9-1. Typical half-mask, twin-cartridge APR. (Courtesy of Scott Aviation)

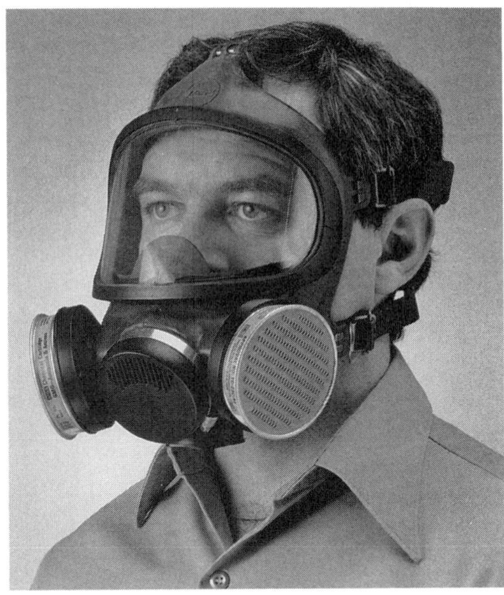

FIGURE 9-2. Typical full-facepiece, twin-cartridge APR. (Courtesy of Mine Safety Appliances Company)

breathing. The purifying materials are contained in disposable cartridge or canister-type purification elements (see Figure 9-4) which must be replaced after a given period of use. Canisters are appreciably larger than cartridges, and are usually worn on a belt or harness and connected to the facepiece by a breathing tube, whereas cartridges are typically mounted directly to the respirator facepiece (compare Figures 9-2 and 9-3).

APRs provide effective protection in atmospheres which contain relatively low concentrations of known contaminants and have near-normal oxygen levels (>19.5 percent). However, if any of the following conditions are present in the operations area, atmosphere-supplying respirators must be used:

- Unidentified breathing hazards
- Excessive concentrations of contaminants
- Extremely toxic contaminants
- Contaminants with poor warning properties (as discussed below)
- Deficient oxygen levels

One or more of these restricting factors is typically present and prevents the use of APRs during emergency response.

FIGURE 9-3. Typical full-facepiece, canister-type APR. (Courtesy of Mine Safety Appliances Company)

Atmosphere-Supplying Respirators

Atmosphere-supplying respirators provide the user with breathing air from outside the contaminated work area. Two types of atmosphere-supplying equipment are commonly used:

- *Supplied-air respirators (SARs)*, or airline respirators, supply breathing air to the worker through an airline connected to a compressor and a purification unit or a bank of air tanks located outside the contaminated area (see Figure 9-6).
- *Self-contained breathing apparatus (SCBA)* supplies breathing air from a tank worn on the user's back (see Figure 9-7). As discussed below, this type of respirator is most frequently used for emergency response.

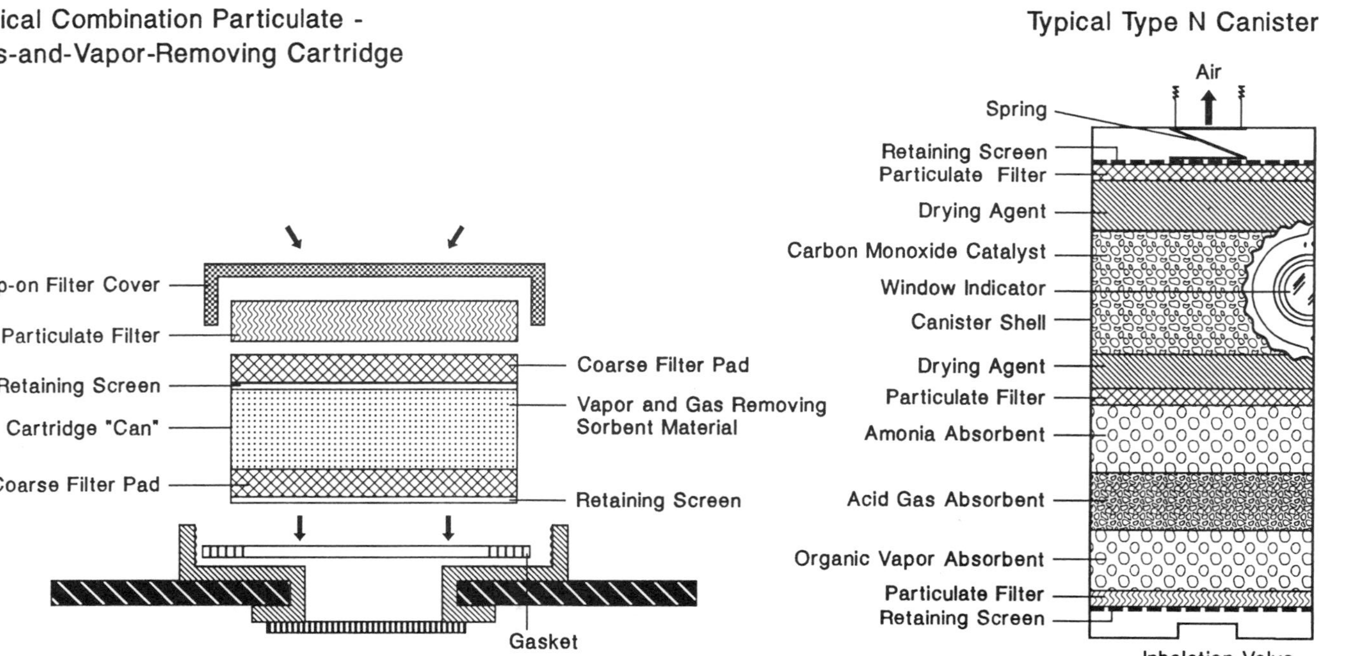
Typical Combination Particulate -
Gas-and-Vapor-Removing Cartridge
Snap-on Filter Cover
Particulate Filter
Retaining Screen
Cartridge "Can"
Coarse Filter Pad
Coarse Filter Pad
Vapor and Gas Removing
Sorbent Material
Retaining Screen
Gasket
Typical Type N Canister
Air
Spring
Retaining Screen
Particulate Filter
Drying Agent
Carbon Monoxide Catalyst
Window Indicator
Canister Shell
Drying Agent
Particulate Filter
Amonia Absorbent
Acid Gas Absorbent
Organic Vapor Absorbent
Particulate Filter
Retaining Screen
Inhalation Valve

FIGURE 9-4. Air-purifying cartridge and canister components.

Mode of Operation

Based on the pressure generated within the facepiece during use, respirators can be classified as operating in either a negative-pressure or a positive-pressure mode of operation.

Negative-Pressure Respirators

Negative-pressure respirators require the wearer to inhale, and generate a negative pressure (or vacuum) within the facepiece, in order to receive breathing air. Thus, a poor facial fit (or any other source of leakage) will allow large volumes of contaminated air to enter the facepiece during inhalation. For this reason, a good fit is absolutely critical when using any respirator operating in the negative-pressure mode. Respirators which operate in this mode include APRs (excluding powered APRs) and SARs or SCBAs which operate in the "demand" mode. Powered air-purifying respirators (PAPRs) are APRs which are fitted with a motor designed to feed a continuous flow of purified air to the facepiece, thus creating a positive pressure within the facepiece.

Positive-Pressure Respirators

Positive-pressure respirators are designed to maintain a slight positive pressure within the facepiece at all times so that any leaking air will, theoretically, move from the inside out. This will aid in preventing contaminants from entering the facepiece. Thus, respirators which operate in the positive-pressure mode have an appreciably higher protection factor than equivalent respirators operating in the negative-pressure mode. Only those SCBAs and SARs which operate in the positive-pressure mode should be used during emergency response operations. Two positive-pressure designs are currently used:

- *Continuous-flow respirators* maintain positive pressure by feeding a continuous stream of breathing air to the facepiece. All PAPRs and some SARs are of the continuous flow design.
- *Pressure-demand respirators* are designed to maintain a slight positive pressure within the facepiece. Breathing air flows into the facepiece only during inhalation, so that air consumption is much lower than with a continuous flow respirator.

It should be noted that during times of peak inhalation while performing strenuous tasks, responders can temporarily "overbreathe" the positive pressure within the facepiece of a respirator operating in the positive-pressure mode. If the face-to-facepiece seal is poor, harmful concentrations of contaminants may enter the facepiece during these intervals of negative pressure. Thus, a good facepiece fit is vital even when using positive-pressure respirators.

Examples of Respiratory Protective Equipment Classifications

Based on facepiece type, method of protection, and mode of operation utilized, respiratory protective equipment may be classified as represented in the following examples:

Half-mask, twin cartridge APRs (Figure 9-1)
Full facepiece, canister type APRs (Figure 9-3)
Half-mask, demand supplied-air (or airline) respirators
Full facepiece pressure-demand SCBAs (Figure 9-7)

Selection of Respiratory Protective Equipment

Each type of respiratory protective equipment has certain advantages and limitations. Since the protective capabilities of different types of respirators vary significantly, the selection process must ensure that the respirator selected will offer adequate protection against atmospheric contaminants. As a minimum, the assigned protection factor of the respirator (as described below) must be adequate. However, a strictly numerical approach based on protection factors may ignore other important considerations. Specific requirements and limitations imposed by conditions in the operations area and the specific task(s) to be performed must also be considered. Thus, careful consideration of a variety of factors is required to determine which type of respiratory protective equipment is appropriate for a specific situation.

Selection Considerations for APRs

Advantages of APRs

APRs offer a number of advantages: They are light in weight, relatively inexpensive, relatively simple to maintain, and place minimal mobility restrictions on the user. Thus, as a general rule, APRs are the preferred type of respirator for any situation in which they will provide adequate protection.

Disadvantages of APRs

A number of disadvantages of APRs must also be considered. For the most part, these disadvantages are related to work area conditions which preclude the use of APRs. These conditions are incorporated in Figure

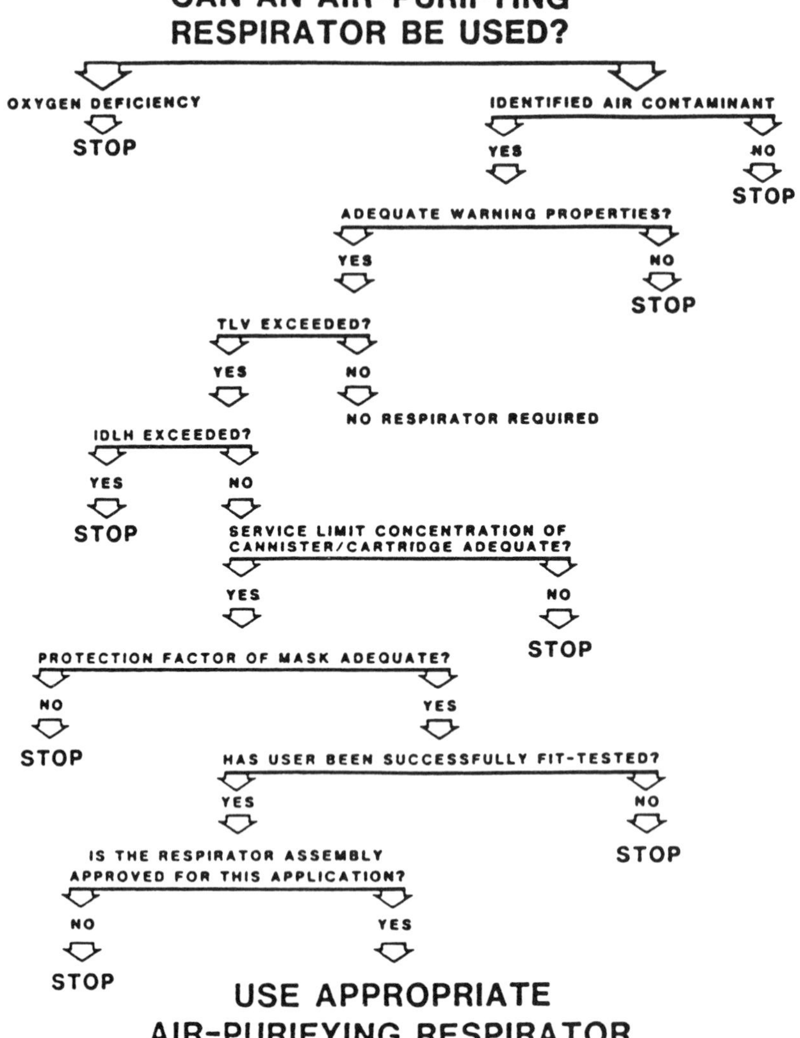

FIGURE 9-5. Flow chart incorporating use limitations of APRs.

9-5, which can be utilized in determining if the use of APRs is safe in a specific situation. Disadvantages of APRs are as follows:

- All potential atmospheric hazards in the operations area must be identified prior to entry in order to determine that APRs will offer adequate protection.

- APRs cannot be used if contaminants which are highly toxic in small concentrations (such as hydrogen cyanide) are present.
- APRs cannot be used in oxygen-deficient atmospheres (i.e., atmospheres containing less than 19.5 percent oxygen).
- APRs cannot be used if contaminant concentrations are excessively high (i.e., in excess of IDLH levels).
- Selection of purification elements must be hazard specific, since even "universal" elements are not effective against all potential contaminants. It should be noted that a system of color coding (as shown in Table 9-1) can be used to ensure that appropriate cartridges or canisters are used.
- An incomplete face-to-facepiece seal may allow a significant amount of contaminant to enter the facepiece, since most APRs work in the negative-pressure mode.
- Canisters and cartridges have a finite service life and must be discarded before the service life is exceeded and "breakthrough" occurs.
- Breathing through APRs requires greater than normal effort. However, PAPRs can sometimes be used to reduce this problem.
- Conditions in the operations area (e.g., high relative humidity) may reduce the effectiveness of some purification materials (e.g., sorbents) used in cartridges and canisters.

Other Considerations for Using APRs

The use of APRs can be considered safe only when the user has some way of knowing when the end of the canister or cartridge service life has been reached. Good warning properties allow the APR user to smell, taste, or experience irritation from concentrations of the contaminant *below appropriate exposure limits* in the event of breakthrough. As a general rule, APRs should be used only for contaminants having good warning properties.

In using APRs, the wearer should follow specific cartridge or canister replacement procedures. These purification elements should not be used if the manufacturer's expiration date has passed. They should be used immediately once the package seal is broken and discarded after completion of the response operation, at the end of service life, or when breakthrough begins to occur, *whichever comes first*.

APRs cannot be used if the concentration of air contaminants in the work area exceeds the *service limit concentration* of the cartridges or canisters used. Depending upon the specific contaminants involved, service limit concentrations vary from 10 ppm (0.001 percent) to 1,000 ppm (0.1 percent) for cartridges and from 5,000 ppm (0.5 percent) to 30,000

TABLE 9-1. Color Codes For Respirator Cartridges and Canisters

Atmospheric Contaminants To Be Protected Against	Color Assigned
Acid gases	White
Hydrocyanic acid gas	White with ½-inch green stripe completely around the canister near the bottom
Chlorine gas	White with ½-inch yellow stripe completely around the canister near the bottom
Organic vapors	Black
Ammonia gas	Green
Acid gases and ammonia gas	Green with ½-inch white stripe completely around the canister near the bottom
Carbon monoxide	Blue
Acid gases and organic vapors	Yellow
Hydrocyanic acid gas and chloropicrin vapor	Yellow with ½-inch blue stripe completely around the canister near the bottom
Acid gases, organic vapors, and ammonia gases	Brown
Radioactive materials, excepting tritium and noble gases	Purple (magenta)
Particulates (dusts, fumes, mists, fogs, or smokes) in combination with any of the above gases or vapors	Canister color for contaminant, as designated above, with ½-inch gray stripe completely around the canister near the top.
All of the above atmospheric contaminants	Red with ½-inch gray stripe completely around the canister near the top.

Gray shall not be assigned as the main color for a canister designed to remove acids or vapors.

Note: Orange shall be used as a complete body or stripe color to represent gases not included in this table. The user will need to refer to the canister label to determine the degree of protection the canister will afford.

Source: 29 CFR 1910.134.

ppm (3 percent) for canisters. Applicable service limit concentrations should be clearly printed on cartridge or canister labels. The assigned protection factor of the facepiece type with which the cartridges or canister is used is also a critical consideration in determining whether or not an APR will provide adequate protection. See the section on respirator protection factor below.

The Role of APRs in Emergency Response

The role of APRs in the area of emergency response tends to be highly restricted. Given the large number of restrictions on the use of APRs, the process of determining that APRs will provide adequate protection in a specific situation is necessarily complicated and time-consuming. The time-critical nature of emergency response will frequently not allow for this process. Also, hazard levels may drastically increase unexpectedly, for example, due to container failure during entries. In some instances, entry into hazardous atmospheres may be required before information such as the identity and concentration of contaminants can be gathered. For such entries, only positive-pressure SCBA is allowable under OSHA regulations. Because of these restrictions, APRs are not typically used for primary response operations.

APRs may provide adequate protection for personnel involved in secondary response operations, such as decontamination, which are carried out in areas of reduced hazard on site. However, from a strictly legal standpoint, it should be determined that all restrictions on the use of APRs are met by conditions in the work area before APRs are assigned.

Selection Considerations for Supplied-Air (Airline) Respirators

Advantages of SARs

The primary advantage of SARs is that they allow extended work periods (in comparison to SCBAs) in atmospheres requiring atmosphere-supplying respiratory protection. SARs are also much less cumbersome than SCBAs.

Disadvantages of SARs

Disadvantages associated with SARs are primarily related to the airline. For example, worker mobility is restricted by the airline, since worker's must retrace their previous steps to exit a work area. Also, the maximum allowable airline length is 300 feet. The airline may be cut, torn, caught, or entangled, trapping the user and/or cutting off the air supply. The airline may also be contaminated or permeated by chemicals.

SAR systems are expensive to purchase. They also require detailed maintenance.

Other Considerations for Using SARs

Only those SARs which operate in the positive-pressure mode should be used during hazardous materials response operations. Work in IDLH

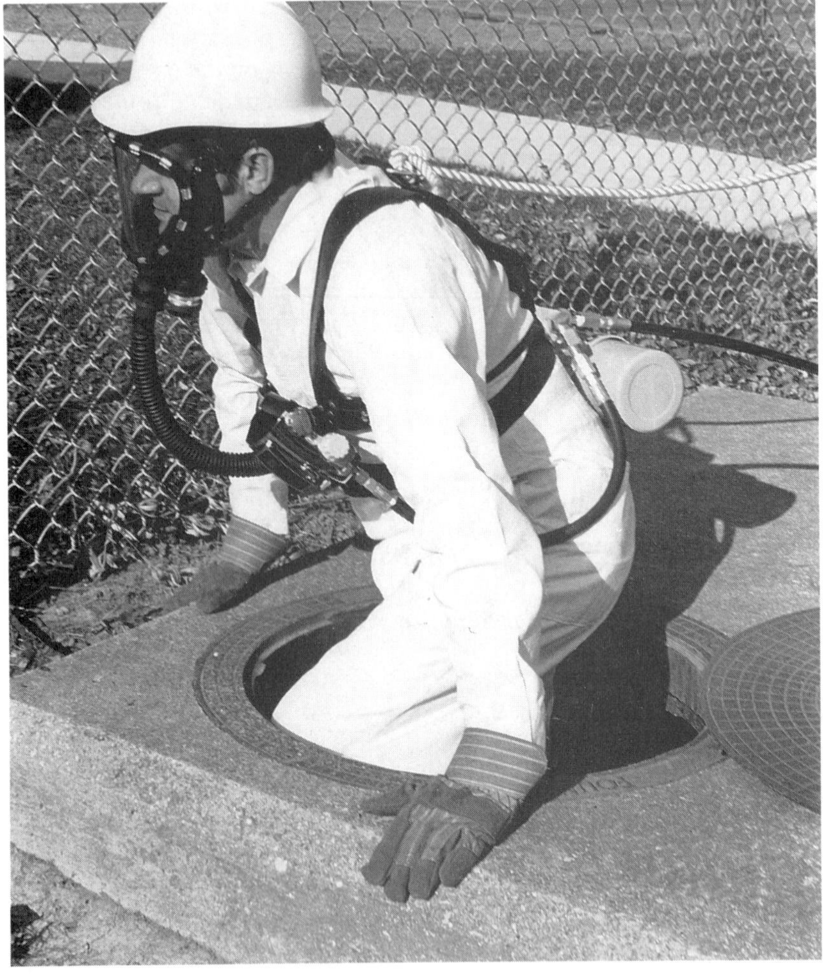

FIGURE 9-6. SAR with escape SCBA. (Courtesy of Scott Aviation)

conditions requires an escape air supply of at least 5 minutes' duration when using SARs (see Figure 9-6).

Air compressors and purification units used with SARs must be designed specifically to supply breathable air of at least grade D quality. Airlines must contain couplings located no farther than 100 feet apart, and all airline couplings must be incompatible with couplings on hoses containing any substance other than breathable air on site.

The Role of SARs in Emergency Response

The role of SARs in the area of emergency response tends to be somewhat restricted. Due to the mobility restrictions involved, SARs should not be used for entry into unknown conditions. However, for situations in which contaminants have been (1) identified and (2) determined to be present in excess of IDLH levels, SARs may be a valuable asset for response operations because of the extended air supply they offer. In using SARs, safety considerations include (1) the permeation resistance of the airline to the chemicals present and (2) the stress levels which the extended duration of entry made possible by SARs may place on response personnel.

Selection Considerations for SCBA Respirators

Advantages of SCBAs

SCBAs offer the primary advantage of providing atmosphere-supplying respiratory protection without the airline-related problems discussed above. When operated in the positive-pressure mode, SCBAs are considered to offer the highest level of respiratory protection presently available.

Disadvantages of SCBAs

Disadvantages of SCBAs are primarily related to the air tank. The presence of the tank makes SCBA units heavy and cumbersome. Passage through some small openings (such as manholes) while wearing SCBA may be impossible. Also, the duration of the air supply is limited. For example, when using an open-circuit SCBA system (as described below), the rated service life or breathing time is approximately 30 minutes for a low-pressure system or 60 minutes for a high-pressure system. However, it should be noted that the actual breathing time during use is typically significantly less than the rated service life for SCBA.

SCBAs are expensive and require detailed maintenance procedures. Air tanks must be hydrostatically tested periodically (at 5-year intervals for steel tanks and at 3-year intervals for aluminum-cored, fiberglass-wound tanks).

Other Considerations for Using SCBAs

Only those SCBAs which operate in the positive-pressure mode should be used for emergency response activities.

The most commonly used SCBAs operate as *open-circuit* systems in which exhaled air exits the system. Extended work times may be achieved by utilizing *closed-circuit SCBAs* in which exhaled air is purified, enriched with oxygen, and rebreathed. However, closed-circuit

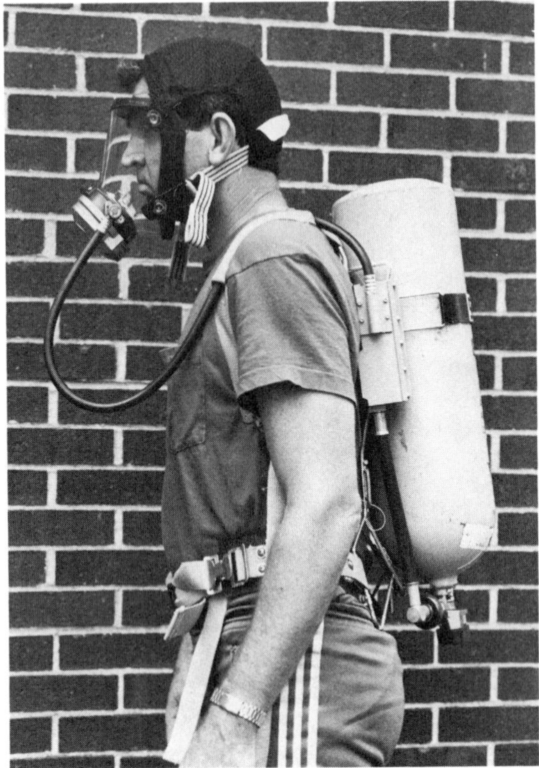

FIGURE 9-7. SCBA. (Courtesy of Scott Aviation)

SCBAs tend to generate heat during the purification process. Thus, heat stress may be a greater problem when using the rebreathers compared to the conventional, open-circuit SCBAs.

Hybrid SCBA/SAR combination systems are currently available which allow users to enter a contaminated area using the SCBA system and then plug into an airline for an extended work period using the SAR system. Air remaining in the tank is then saved for exiting the hazard area. When this type of system is used, no more than 20 percent of the SCBA air supply should be used during entry.

An interesting variation on the hybrid SCBA/SAR concept is the "quick-fill" type of system (Figure 9-8). This system allows the user to periodically replenish the SCBA air supply by briefly plugging into an airline located within the immediate work area. Thus, an extended work period can be achieved while avoiding the airline-related problems discussed previously. A variant of this type of system can be used during emergency operations to provide escape breathing air for a responder whose SCBA air supply becomes depleted during an entry. This is achieved using a short section of hose to "transfill" or transfer breathing air from a charged SCBA to the depleted unit (Figure 9-9).

Some SCBA units are designed to be used only for escape from atmospheres which suddenly become hazardous. These units are compact and can be donned quickly (Figure 9-10). Escape-only SCBA units have a very limited air supply, which is fed to the wearer in a continuous flow after the unit is donned and activated. These units are not suitable for entry into a hazardous atmosphere.

The Role of SCBA in Emergency Response

Because of the high protection factor and lack of airline-related mobility restrictions, positive-pressure SCBA tends to be the mainstay of respira-

FIGURE 9-8. Hybrid system for quick filling of SCBA air tanks. (Courtesy of Mine Safety Appliances Company)

FIGURE 9-9. Transfer of breathing air from a charged SCBA to a depleted SCBA using the "transfill" system. (Courtesy of Mine Safety Appliances Company)

tory protection for emergency response personnel. For this reason, OSHA regulations contained in 29 CFR 1910.120 require that positive-pressure SCBA be used during emergency response operations in hazardous areas until such time as the IC determines, based on air monitoring results, that some less protective type of respiratory protective equipment is appropriate.

Importance of Respirator Fit and Fit Testing

The face-to-facepiece fit of a respirator is of the utmost importance if the wearer is to be adequately protected. A poor fit will allow dangerously high volumes of contaminated air to enter the facepiece. Thus, fit tests must always be conducted as part of the respirator assignment procedure. Methods of fit testing are as follows.

- *Qualitative fit tests* are simple tests designed to determine whether or not an acceptable fit has been achieved. Qualitative fit tests may be conducted as follows:
  - *Negative-pressure tests* are conducted by blocking the inhalation pathway of the facepiece, inhaling gently, and holding the breath for 10 seconds while checking for leakage (Figure 9-11). If the fit is acceptable, the respirator should be pulled back toward the face by the vacuum generated through inhalation and remain there as the wearer holds his breath. Care should be taken to cover the inhalation pathway gently so as not to distort the normal fit of the facepiece.

- *Positive-pressure tests* can be performed by blocking the facepiece exhalation valve and gently exhaling (Figure 9-12). Failure to generate a positive pressure inside the facepiece indicates a poor fit. It should be noted that the exhalation port construction of some respirators makes it impossible to perform this test.
- *Irritant smoke, odorous vapor, and sweetener tests* are performed by exposing the wearer to irritants (such as stannic chloride) or substances which have distinctive odors or tastes (such as banana

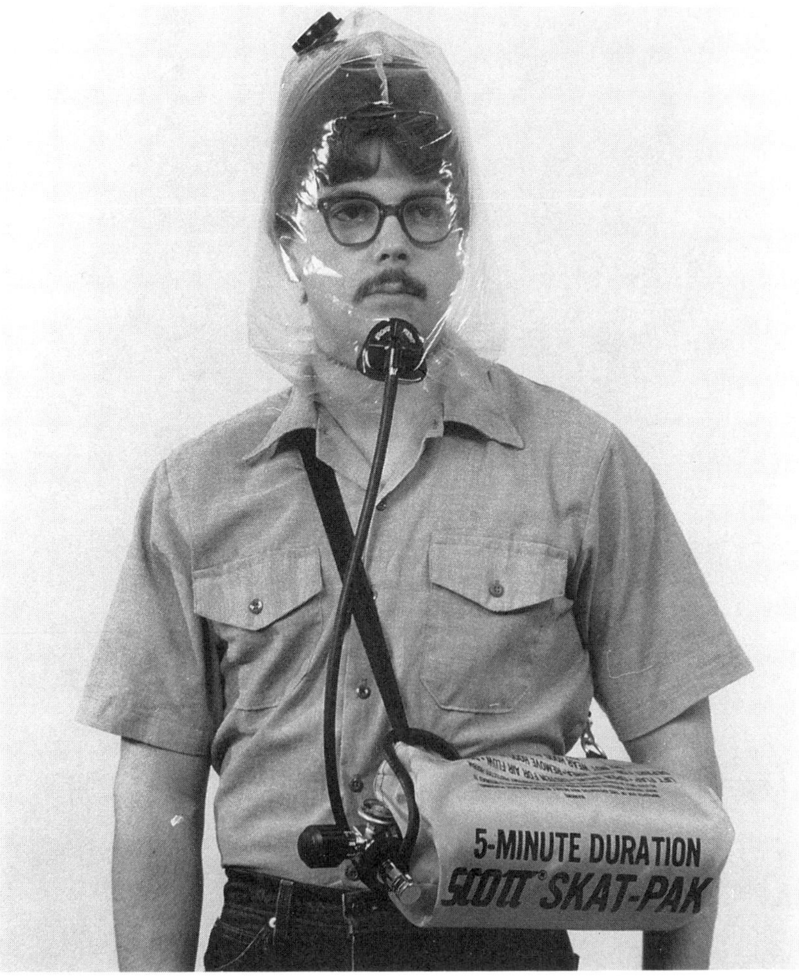

FIGURE 9-10. Escape-only SCBA. (Courtesy of Scott Aviation)

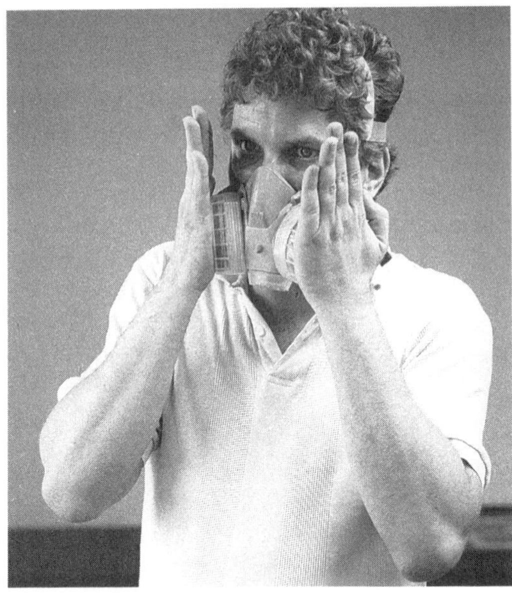

FIGURE 9-11. Negative-pressure fit check technique.

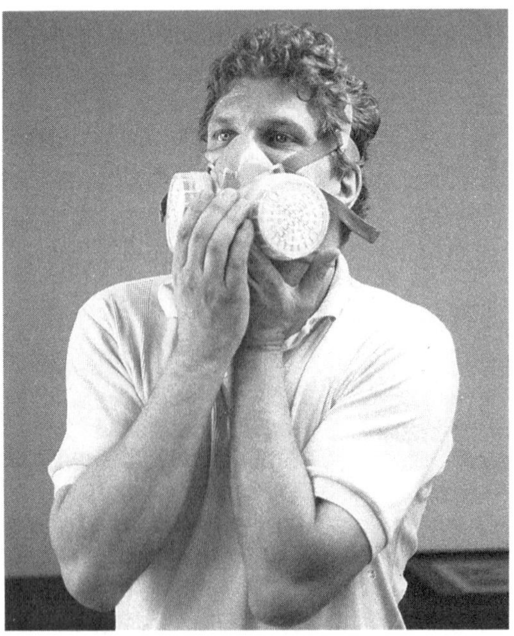

FIGURE 9-12. Positive-pressure fit check technique.

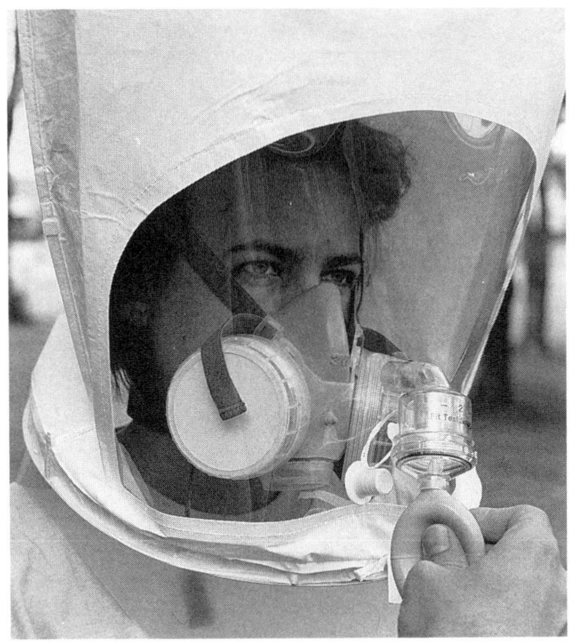

FIGURE 9-13. Qualitative fit testing using the saccharin mist protocol. (Courtesy of 3M Company)

oil or saccharin mist). If the facepiece fit is good, the wearer should experience no reactions or sensations related to the substance used. Qualitative fit test apparatus can be obtained in kit form from respirator manufacturers (Figure 9-13) or assembled from commonly available items. Specific fit testing protocols (NIOSH, 1987) should be conducted by a person with the appropriate expertise in occupational safety and health.

- *Quantitative fit tests* are complicated tests designed to produce a numerical value (or fit factor) indicating the degree of fit. Quantitative testing is typically performed by placing the wearer in an enclosure containing a known concentration of a contaminant (Figure 9-14). A sample is drawn from within the facepiece and analyzed to determine the concentration of the contaminant it contains. The fit factor is then calculated as follows:

$$\text{Fit Factor} = \frac{\text{Airborne Concentration of Contaminant}}{\text{Concentration of Contaminant within Facepiece}}$$

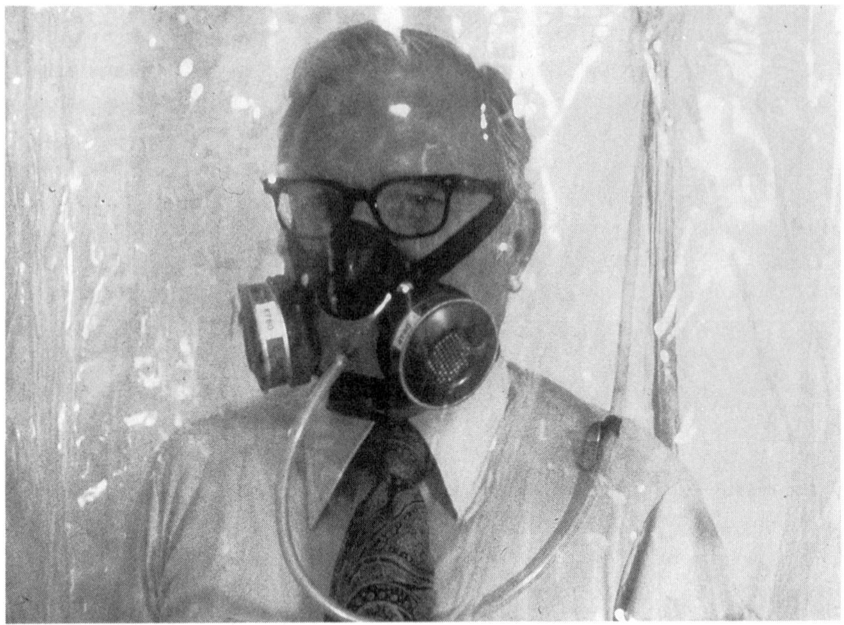

FIGURE 9-14. Quantitative fit testing apparatus.

Respirator Protection Factor

The use of quantitative fit testing to determine an actual fit factor for a given respirator is frequently impractical outside of a laboratory setting. However, protection factors have been assigned to the different types of respirators by the American National Standards Institute (ANSI). Examples of assigned protection factors are shown in Table 9-2.

TABLE 9-2. Examples of Assigned Protection Factors

Respirator Type	Protection Factor
Half-mask APRs	10
Full-facepiece APRs	50
SAR with full facepiece in positive-pressure mode	2,000
SCBA with full facepiece in positive-pressure mode	10,000

Source: NIOSH, 1987.

Assigned protection factors can be utilized to assess the effectiveness of respiratory protective equipment, as illustrated by the following equation:

$$\frac{CCwa}{PF} = CCfp$$

where

CCwa = contaminant concentration in the work area
PF = protection factor assigned to the type of respirator under consideration
CCfp = contaminant concentration within the facepiece

Obviously, if the contaminant concentration within the facepiece is in excess of the exposure limit for the contaminant of concern, the protection provided the worker is inadequate and some type of respiratory protective equipment providing a higher protection factor should be used.

CHEMICAL PROTECTIVE CLOTHING AND ACCESSORIES

Numerous substances likely to be encountered during emergency response operations pose a threat to the skin of personnel involved in response activities. These substances may directly attack the skin or pass through the skin to reach and attack other organs of the body. To prevent injury due to contact with these substances, it is necessary to place a barrier, in the form of chemical protective clothing (CPC) and related accessories, between the responder and the substances to which he or she is potentially exposed during response operations. This section serves as an introduction to the topic of CPC and accessories.

Selection of CPC

Selection Considerations

CPC is virtually worthless if not properly selected. For example, two gloves which appear to be typical rubber gloves to the casual observer may actually be composed of different materials which have completely different protective capabilities. Given the complexities involved and the potentially serious consequences of improper selection, it is obvious that selection is a task which should be undertaken only by personnel with appropriate knowledge, training, and experience. Proper selection requires full consideration of the following factors:

- *Specific chemical contaminants likely to be encountered* should be identified, if possible, prior to selection, since no single protective material is effective against all potential chemical assaults.
- The *performance characteristics of available protective clothing* in resisting chemical attack and physical damage must be evaluated.
- *Site- and/or task-specific requirements and limitations* must also be considered. For example, the selection process must take into account
  - The physical state of contaminants (e.g., liquid versus vapor)
  - The exposure time required for a given task (e.g., patching a leaking tank)
  - The likelihood of direct exposure, such as through chemical spraying or splashing, during a given procedure
  - The degree of stress (particularly heat stress) placed on the wearer by a given article of protective clothing

The information presented here is intended to serve as an introduction to the basics of CPC selection.

Attacks on CPC

In order to provide adequate protection for the wearer, an article of CPC must be sufficiently resistant to attacks by chemicals and physical agents present in the work aera. These attacks can be classified as follows:

- *Permeation* refers to the process by which a solid, liquid, or gaseous chemical dissolves into or passes through a chemical protective material at the molecular level (i.e., in the vapor phase).
- *Penetration* refers to the bulk movement of liquids through pores or small flaws in an article of CPC. Penetration may occur through imperfect seams, zippers, or pinholes in an article of CPC.
- *Degradation* refers to the loss of chemical resistance or physical competence of a protective material. This may occur due to chemical exposure or physical wear and abrasion while the garment is in use.

Resistance to Chemical Attacks

Chemical resistance of CPC is typically reported in the following terms:

- *Breakthrough time* is the time elapsing between the introduction of a chemical to the outer surface of a protective material and the initial detection of that chemical on the inner surface of the material.

- *Permeation rate* is the rate at which a permeating chemical moves through a given material after breakthrough, as determined under a set of test conditions. Permeation rate is reported in terms of the mass of contaminant passing through a specific area of material during a specific length of time (e.g., micrograms per square centimeter per minute).

Resistance to degradation may also be reported. For example, observation of subjective evidence of degradation (e.g., discoloration, swelling, stiffness, delamination, or cracking) of CPC material after chemical exposure may be noted. Objective factors such as weight change, loss of physical strength, and reduced permeation resistance may also serve as evidence of degradation.

Basic Principles of CPC Selection

In selecting and using CPC for emergency response operations, the following basic principles should be kept in mind at all times:

- No single chemical protective material offers protection against all chemicals. Selection should be chemical specific if the identities of chemicals which may be involved in accidental releases are known or anticipated. Data presented in Table 9-3 and Figure 9-15 offer excellent examples of the importance of chemical-specific selection.
- No protective material currently available is truly impermeable or able to provide an effective barrier to prolonged exposure. Breakthrough is simply a matter of time. The key question is whether or not a significant exposure will occur due to permeation of CPC before a given task can be completed.
- CPC should be selected which offers the widest range of protection available against the specific chemicals known or expected to be encountered. Materials which offer the longest breakthrough times and lowest permeation rates for the chemicals of concern are most desirable.
- As a general rule, CPC is not designed to provide thermal protection, and many CPC materials are in fact highly flammable. If CPC is worn in situations where flammable atmospheres may be encountered, special precautions (such as the use of aluminized oversuits) are required.

CPC Materials and Technologies

In the past, CPC technologies focused almost solely on reusable fabrics such as butyl rubber, Viton, urethane, nitrile rubber, neoprene, polyvinyl chloride, and others. Each of these materials offers excellent protection against certain chemicals. In recent years, manufacturers have begun to offer reusable CPC constructed of two or more layers of different materials laminated together, thus offering a widened range of protection. Reusable CPC materials are generally quite durable and offer adequate protection if properly selected. However, the reusables are expensive to purchase, require complete decontamination after every use, and may present matrix permeation problems (as discussed in the section on use of PPE).

In recent years, disposable items of CPC have come to be used commonly in certain situations. Tyvek is probably the most popular of the disposable materials, offering adequate protection against particulate contaminants in many instances. Materials constructed of Tyvek coated with polyethylene or Saran offer protection against certain liquids, gases, and vapors, as well as particulate contaminants. The disposables are relatively inexpensive to purchase and require no decontamination after use. However, they typically have low durability relative to the reusables.

The most recent additions to the disposable CPC market have been constructed of laminated film-based materials. Examples include the "Responder" by Life Guard, Inc., and the "Chemrel" suit by Chemron, Inc. These new, high-tech materials offer protection against a wide range of chemicals for an extended period of time and are quite inexpensive relative to reusable materials. For this reason, they have generated a significant amount of interest within the hazmat field.

Availability of Information on Performance Characteristics of CPC

Information which may be used in selecting CPC is typically available in two forms:

- *Qualitative recommendations* rate protective materials in nonstandard, subjective ways using terms such as "excellent," "good," "fair," "poor," "not recommended" for protection against specific chemicals (Figure 9-15). These recommendations may be based on subjective evaluations of resistance to degradation or on permeation resistance.
- *Quantitative chemical resistance data* report actual permeation rates and/or breakthrough times for a given protective material under attack by specific chemicals (Table 9-3).

Sources of information for use in selecting CPC include the following:

- *Various published reference documents*, such as the American Conference of Governmental Industrial Hygienists' "*Guidelines for the*

Selection of Chemical-Protective Clothing'' and Forsberg and Mansdorf's ''*Quick Selection Guide to Chemical Protective Clothing*'' (see the Reference section at the end of this chapter).

- *Various computer databases*, such as Hyper CPC Stacks, CPC base, and GlovES+ (see Chapter 16).
- *Vendor literature*, such as shown in Table 9-3 and Figure 9-15, which is published in various forms by manufacturers of CPC.
- *Other sources*, such as the *Emergency Action Guides* by the Association of American Railroads, which provide guidance on selecting CPC, in addition to numerous other topics related to emergency response.

Problems with Information Available on CPC

Information currently available on the performance characteristics of CPC has several deficiencies which may complicate the selection process. Each of the following limitations must be carefully considered when selecting CPC suitable for usage during a hazardous materials incident.

- Permeation rate and breakthrough time have traditionally been determined by testing small swatches of protective material while ignoring the performance of the suit or article of clothing as a whole. For example, the chemical resistance of seams, zippers, and visors is usually not considered in test results of this type. Also, boots and gloves used with a suit may be constructed of a different material than the suit itself.
- For the most part, protective materials are currently tested using single-chemical assaults. However, hazmat operations may involve exposure to mixtures, which are usually much more aggressive in attacking chemical protective materials than a single component. Data on resistance to multicomponent chemical attacks are currently inadequate.
- Permeation rates and breakthrough times published for a given material are determined in a carefully controlled experimental setting. However, performance characteristics of CPC in actual use during emergency response operations may vary widely with variations in factors such as type and concentration of chemicals, ambient temperature, and humidity.
- Procedures for making qualitative recommendations are not standardized within the CPC industry and may vary widely from one manufacturer to another. For example, one manufacturer may assign a rating of ''excellent'' to a material which is resistant to a given chemical for 2 hours, while another manufacturer would require a material

Chemical resistance offered by Fyrepel Fire/Chemical Suits.

Chemical	PVC	Viton	Butyl
Acetaldehyde	I	P	E
Acetic Acid	P	P	G
Acetone	P	P	E
Acrylonitrile	P	F	G
Aluminum Chloride	GF	E	E
Ammonium Hydroxide	E	G	E
Amyl Acetate	P	P	E
Amyl Alcohol	E	G	E
Aniline	P	G	G
Animal Fats	GF	E	G
Benzaldehyde	P	P	E
Benzene	FP	G	P
Benzyl Alcohol	P	E	G
Benzyl Chloride	P	E	F
Butane	GF	E	P
Butyl Acetate	P	P	G
Butyl Alcohol	GF	E	G
Butyraldehyde	P	P	G
Calcium Hypochlorite	E	E	E
Carbolic Acid	E	E	G
Carbon Tetrachloride	P	E	P
Castor Oil	GF	E	G
Chlorine (Dry)	GF	G	F
Chlorine (Wet)	P	G	F
Chloroacetone	P	P	G
Chlorobenzene	P	E	P
Chloroform	P	E	P
Chromic Acid	E	E	P
Citric Acid	E	E	E
Cottonseed Oil	E	E	F
Creosote	FP	E	P

Chemical	PVC	Viton	Butyl
Fluorine	I	G	F
Formaldehyde	E	P	E
Formic Acid	GF	F	E
Freon 11	FP	G	P
Freon 12	E	E	G
Freon 21	I	P	P
Freon 22	E	P	E
Furfural	P	P	G
Gasoline	FP	E	P
Glycerin	E	E	E
Hexane	GF	E	P
Hydraulic Fluid—Petroleum Base	I	E	P
Hydraulic Fluid—Ester Base	F	F	G
Hydrobromic Acid	E	E	E
Hydrochloric Acid 37%	E	E	E
Hydrofluoric Acid	E	G	G
Hydrogen Peroxide	E	E	G
Hydroquinone	E	G	P
Isobutyl Alcohol	GF	E	E
Iso-Octane	GF	E	P
Isopropyl Alcohol	E	E	E
Kerosene	GF	E	P
Lactic Acid	E	E	E
Lard	E	E	F
Linseed Oil	GF	E	G
Lubricating Oils (Petroleum)	GF	E	P
Maleic Acid	E	E	P
Methyl Acetate	P	P	G

Chemical	PVC	Viton	Butyl
Octyl Alcohol	E	E	G
Oleic Acid	E	G	P
Olive Oil	E	E	G
Oxalic Acid	E	E	E
Paint Remover	F	G	P
Pentane	I	E	P
Perchloric Acid	I	E	G
Perchloroethylene	P	E	P
Phenol	FP	E	G
Phosphoric Acid	I	E	E
Phosphoric Acid (20%, 45%)	E	E	E
Pickling Solution	E	F	F
Picric Acid	E	E	G
Pine Oil	GF	E	P
Plating Solutions—Chrome	I	E	E
Potassium Hydroxide (50%)	E	F	E
Printing Ink	I	I	I
Propane	FP	E	P
Propyl Acetate	FP	P	G
Propyl Alcohol	E	E	E
Propylene	E	E	P
Skydrol 500	I	F	G
Sodium Hydroxide (50%)	GF	G	E
Sodium Hypochlorite	GF	E	E
Soybean Oil	E	E	F
Stearic Acid	E	I	G
Stoddard Solvent	FP	E	P
Styrene	P	G	P
Sulfuric Acid (Diluted)	E	E	G
Sulfuric Acid (Concentrated)	E	E	G
Tannic Acid	E	E	E

Cutting Oil	I	E	F
Cyclohexane	GF	E	P
Cyclohexanol	FP	E	P
Diacetone Alcohol	FP	P	G
Dibenzyl	F	P	G
Dibutyl Phthalate	P	F	F
Diethylamine	GF	P	G
Di-Isobutyl Ketone	I	I	E
Di-Isocyanate	I	I	I
Dimethyl Formamide	P	P	E
Dioxane	P	P	G
Epoxy Resins	I	P	E
Ethyl Acetate	P	P	G
Ethyl Alcohol	GF	G	E
Ethyl Ether	P	P	G
Ethyl Formate	GF	E	G
Ethylene Dichloride	P	G	F
Ethylene Glycol	E	E	E
Ethylene Trichloride	FP	E	F
Methyl Alcohol	E	P	E
Methyl Bromide	I	E	F
Methyl Cellosolve	E	P	G
Methylene Chloride	P	G	F
Methyl Ethyl Ketone (M.E.K.)	P	P	E
Methyl Formate	GF	I	G
Methyl Isobutyl Ketone	P	P	G
Methylamine	I	I	G
Methyl Methacrylate	P	P	F
Mineral Oil	E	E	P
Monoethanolamine	I	P	G
Morpholine	I	I	G
Muriatic Acid	I	I	E
Naptha	FP	E	P
Nitric Acid—Concentrated	P	G	F
Nitric Acid—Diluted	GF	G	F
Nitric Acid—Red Fuming	P	F	P
Nitrobenzene	P	G	P
Nitromethane	I	P	G
Tetrahydrofuran	P	P.	G
Toluene	P	E	P
Toluene Di-Isocyanate	FP	P	G
Trichlorethylene	P	E	P
Tricresyl Phosphate	FP	G	E
Triethanol Amine	GF	P	G
Trinitrotoluene	FP	G	P
Tung Oil	FP	E	F
Turbine Oil	FP	E	P
Turpentine	FP	E	P
Vegetable Oil	FP	E	E
Vinyl Chloride	P	G	I
Xylene	P	E	P

E — No Effect
G — Minor Effect
F — Moderate Effect
P — Severe Effect
I — Insufficient Data Available

Best to have around you around danger.

PROTECTIVE WEAR

A Division of Lakeland Industries, Inc.

WARRANTY—Goods sold by Fyrepel shall be free from defects in workmanship and materials. No use warranty is implied or intended. Technical advice on typical applications is offered without obligation or liability.

Fyrepel Products Inc. Box 518, Newark, Ohio 43055
614/344-0391

FIGURE 9-15. Typical vendor literature showing qualitative data on chemical resistance of CPC. (Courtesy of Lakeland Industries, Inc.)

TABLE 9-3. Typical Vendor-Provided Quantitative Data on Chemical Resistance of Chemical Protective Clothing

Chemical Permeation Resistance Chart	VAUTEX 23 mils thick (19.1 oz/yd^2)						BETEX 19 mils thick (15.9 oz/yd^2)					
	Breakthrough Time In Minutes			Permeation Rate (Micrograms Per cm^2 Per Minute)			Breakthrough Time In Minutes			Permeation Rate In Micrograms Per cm^2 Per Minute		
	Test #			Test #			Test #			Test #		
Chemical	1	2	3	1	2	3	1	2	3	1	2	3
Acetone	15	15	15	33.4	24.8	36.7	135	135	135	12.9	10.8	7.5
Acetonitrile†	20	20	20	4.0	3.3	3.8	165	165	165	.04	.05	.05
Acrylonitrile	<5	<5	<5	24	21	23	125	125	125	0.18	0.22	0.10
Adiponitrile	240+	240+	240+	NA	NA	NA	240+	240+	240+	NA	NA	NA
Allyl alcohol	240+	240+	240+	NA	NA	NA	240+	240+	240+	NA	NA	NA
Ammonia gas	240+	240+	240+	NA	NA	NA	240+	240+	240+	NA	NA	NA
Ammonia liquid splash	5	5	5	.121	.025	.096	5	5	5	.117	.076	.022
	60	60	60	.020	.010	.014	60	60	60	.004	.007	.002
	240	240	240	.014	.000	.005	240	240	240	.000	.001	.000
Aniline	240+	240+	240+	NA	NA	NA	240+	240+	240+	NA	NA	NA
Benzene	65	65	165	.05	.08	.09	<5	<5	<5	144	147	147
Carbon disulfide†	240+	240+	240+	NA	NA	NA	<5	<5	<5	5.8	6.0	5.5
Chlorine gas	240+	240+	240+	NA	NA	NA	240+	240+	240+	NA	NA	NA
Dichloromethane†	20	20	20	50.1	52.6	55.1	<5	<5	<5	193	310	259
Diethylamine	35	35	35	53.7	50.1	51.9	<5	<5	<5	78.8	102.1	55.5
Dimethyl formamide†	30	30	30	6.8	7.7	8.4	240+	240+	240+	NA	NA	NA

Ethyl acetate†	15	15	15	57.2	59.1	59.1	30	30	30	5.5	7.0	5.7
Ethyl acrylate	<5	<5	<5	42.6	49.6	46.4	25	25	25	4.4	4.7	4.4
Ethylene dibromide	240+	240+	240+	NA	NA	NA	15	15	15	24.4	23.3	21.8
Formaldehyde	240+	240+	240+	NA	NA	NA	240+	240+	240+	NA	NA	NA
Gasoline	240+	240+	240+	NA	NA	NA	25	25	25	300	212	287
n-Hexane†	240+	240+	240+	NA	NA	NA	<5	<5	<5	36.6	36.6	36.6
Hexamethylene diamine	240+	240+	240+	NA	NA	NA	240+	240+	240+	NA	NA	NA
Hydrazine	240+	240+	240+	NA	NA	NA	240+	240+	240+	NA	NA	NA
Hydrochloric acid	240+	240+	240+	NA	NA	NA	240+	240+	240+	NA	NA	NA
Hydrofluoric acid	240+	240+	240+	NA	NA	NA	240+	240+	240+	NA	NA	NA
Hydrogen cyanide gas	80	80	80	0.7	0.7	0.6	240+	240+	240+	NA	NA	NA
Isobutylamine	45	45	45	15.4	16.0	13.2	50	50	50	6.9	6.8	6.4
Methyl alcohol† (methanol)	240+	240+	240+	NA	NA	NA	240+	240+	240+	NA	NA	NA
Methyl ethyl ketone	<5	<5	<5	62.4	68.6	64.6	30	30	30	12.4	12.2	12.6
Nitric acid	240+	240+	240+	NA	NA	NA	240+	240+	240+	NA	NA	NA
Nitrobenzene†	240+	240+	240+	NA	NA	NA	240+	240+	240+	NA	NA	NA
Nitrogen tetroxide†	15	15	15	2*	<1*	<1*	15	15	15	0*	<1*	<1*
	30	30	30	7*	<1*	<1*	30	30	30	>40*	<1*	>40*
	45	45	45	>40*	>40*	<7*	45	45	45	>40*	>40*	

NA = Not Applicable if breakthrough time exceeds 240 minutes.
† Listed in ASTM F1001.
* Concentrations observed expressed in parts per million. Permeation rates not available.
Complete degradation occurred after reaching 40 ppm.

Source: Mine Safety Appliances Company.

TABLE 9-4. Protective Clothing For Chemical Hazards

Type of Clothing or Accessory	Description	Type of Protection	Use Considerations
Fully encapsulating suit	One-piece garment. Boots and gloves may be integral, attached and replaceable, or separate.	Protects against splashes, dust, gases, and vapors.	Does not allow body heat to escape. May contribute to heat stress in wearer, particularly if worn in conjunction with a closed-circuit SCBA; a cooling garment may be needed. Impairs worker mobility, vision, and communication.
Nonencapsulating suit	Jacket, hood, pants, or bib overalls and one-piece coveralls.	Protects against splashes, dust, and other materials but not against gases and vapors. Does not protect parts of head or neck.	Do not use where gas-tight or pervasive splashing protection is required. May contribute to heat stress in wearer. Tape-seal connections between pant cuffs and boots and between gloves and sleeves.
Aprons, leggings, and sleeve protectors	Fully sleeved and gloved apron. Separate coverings for arms and legs. Commonly worn over nonencapsulating suit.	Provides additional splash protection of chest, forearms, and legs.	Whenever possible, should be used over a nonencapsulating suit (instead of using a fully encapsulating suit) to minimize potential for heat stress. Useful for sampling, labeling, and analysis operations. Should be used only when there is a low probability of total body contact with contaminants.

TABLE 9-4. (Continued)

Type of Clothing or Accessory	Description	Type of Protection	Use Considerations
Gloves and sleeves	May be integral, attached, or separate from other protective clothing.	Protect hands and arms from chemical contact.	Wear jacket cuffs over glove cuffs to prevent liquid from entering the glove. Tape-seal gloves to sleeves to provide additional protection.
	Overgloves.	Provide supplemental protection to the wearer and protect more expensive undergarments from abrasions, tears, and contamination.	
	Disposable gloves.	Should be used whenever possible to reduce decontamination needs.	
Safety boots	Boots constructed of chemical-resistant material.	Protect feet from contact with chemicals.	
	Boots constructed with some steel materials (e.g., toes, shanks, insoles).	Protect feet from compression, crushing, or puncture by falling, moving, or sharp objects.	All boots must at least meet the specifications required under OSHA 29 CFR Part 1910.136 and should provide good traction.
	Boots constructed from nonconductive, spark-resistant materials or coatings.	Protect the wearer against electrical hazards and prevent ignition of combustible gases or vapors.	
Disposable shoe or boot covers	Made of a variety of materials. Slip over the shoe or boot.	Protect safety boots from contamination. Protect feet from contact with chemicals.	Covers may be disposed of after use, facilitating decontamination.

Source: Adapted from NIOSH, 1985.

to be resistant to the same chemical for 8 hours in order to receive the same rating.

Types of CPC and Accessories

Numerous types of CPC and related accessories are available to protect personnel involved in emergency response activities. As in all cases involving PPE, chemical protective garments must be appropriate for the hazards and workplace conditions in which they are used.

Chemical Protective Clothing

Table 9-4 presents a summary of general information on various types of protective clothing, some of which are discussed below.

Totally Encapsulating Chemical Protective Suits

Totally encapsulating chemical protective (TECP) suits (Figures 9-16 and 9-17) completely enclose the wearer. These suits are designed to seal out gases and vapors, in addition to protecting against liquid splashes and particulate contaminants. TECP suits must be capable of maintaining a positive internal pressure so that any leakage will theoretically occur from the inside out.

TECP suits have traditionally been constructed of reusable materials, but have more recently been produced using disposable materials as well. It should be noted that problems in maintaining adequate positive pressure have been encountered with some disposable TECP suits.

Nonencapsulating Suits

Nonencapsulating chemical protective (NECP) suits cover most of the wearer's skin, thus providing good protection from splashes and/or particulate contaminants (Figures 9-18, 9-19, 9-20, and 9-21). Protection can be enhanced by sealing cuffs to boots and gloves with tape. However, it is not possible to convert a NECP suit into a TECP suit through the use of tape.

Disposable Overgarments

Disposable overgloves and overboots or boot covers are recommended to minimize exposure of the underlying chemical-protective garments and enhance personal protection. For example, a cheap glove can be worn over an expensive glove so that the cheaper outer glove will bear the brunt of wear and gross contamination. Thus, the expensive inner glove is protected. At the time of decontamination, the outer glove can simply be disposed of so that problems related to decontamination are minimized.

TABLE 9-5. Protective Clothing For Nonchemical Hazards

Type of Clothing or Accessory	Description	Type of Protection	Use Considerations
Fire fighters' protective clothing	Gloves, helmet, running or bunker coat, running or bunker pants (NFPA No. 1971, 1972, 1973), and boots.	Protects against heat, hot water, and some particles. Does not protect against gases and vapors, or chemical permeation or degradation.	Decontamination is difficult. Should not be worn in areas where protection against gases, vapors, chemical splashes, or permeation is required.
Fire entry suit	Constructed of multiple layers of insulated, flame-retardant material with an aluminized outer layer.	Protects against heat during short-term entry into total flame environments.	Not designed to provide protection against chemical permeation, penetration, or degradation.
Proximity suit	Constructed of an inner shell of insulated, flame-retardant material with an aluminized outer shell.	Protects against short-duration exposure to heat in close proximity to areas of active combustion.	Not designed to provide protection against chemical permeation, penetration, or degradation.
Flash protective suit (flash cover)	Oversuit of Nomex or Kevlar with an aluminized outer surface	Worn over CPC, provides limited protection to allow escape in the event of flash fire during entry. Not suitable for use in situations involving active combustion.	Adds bulk and may exacerbate heat stress problems and impair mobility.
Blast and fragmentation suit	Blast and fragmentation vest and clothing, bomb blankets, and bomb carriers.	Provides some protection against very small detonations. Bomb blankets and baskets can help redirect a blast.	Does not provide hearing protection.
Radiation-contamination protective suit	Various types of protective clothing designed to prevent contamination of the body by radioactive particles.	Protects against alpha and beta particles. *Does NOT protect against gamma radiation.*	Designed to prevent skin contamination.

Source: Adapted from NIOSH, 1985.

Likewise, cheap disposable chemical protective suits can be used as oversuits to be worn over expensive reusable chemical protective suits and then discarded at the time of decontamination.

Protective Clothing for Nonchemical Hazards

Situations which pose potential hazards in addition to chemical exposure may arise during emergency response operations. Examples of such situations are fires or potential explosions on site. Types of protective clothing which may be required in responding to such situations are described in Table 9-5.

Accessories

In addition to the various articles of protective clothing previously described, numerous accessory items may be used to enhance protection and allow for safer operations during emergency response. Some of these accessories are described in Table 9-6 and shown in Figure 9-24.

CLASSIFICATION OF PROTECTIVE ENSEMBLES

Individual components of respiratory protective equipment, CPC, and various accessories may be assembled into personal protective ensembles providing protection as demanded by site-specific hazards. Systematic approaches to the classification of PPE ensembles include (1) EPA levels of protection and (2) NFPA certification standards for CPC. Both of these systems are described in this section.

EPA Levels of Protection

EPA has created recommendations for levels of protection for PPE ensembles. These recommendations are shown in Tables 9-7 through 9-10 and are briefly discussed below.

Level A Protection

Level A protection provides the highest available degree of both respiratory protection and skin and eye protection. Thus, the TECP suit can be considered the definitive protective item of the Level A ensemble (Figure 9-16). Information on Level A protection is included in Table 9-7.

Level A protection is required for entries into atmospheres containing high concentrations of unidentified contaminants (Figure 9-17). It is also required for entries into atmospheres containing identified airborne contaminants which are known to pose a high degree of hazard to the skin.

TABLE 9-6. Accessory Items For Protective Ensembles

Type of Clothing or Accessory	Description	Type of Protection	Use Considerations
Flotation gear	Life jackets or work vests. (Commonly worn underneath CPC to prevent flotation gear degradation by chemicals.)	Adds 15.5 to 25 pounds (7 to 11.3 kilograms) of buoyance to personnel working in or around water.	Adds bulk and restricts mobility. Must meet USCG standards (46 CFR Part 160).
Cooling garment	One of three methods: 1. A pump circulates cool dry air throughout the suit or portions of it via an air line. Cooling may be enhanced by use of a vortex cooler, refrigeration coils, or a heat exchanger. 2. A jacket or vest having pockets into which packets of ice are inserted. 3. A pump circulates chilled water from a water/ice reservoir and through circulating tubes, which cover part of the body (generally the upper torso only).	Removes excess heat generated by worker activity, the equipment, or the environment.	1. Pumps circulating cool air require 10 to 20 ft^3 (0.3 to 0.6 m^3) of respirable air per minute, so it is often uneconomical for use. 2. Jackets or vests pose ice storage and recharge problems. 3. Pumps circulating chilled water; poses ice storage problems. The pump and battery add bulk and weight.
Safety helmet (hard hat)	For example, a hard plastic or rubber helmet.	Protects the head from blows.	Helmet shall meet OSHA Standard 29 CFR Part 1910.135.

TABLE 9-6. (Continued)

Type of Clothing or Accessory	Description	Type of Protection	Use Considerations
Helmet liner		Insulates against cold. Does not protect against chemical splashes.	
Hood	Commonly worn with a helmet.	Protects against chemical splashes, particulates, and rain.	
Faceshield	Full-face coverage, 8-inch minimum.	Protects against chemical splashes. Does not protect adequately against projectiles.	Faceshields and splash hoods must be suitably supported to prevent them from shifting and exposing portions of the face or obscuring vision. Provide limited eye protection.
Splash hood		Protects against chemical splashes. Does not protect adequately against projectiles.	
Safety glasses		Protect eyes against large particles and projectiles.	
Goggles		Depending on their construction, goggles can protect against vaporized chemicals, splashes, large particles, and projectiles (if constructed with impact-resistant lenses).	

TABLE 9-6. (Continued)

Type of Clothing or Accessory	Description	Type of Protection	Use Considerations
Earplugs and muffs		Protect against physiological damage and psychological disturbance.	Must comply with OSHA Standard 29 CFR Part 1910.95. Can interfere with communication. Use of earplugs should be carefully reviewed by a health and safety professional because chemical contaminants could be introduced into the ear.
Headphones	Radio headset with throat microphone.	Provide some hearing protection while enabling communication.	Highly desirable, particularly if emergency conditions arise.
Personal dosimeter		Measures worker exposure to ionizing radiation and to certain chemicals.	
Personal locator beacon	Operated by sound, radio, or light.	Enables emergency personnel to locate victim.	
Two-way radio		Enables field workers to communicate with personnel in the cold zone.	
Knife		Allows a person in a fully encapsulating suit to cut his way out of the suit in the event of an emergency or equipment failure.	Should be carried and used with caution to avoid puncturing the suit.

Source: Adapted from NIOSH, 1985.

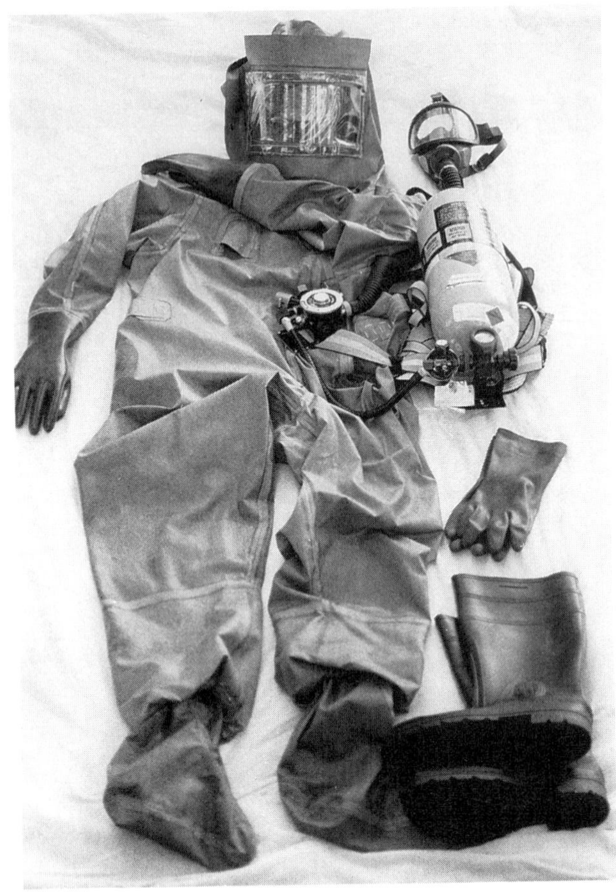

FIGURE 9-16. Totally encapsulating chemical protective suit, SCBA, and other components of typical Level A ensemble of PPE.

Such contaminants may be present as gases, vapors, or particulates and may attack the skin directly on contact or pass through the skin to attack other organs. Level A protection may also be required if liquids hazardous to the skin are likely to be encountered in an operation involving major splashing or high-pressure spraying (e.g., valve repair).

Level B Protection

Level B protection provides the same degree of respiratory protection as Level A but a lesser degree of skin protection. Thus, the Level B ensemble is quite similar to the Level A ensemble, except for the use of a

TABLE 9-7. Equipment and Use Considerations For Level A Protection

Equipment	Protection Provided	Should Be Used When:	Limiting Criteria
Recommended • Pressure-demand, full-facepiece SCBA or pressure-demand SAR with escape SCBA. • Fully encapsulating, chemical-resistant suit. • Inner chemical-resistant gloves. • Chemical-resistant safety boots/shoes. • Two-way radio communications. Optional • Cooling unit. • Coveralls. • Long cotton underwear. • Hard hat. • Disposable gloves and boot covers.	The highest available level of respiratory, skin, and eye protection.	• The chemical substance has been identified and requires the highest level of protection for skin, eyes, and the respiratory system based on either: — measured (or potential for) high concentration of atmospheric vapors, gases, or particulates or — site operations and work functions involving a high potential for splash, immersion, or exposure to unexpected vapors, gases, or particulates of materials that are harmful to skin or capable of being absorbed through the intact skin. • Substances with a high degree of hazard to the skin are known or suspected to be present, and skin contact is possible. • Operations must be conducted in confined, poorly ventilated areas until the absence of conditions requiring Level A protection is determined.	• Fully encapsulating suit material must be compatible with the substances involved.

Source: Adapted from NIOSH, 1985.

FIGURE 9-17. Use of Level A protection.

nonencapsulating suit such as a two-piece splash suit (Figure 9-18). Information on Level B protection is included in Table 9-8. Level B is used whenever the conditions for the use of APRs are not met and threats to the skin are non-IDLH and in the form of liquid splashes or particulate contaminants (Figure 9-19).

Level C Protection

Level C protection involves the same degree of skin protection as level B but a lesser degree of respiratory protection. The Level C ensemble is quite similar to the Level B ensemble, except that respiratory protection is provided by an APR rather than an SCBA (Figure 9-20). Information on Level C equipment is presented in Table 9-9. Level C should be used only in work areas where it can be determined that the use of APRs is safe (Figure 9-21). For this reason, the role of Level C in emergency response tends to be highly restricted (see the section on APRs).

Level D Protection

Level D PPE provides protection only against normal workplace safety hazards. Hard hats, steel-toed boots, safety glasses, and cotton work clothes are basic components of the Level D ensemble (Figure 9-22 and Table 9-10). Level D should be used only in work areas such as the cold zone, where both respiratory and skin hazards are absent (Figure 9-23).

An important aspect of PPE is protection from physical safety hazards, as well as the chemical-related hazards typical of the emergency operations area. All levels of protection should incorporate the basic safety equipment (as represented by the Level D ensemble) needed to ensure responder safety from all site hazards.

Modified Levels of Protection

In considering levels of protection, it should be noted that the four levels presented here are highly generalized and should be fine-tuned to provide

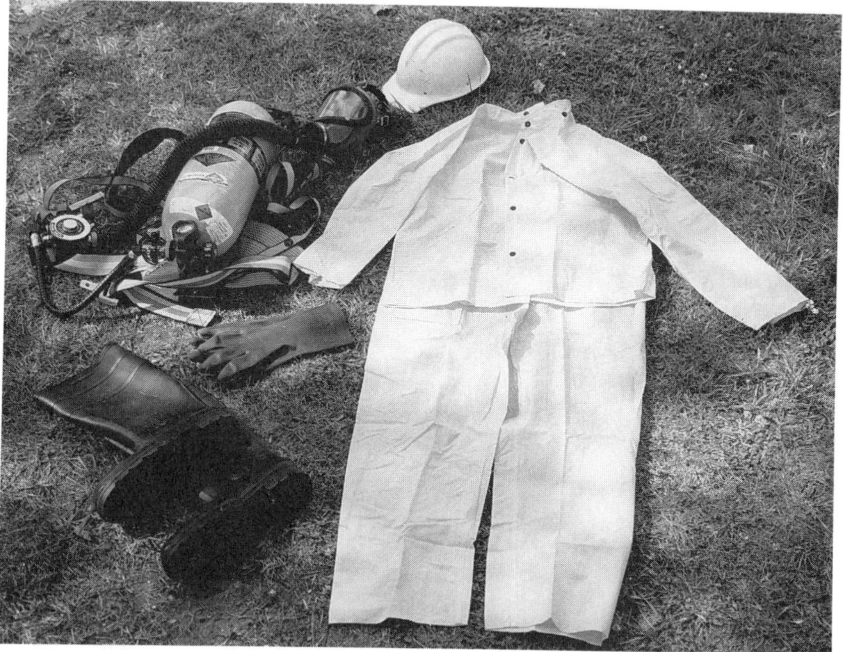

FIGURE 9-18. Nonencapsulating chemical protective suit, SCBA, and other components of a typical Level B ensemble of PPE.

TABLE 9-8. Equipment and Use Considerations For Level B Protection

Equipment	Protection Provided	Should Be Used When:	Limiting Criteria
Recommended • Pressure-demand, full-facepiece SCBA or pressure-demand SAR with escape SCBA. • Chemical-resistant clothing (overalls and long-sleeved jacket; hooded, one- or two-piece chemical splash suit; disposable chemical-resistant one-piece suit). • Inner and outer chemical-resistant gloves. • Chemical-resistant safety boots/shoes. • Hard hat. • Two-way radio communications. Optional • Coveralls. • Disposable boot covers. • Faceshield. • Long cotton underwear.	The same level of respiratory protection but less skin protection than Level A. It is the minimum level recommended for initial site entries until the hazards have been further identified.	• The type and atmospheric concentration of substances have been identified and require a high level of respiratory protection but less skin protection. This involves atmospheres: — with IDLH concentrations of specific substances that do not represent a severe skin hazard or — that do not meet the criteria for use of APRs. • Atmosphere contains less than 19.5 percent oxygen. • Presence of incompletely identified vapors or gases is indicated by direct-reading organic vapor detection instrument, but vapors and gases are not suspected of containing high levels of chemicals harmful to skin or capable or being absorbed through the intact skin.	• Use only when the vapor or gases present are not suspected of containing high concentrations of chemicals that are harmful to skin or capable of being absorbed through the intact skin. • Use only when it is highly unlikely that the work being done will generate either high concentrations of vapors, gases, or particulates or splashes of material that will affect exposed skin.

Source: Adapted from NIOSH, 1985.

FIGURE 9-19. Use of Level B protection.

FIGURE 9-20. Nonencapsulating chemical protective suit, APR, and other components of a typical Level C ensemble of PPE.

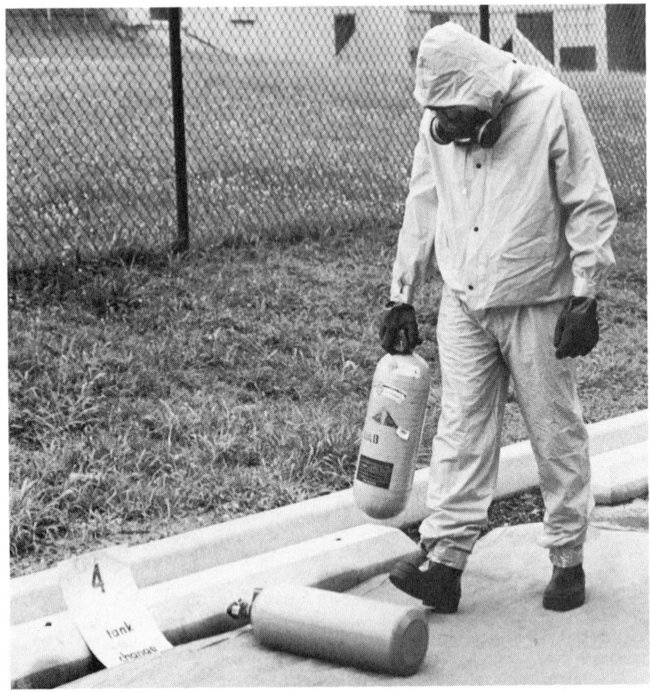

FIGURE 9-21. Use of Level C protection.

the specific degree of protection required for a particular task. Any number of modified levels of protection may be used in emergency response.

Accessory items, as described in Table 9-6 and shown in Figure 9-24, can be used to enhance the protection provided by various personal protective ensembles.

The classification of PPE represented by the four levels of protection completely ignores the topic of CPC flammability. This is significant to the responder, since many chemical protective materials burn readily if ignited (Figure 9-25). The NFPA has proposed a more specific classification of CPC (as described in the following section) which takes into consideration additional factors, including the degree of flammability of various chemical protective materials.

For all entries during which flammable atmospheres may be encountered, a flash cover should be worn over CPC, as shown in Figure 9-26. Flash covers provide limited protection against radiant heat in order to allow the entrant to escape from the hazardous area in the event that flash fire occurs during entry. However, flash covers are not designed to provide adequate thermal protection for entry into areas where active combustion is occurring. More specialized thermal protective equipment is required for such entries.

NFPA Certification Standards for Chemical Protective Garments

NFPA has developed three performance-based standards which require that chemical protective suits provide a specified minimum level of pro-

TABLE 9-9. Equipment and Use Considerations For Level C Protection

Equipment	Protection Provided	Should Be Used When:	Limiting Criteria
Recommended • Full-facepiece, air-purifying, canister-equipped respirator. • Chemical-resistant clothing (overalls and long-sleeved jacket; hooded, one- or two-piece chemical splash suit; disposable chemical-resistant one-piece suit). • Inner and outer chemical-resistant gloves. • Chemical-resistant safety boots/shoes. • Hard hat. • Two-way radio communications. Optional • Coveralls. • Disposable boot covers. • Faceshield. • Escape mask. • Long cotton underwear.	The same level of skin protection as Level B but a lower level of respiratory protection.	• The atmospheric contaminants, liquid splashes, or other direct contact will not adversely affect any exposed skin. • The types of air contaminants have been identified and their concentrations measured, and a canister is available that can remove the contaminant. • All criteria for the use of APRs have been met.	• Atmospheric concentration of chemicals must not exceed IDLH levels. • The atmosphere must contain at least 19.5 percent oxygen.

Source: Adapted from NIOSH, 1985.

FIGURE 9-22. Typical components of a Level D personal protective ensemble.

TABLE 9-10. Equipment and Use Considerations For Level D Protection

Equipment	Protection Provided	Should Be Used When:	Limiting Criteria
Recommended • Coveralls. • Safety boots/ shoes. • Safety glasses or chemical splash goggles. • Hard hat. Optional • Gloves. • Escape mask. • Faceshield.	No respiratory protection. Minimal skin protection.	• The atmosphere contains no known hazard. • Work functions preclude splashes, immersion, or the potential for unexpected inhalation of or contact with hazardous levels of any chemicals.	• This level should not be worn in the hot zone. • The atmosphere must contain at least 19.5 percent oxygen.

Source: Adapted from NIOSH, 1985.

FIGURE 9-23. Use of Level D protection.

tection in order to be certified for use. This process involves a comprehensive approach requiring whole-ensemble evaluation, as opposed to merely specifying factors such as the type of suit, barrier fabric, and methods of construction.

NFPA 1991: Vapor-Protective Suits for Hazardous Chemical Emergencies

For certification under NFPA 1991, vapor-protective suits must be able to pass a pressure test conducted in accordance with ASTM F 1052, Practice for Pressure Testing of Gas-Tight Totally Encapsulated Chemical Protective Suits. Suits must also be able to pass a "shower test" for penetration resistance as specified by NFPA.

The primary material of which a suit is constructed (i.e., the barrier fabric), as well as visors, boots, and gloves, must be able to resist breakthrough for at least 1 hour when tested for permeation by each of 15 chemicals specified in ASTM Standard Guide F 1001, plus ammonia and chlorine gas. The NFPA test battery thus consists of 17 chemicals which are considered to be representative of the major chemical classes encoun-

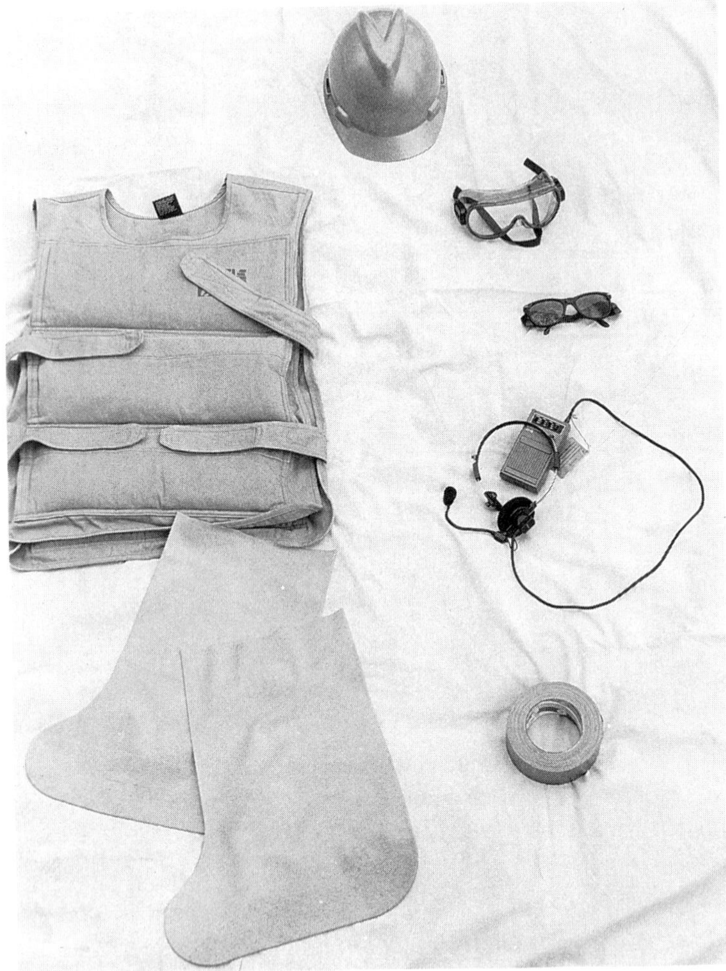

FIGURE 9-24. Accessory items of PPE, including (clockwise from top) hard hat, splash goggles, safety glasses, voice-actuated two-way radio, duct tape, disposable boot covers, and cooling vest.

tered during hazmat incidents. The test battery for certification under NFPA 1991 includes the following chemicals:

- Acetone
- Acetonitrile
- Anhydrous ammonia

- Carbon disulfide
- Chlorine
- Dichloromethane
- Diethyl amine
- Dimethyl formamide
- Ethyl acetate
- Hexane
- Methanol
- Nitrobenzene
- Sodium hydroxide
- Sulfuric acid
- Tetrachloroethylene
- Tetrahydrofuran
- Toluene

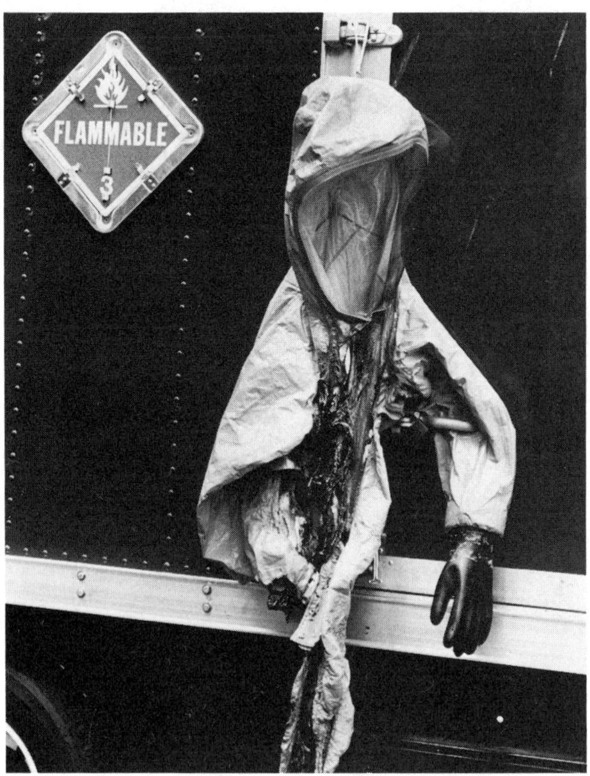

FIGURE 9-25. Effects of fire on CPC.

FIGURE 9-26. Use of flash cover with a TECP suit.

The suit must also offer adequate permeation resistance for any additional chemicals for which the suit is certified. Any ensemble components worn outside the primary suit material must also meet the minimum requirements for chemical resistivity.

All suit components must meet minimum performance-based requirements for certification under NFPA 1991. These components include visors, valves, seams, closures, gloves, and boots. The suits must also meet minimum requirements for durability and flammability resistance.

NFPA 1992: Liquid Splash-Protective Suit for Hazardous Chemical Emergencies

NFPA 1992, like NFPA 1991, is a whole-ensemble, performance-based standard. However, the major focus of 1992 is on protection in chemical splash environments. Thus, no requirements related to gas-tight integrity are applicable under 1992. However, overall suit water penetration resistance must be demonstrated through shower testing.

NFPA 1992 also requires that splash-protective ensembles adequately resist penetration by chemicals. The NFPA battery of challenge chemicals used for this standard was developed by deleting from the ASTM F 1001 chemical assemblage those chemicals known to be skin toxins or known or suspected of causing cancer. The test battery for certification under NFPA 1992 includes the following chemicals:

- Acetone
- Diethylamine
- Ethyl acetate
- Hexane
- Sodium hydroxide
- Sulfuric acid
- Tetrahydrofuran
- Toluene

All ensemble components, including barrier fabric, seams, closures, gloves, boots, visors, and respirator components (if worn outside the CPC), must resist penetration by these chemicals and any additional chemicals for which the suit is certified for at least 1 hour. All suit components must also meet minimum requirements for durability and flammability resistance for certification under NFPA 1992.

NFPA 1993: Support Function Protective Garments for Hazardous Chemical Operations

NFPA 1993, like NFPA 1992, is a whole-ensemble, performance-based standard with a focus on splash protection. However, ensembles certified under NFPA 1993 are intended only for use in support functions, as opposed to use in a hot zone situation. Examples of support functions are decontamination or cleanup operations, in which all activities are carried out in controlled environments where conditions are known. Certification requirements under NFPA 1993 are similar to those under NFPA 1992, with the major exception that flammability resistance is not required for certification under 1993.

SAFE USE OF PPE

Properly selected PPE can provide adequate protection only if properly used. Otherwise, PPE may provide nothing more than a false sense of security to the user. Also, in some instances, the physical stress imposed upon the responder by PPE may be more injurious than the chemicals present. Safe use of respirators, CPC, and various accessories requires attention to a number of specific considerations. This section is intended to make the reader aware of various practical concerns for the safe use of PPE.

PPE Program Requirements

Use of PPE by emergency responders must be covered by a written PPE program, as specified in 29 CFR 1910.120. While using PPE, all personnel must adhere strictly to the provisions of the program. OSHA requires that the PPE program address each of the following topics:

- Selection of PPE
- Use and limitations of equipment
- Work mission duration
- Maintenance
- Storage
- Decontamination and/or disposal after use
- Training
- Proper fitting of PPE
- Donning and doffing procedures
- Inspection procedures
- Limitations during temperature extremes
- Program evaluation

Under 29 CFR 1910.134 (which is incorporated by reference in 29 CFR 1910.120), OSHA requires the employer to implement a written respiratory protection program covering all employees whose job assignments require the use of respirators. The respiratory protection program must meet the following requirements:

- The program must be in writing.
- Respirator selection must be hazard specific.
- Workers must be adequately trained in respirator use.
- Fit testing must be included.
- Only approved equipment may be used.
- Work area surveillance must be conducted to ensure that respiratory protection is appropriate and used as required by the program.
- Respirators must be decontaminated and sanitized after each use.
- Storage areas for respirators must be convenient, clean, and sanitary.
- Respiratory equipment must be regularly inspected, maintained, and repaired as needed.

- Workers must be medically determined to be physically fit to wear a respirator before being assigned to tasks requiring respiratory protection.
- The effectiveness of the program should be evaluated on a regular basis.

The written PPE program must adequately address all the above topics and requirements related to respirators, CPC, and other protective equipment as appropriate for the types of protective gear available to response personnel. Specific considerations which the PPE program should address are discussed in this section or elsewhere in this chapter.

Training in PPE Use

Training in PPE use is required by OSHA in 29 CFR Parts 1910.120 and 1910 Subparts I and Z. Before entry into an area requiring PPE, all responders should be trained sufficiently to

- Establish user familiarity and confidence
- Make the user aware of the capabilities and limitations of the equipment used
- Maximize the protective efficiency of equipment
- Maximize the ability to work efficiently in PPE

An adequate PPE training program should cover all major points presented below. However, PPE training must be largely site specific, since both equipment and hazard-specific considerations of use will vary widely.

Factors Limiting Safe Work Mission Duration

Work mission duration must be estimated before work in PPE actually begins. The factors limiting the length of time that one can safely remain in a contaminated area will now be discussed.

Air Supply Consumption

In situations in which the SCBA is used, limited air supply can be a major problem and, in some instances, a major threat to the safety of responders. Air supply consumption with an SCBA unit may be significantly increased (thus reducing the time on task) by factors such as strenuous work rate, lack of user fitness, and/or large body size of user. Shallow, rapid, irregular breathing patterns, or hyperventilation, can also lead to rapid air consumption. These conditions may result from heat stress, anxiety, lack of acclimatization, or lack of familiarization with the SCBA.

Permeation and Penetration of Protective Clothing or Equipment

Work mission duration cannot safely exceed the length of time during which items of CPC used can be expected to provide adequate protection. Thus, penetration and permeation of CPC (as described previously) are of major concern.

Penetration may occur due to leakage of fasteners or valves on PPE, particularly under extreme temperature conditions (as discussed below).

Permeation may occur due to improper selection of material or prolonged use of equipment in a given atmosphere.

Ambient Temperature Extremes

Ambient temperature extremes can affect responder safety and safe work duration in a number of ways. For example, the effectiveness of PPE may be reduced as hot or cold temperatures affect

- *Valve operation* on suits and/or respirators
- *Durability and flexibility* of CPC materials
- *Integrity of fasteners* on suits
- *Concentration* of airborne contaminants
- *Breakthrough time and permeation rate* of chemicals

In many instances, heat stress is the most immediate hazard to the wearer of PPE and the greatest limitation on work mission duration. The coolant supply will directly affect mission duration when cooling devices are required to prevent heat stress. Specific methods of dealing with heat stress are discussed below.

Personal Factors Affecting Respirator Use

A number of personal use factors may diminish the effectiveness of respirators. For example, any facial hair or long hair which comes between the respirator's facial seal gasket and the wearer's skin will prevent a good respirator fit. Beards are not allowable for response team members required to use respirators. Facial features such as scars, hollow cheeks, deep skin creases, or missing teeth may also prevent a good respirator fit. Chewing gum and tobacco should be prohibited during respirator use.

Likewise, the temple pieces on conventional eyeglasses interfere with respirator fit. However, spectacle kits are available which can be used to mount the corrective lenses within the facepiece. Contact lenses may not currently be worn with a full-facepiece respirator under OSHA regulations (29 CFR 1910.134). However, the American National Standards Institute (ANSI, 1988) has proposed that the use of contact lenses be allowed with full-facepiece respirators, provided that the wearer can demonstrate the ability to do so without problems.

Donning PPE

In donning an ensemble of PPE, an established routine should be followed. All equipment should be inspected as part of the donning procedure. Donning and doffing of PPE should always be done with the aid of an assistant. A field check (i.e., positive- and negative-pressure tests) for respirator fit should *always* be performed as part of the donning procedure. After donning, all ensemble components should be checked for proper fit, proper functioning, and relative comfort before entering a hazardous area.

Inspection of PPE

PPE should be fully inspected before each use. PPE inspection checklists such as the following may be used.

Inspecting CPC

- General Inspection Procedure (applicable to all items of CPC)
  - Determine that the clothing material is correct for the specified task at hand.
  - Inspect visually for imperfect seams, nonuniform coatings, tears, and malfunctioning closures.
  - Hold CPC up to a light and check for pinholes.
  - Flex the product and observe for cracks and other signs of shelf deterioration.
  - If the product has been used previously, inspect it inside and out for signs of chemical attack, such as discoloration, swelling, and stiffness.
- Inspecting Fully Encapsulating Suits
  - Check the operation of pressure relief valves.
  - Inspect the fitting of wrists, ankles, and neck.
  - Check the faceshield, if so equipped, for cracks, crazing, and/or fogginess.
  - TECP suits require periodic pressure testing or whole suit in-use testing (as described in Appendix A of 29 CFR 1910.120).

- Inspecting Gloves
  - Before use, check for pinholes. Blow into the glove, and then roll the gauntlet toward the fingers and holt it underwater. No air should escape.

Inspecting Respirators

- General Procedures (applicable to all types of respirators)
  - Check the material condition of the harness, facial seal, and breathing tube (if so equipped) for pliability, signs of deterioration, discoloration, and damage.
  - Check faceshields and lenses for cracks, crazing, and fogginess.
  - Check inhalation and exhaust valves for proper operation.
- Inspecting APRs
  - Inspect APRs:
    Before each use to be sure that they have been adequately cleaned.
    After each use during cleaning.
    At least monthly if in storage for emergency use.
  - Examine cartridges or canisters to ensure that
    They are the proper type for the intended use.
    The expiration date has not passed.
    They have not been opened or used previously.
- Inspecting SCBAs
  - Inspect SCBAs
    Before and after each use.
    At least monthly when in storage.
    Every time they are cleaned.
  - Check the air supply.
  - Check all connections for tightness.
  - Check for proper setting and operation of regulators and valves (according to manufacturers' recommendations).
  - Check the operation of alarms.
- Inspecting SARs
  - Inspect SARs
    Daily when in use.
    At least monthly when in storage.
    Every time they are cleaned.
  - Inspect the airline prior to each use, checking for cracks, kinks, cuts, frays, and weak areas.
  - Check for proper setting and operation of regulators and valves (according to manufacturers' recommendations).
  - Check the escape air supply (if applicable).
  - Check all connections for tightness.

In-Use Monitoring of PPE

While working in PPE, responders should constantly monitor equipment performance. If indications of possible in-use equipment failure are noted, personnel should exit the contaminated area immediately and investigate.

Degradation of ensemble components during use may be indicated by discoloration, swelling, stiffening, and softening of materials. Likewise, any tears, punctures, or splits at seams or zippers should be noted, as should any unusual residues on items of PPE (see Chapter 10).

Perception of odors; irritation of skin, eyes, and/or the respiratory tract; and general discomfort may be indications of equipment failure. Also, symptoms commonly associated with chemical exposure and oxygen deficiency may be the end result of equipment failure. These symptoms include rapid pulse, nausea, chest pain, difficulty in breathing, or undue fatigue. Restrictions of movement, vision, or communication may also result from equipment failure.

Doffing PPE

Like donning, doffing of PPE should be done according to an established routine. Furthermore, doffing routines should be well integrated with decontamination and disposal procedures for used PPE.

Storage of PPE

Storage is an important aspect of PPE use for emergency responders, since hazmat equipment is typically not used on a regular basis. Improper storage may lead to damage due to contact with dust, moisture, sunlight, damaging chemicals, extreme temperatures, and physical abrasion.

The following considerations should be observed in storing CPC:

- Potentially contaminated clothing should be stored in an area separate from street clothing.
- Potentially contaminated clothing should be stored in a well-ventilated area, with good air flow around each item if possible.
- Different types and materials of clothing and gloves should be stored separately to prevent issuing the wrong material by mistake.
- Protective clothing should be folded or hung in accordance with manufacturers' recommendations.

The following considerations should be observed in storing respirators:

- SCBAs, SARs, and APRs should be dismantled, washed, and disinfected after each use.

- SCBAs should be stored in storage containers supplied by the manufacturer.
- APRs should be stored individually in their original cartons or carrying cases.
- All respirator facepieces should be sealed inside a plastic bag for storage.

Reuse of CPC

Items of CPC must be completely decontaminated prior to reuse. Otherwise, the items cannot be considered safe to use.

In some instances, contaminants may permeate the CPC material and may be difficult or impossible to remove. Such contaminants may continue to diffuse through the CPC material toward the inner surface during storage, posing the threat of direct skin contact to the next wearer.

If matrix permeation is possible, the article of CPC should be hung in a warm, well-ventilated place to allow the item to release the permeated contaminant.

Extreme care should be taken to ensure that permeation and degradation have not rendered CPC unsafe for reuse. It should also be noted that permeation and degradation may occur without any visible indications.

Maintenance of PPE

Effective maintenance is vital to the proper functioning of PPE. Thus, the employer's PPE program should include specific maintenance schedules and procedures in accordance with manufacturers' recommendations for all reusable items of PPE.

Maintenance can generally be divided into three levels:

- Level 1: User or wearer maintenance, requiring a few common tools or no tools at all.
- Level 2: Shop maintenance, which can be performed by the employer's maintenance shop.
- Level 3: Specialized maintenance, which can be performed only by the factory or an authorized repair person.

In some instances, it may be advisable for an employer to send selected employees through a manufacturer's training course in order to establish an in-house PPE maintenance program.

Heat Stress and Other Physiological Factors

Heat stress and other physiological factors directly affect the ability of personnel to operate safely and effectively while wearing PPE and thus are important considerations in selecting equipment and planning response operations. Factors that may predispose a worker to reduced work tolerance are

- Poor physical condition or obesity
- Alcohol and drug use (including prescription drugs)
- Dehydration or sunburn
- Old age
- Infection, illness, or disease
- Lack of acclimatization
- Work environments with elevated temperatures
- Work environments requiring burdensome amounts of PPE
- High workloads

Heat Stress and PPE

The body has several mechanisms for the control of internal temperature, as discussed in Chapter 18. A heat-induced illness (heat strain) can result any time these mechanisms are compromised or overloaded.

The use of PPE, especially impermeable clothing, affects the body's ability to cool itself in several ways. First, the blood carries heat from deep inside the body to the skin. There, the natural principle of convection causes the heat to be released to the lower-temperature air outside the body. Impermeable clothing, especially totally encapsulating ensembles, prevent the heated air from leaving the skin surface, thus eliminating this natural heat transfer.

Also, evaporated sweat from the body quickly saturates the air near the body, thus preventing this vital cooling mechanism. The body continues to produce sweat, which drips from the body wasted; dehydration and loss of important electrolytes can result.

PPE, including CPC and respiratory protection, can add weight and reduce mobility, thereby causing the responder to work harder. This increased work results in increased metabolic heat and increased demand for oxygen as energy is produced. The heat produced adds to the load on the cooling system of the body. Because the increased demand for oxygen usually occurs at a time when much of the body's blood is at the skin surface and not available for the heart to pump, the heart must pump more frequently to move enough blood through the lungs to receive oxygen. Thus, a higher pulse rate usually accompanies heat stress.

Given these examples of how use of PPE can promote heat stress, the need for special attention to potential heat conditions is obvious. Generally speaking, the use of ambient temperature, humidity, and wind speed as indicators of heat conditions has limited use at best, since the PPE essentially creates its own internal environment. Therefore, heat stress monitoring and control activities should focus on the individuals using the equipment and not on environmental parameters. For instance, medical monitoring using such parameters as heart rate, oral temperature, and body weight loss, as described in Chapter 18, should be instituted whenever the work area temperature exceeds 70°F.

Heat Injury Prevention

Some encapsulating suit manufacturers have begun designing temperature control capabilities into their suits. Such designs include tubing for distribution of cooled air or water throughout the suit. The air distribution systems typically require the use of an airline and large quantities of respirable air. Water cooling systems require ice storage or refrigeration units and a pump, which add bulk and weight to the suit. Therefore, these suits are often uneconomical or impractical for emergency response work.

Another cooling option is the use of cooling garments or vests containing ice packs. These garments cool the wearer and do not require airlines or water cooling systems. However, they are fairly heavy and require an on-site source of ice or the ability to freeze coolant gel packs. Also, once the cooling medium has lost its cooling effect (e.g., the ice has melted), the cooling garment becomes an insulator which holds heat close to the body and compounds the heat stress problem. Therefore, the cooling medium must be regularly checked and replaced during the use of the garment.

SUMMARY

In many instances, PPE will constitute the responder's only line of defense against the dangers associated with a hazardous materials incident. An appropriate ensemble of protective gear can provide adequate protection against chemical exposure and physical safety hazards related to the incident. However, the protective equipment involved must be properly selected, cared for, and used so that its protective potential can be maximized. For this reason, a working knowledge of PPE is vital for all responders who may be required to enter potentially hazardous areas during incident mitigation.

REFERENCES

ANSI 1973. *Identification of Air Purifying Respirator Canisters and Cartridges*, K13.1. New York: American National Standards Institute.

ANSI 1980. *Practices for Respiratory Protection*, Z88.2. New York: American National Standards Institute.

ANSI 1988. *Proposed Revisions to ANSI Z88.2*. 1980. New York: American National Standards Institute.

Bureau of Explosives, Association of American Railroads. 1990. *Emergency Action Guide*. Washington, DC: Association of American Railroads.

Forsberg, K., and S.Z. Mansdorf. 1989. *Quick Selection Guide to Chemical Protective Clothing*. New York: Van Nostrand Reinhold.

Henry, M.F., ed. 1989. *Hazardous Materials Response Handbook*. Quincy, MA: National Fire Protection Association.

Langley, J. June 1988. "Choosing Adequate Personal Protective Clothing for the Haz Mat Team." *Hazardous Waste Management Magazine*, 10–11.

NFPA. 1990. *Liquid Splash-Protective Suits for Hazardous Chemical Emergencies*. NFPA Standard 1992. Quincy, MA: National Fire Protection Association.

NFPA. 1990. *Support Function Protective Garments for Hazardous Chemical Operations*. NFPA Standard 1993. Quincy, MA: National Fire Protection Association.

NFPA. 1990. *Vapor-Protective Suits for Hazardous Chemical Emergencies*. NFPA Standard 1991. Quincy, MA: National Fire Protection Association.

NIOSH, 1986. *Criteria for a Recommended Standard: Occupational Exposure to Hot Environments. Revised Criteria 1986*. NIOSH Publication No. 86-113. Washington, DC: U.S. Government Printing Office.

NIOSH, 1987. *NIOSH Guide to Industrial Respiratory Protection*. NIOSH Publication No. 87-116. Washington, DC: U.S. Government Printing Office.

NIOSH, 1990. *A Guide for Evaluating the Performance of Chemical Protective Clothing*. NIOSH Publication No. 90-109. Washington, DC: U.S. Government Printing Office.

NIOSH/OSHA/USCG/EPA 1985. *Occupational Safety and Health Guidance Manual for Hazardous Waste Site Activities*. NIOSH Publication No. 85-115. Washington, DC: U.S. Government Printing Office.

Noll, G.G., M.S. Hildebrand, and J.G. Yvorra. 1988. *Hazardous Materials: Managing the Incident*. Stillwater, OK; Fire Protection Publications.

OSHA, 1986. *Respiratory Protection*. OSHA Publication No. 3079. Washington, DC: U.S. Government Printing Office.

Rodgers, S.J. "How to Select and Use Personal Protective Garments." September 1988. *Industrial Hygiene News*, 44–55.

Schwope, A.D., P.P. Costas, J.O. Jackson, and D.J. Weitzman. 1985. *Guidelines for the Selection of Chemical-Protective Clothing*, 2nd ed. Cincinnati: American Conference of Governmental Industrial Hygienists, Inc.

U.S. Department of Labor. Title 29 Part 1910. Washington, DC: U.S. Government Printing Office.

10

Decontaminating Personnel and Equipment

Barbara M. Hilyer, M.S., M.S.P.H., and
Lori P. Andrews, P.E.

An in-house hazmat team responded to a methyl ethyl ketone release of approximately 250 gallons which occurred at a paint blending plant. The responders wore butyl rubber boots, which are recommended by the manufacturer for use in working around the chemical. Although it was necessary for team members to walk through pooled liquid to approach the point of release, they did not decontaminate their boots after the incident. When the boots were examined the next day, the soles had been destroyed.

Decontamination is an important part of any hazardous materials incident response. It is not difficult to learn or to do, as long as plans and training are completed well in advance of the need for it. In this chapter we will examine the rationale for effective decontamination and suggest criteria for determining the most efficient way to accomplish decontamination during an incident.

OBJECTIVES

- Be able to define decontamination and give reasons why it should be done
- Understand the physical and chemical bases of adequate decontamination
- Know how to set up a seven-station decon line with appropriate equipment at every station
- Be prepared to work at any station on a decon line with an understanding of the duties and protection at each station

- Know the criteria for locating the decon area
- Recognize differences in suitable decon methods and solutions for people, clothing, tools, and vehicles
- Be aware of special considerations for the decontamination of sick or injured persons

DEFINITION OF AND JUSTIFICATION FOR DECONTAMINATION

"Decontamination" is defined as the removal or neutralization of contaminants that have accumulated on tools, clothing, personnel, and vehicles that have been inside the hot zone where hazardous materials are present. Since the hot zone's boundary is set at a safe distance from the hazard, defined as the distance at which no contamination is present, people or equipment crossing this boundary have the potential to carry contaminants from a "dirty" zone to a "clean" one. The goal of the decontamination process, then, is to remove contaminants from these carriers, confine the contaminants to the decon area, and contain and dispose of contaminated materials in a safe and legal manner.

The overall response goal in any incident is protection of people, both responders and members of the community, from exposure to hazards. Responders must ensure that chemical hazards are not spread outside the contaminated zone and that they themselves do not carry any chemicals away with them, putting themselves and their families at risk from chemical exposure. An effective decontamination plan, efficiently carried out, will eliminate the spread of chemical hazards.

METHODS OF DECONTAMINATION

There are a variety of ways to remove chemicals from people and equipment, and each has a use in certain situations. We can group these decontamination methods into two basic types: physical and chemical.

Physical Methods

Physical decontamination includes all methods which manually separate a chemical from the surface to which it adheres. These are basically scrubbing techniques which force the chemical to lose its grip on the surface. Physical methods are effective for removing the major portions of thick, gooey materials such as mud and sludges, and will entirely remove many contaminants.

Advantages

Physical decontamination has far more advantages than disadvantages, since it utilizes no chemicals and can be used without causing any collateral harm. The advantages of physical decontamination are

- Immediate removal of materials that, if left in contact with chemical-resistant fabrics, will be more likely to permeate (diffuse through) the fabrics.
- Reduction of reliance on chemical solutions that may be harmful to responders, decon station personnel, and suit fabrics.
- Ready availability of water streams and scrubbing tools.
- High effectiveness against many kinds of chemicals.

Disadvantage

The disadvantage of physical decontamination is that it may not completely remove all residues of some chemicals, especially those that are oily. Physical removal should, in these cases, be followed by washing with a chemical solution.

Equipment

Equipment used in physical decontamination includes, but is not limited to, the following five types.

Water Streams

It is almost impossible to decontaminate equipment and personnel thoroughly without a water stream of some sort. This can be delivered by a garden hose, fire hose, shower, or, less effectively, a portable pressure-pump sprayer. The garden hose should be provided with a spray nozzle to increase the pressure of delivery, preferably a nozzle with a trigger grip which can be activated only when needed. If fire hose is used, it should be connected to one or more smaller hoses by means of an adaptor, since the pressure and volume are greater than are needed.

Probably the most effective water spray for personnel decon comes from a specially designed shower, which can be put together with polyvinyl chloride pipe and connectors. The shower can be made more easily portable by the use of quick-release connectors at strategic points so that it can be broken down for transport on a hazmat vehicle. The best showers are designed so that the spray hits the person from several different directions at two or more levels, and they are able to maintain enough pressure to accomplish active physical removal. Some sort of easily activated control mechanism is advisable so that the shower runs only when needed.

FIGURE 10-1. Physical decontamination makes use of scrapers and scrubbers.

Scrapers and Scrubbers

Many types of scrapers and scrubbers (Figure 10-1) can be used in conjunction with a water spray to aid the removal of thick contaminants. For heavy muds and sludges, a smooth-edged scraper such as those sold in tack shops for scraping sweat from horses is appropriate. No scraper should be used on protective clothing which might degrade the surface of the fabric. A brush is handy for scrubbing; those with long handles allow decon personnel to avoid contact with the chemical.

The palms of the gloves and the soles of the outer boots require special attention when scrubbing, as they are generally the most contaminated areas (Figure 10-2). Care should be taken to brush all areas of the suit, including those under the arms and inside the legs.

Dry Brushes

If the contaminant is in dry form, it can be dry-brushed off. After the major portion has been brushed off, the remainder can be rinsed with water; if copious amounts of water are used, the heat from any water reactive materials will be dissipated.

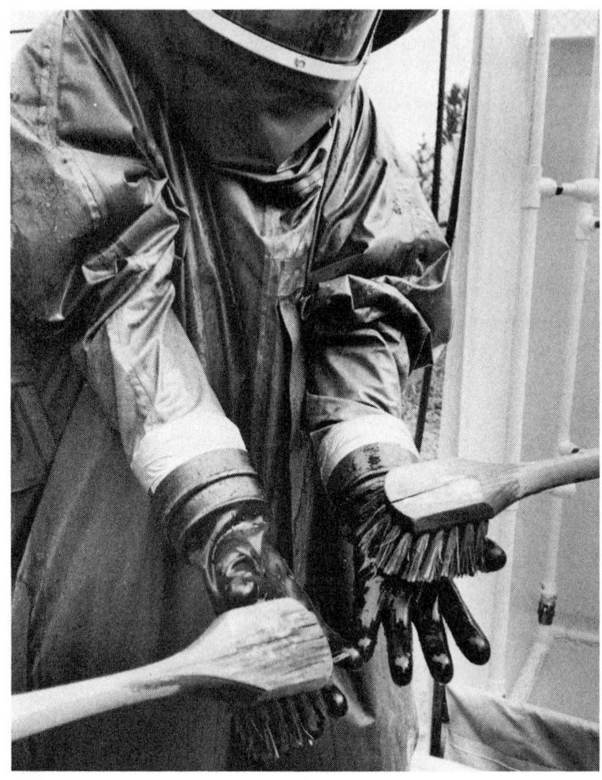

FIGURE 10-2. It is important to scrub the palms of the gloves and the soles of the boots, since they may be more heavily contaminated.

Steam

Steam from a pressure jet adds high temperature to the balance between surface adhesion and separation. High temperature may soften the contaminant as its temperature approaches its boiling point and encourage it to evaporate off the surface. Steam cannot be used on people and is likely to be harmful to protective fabrics, but it is appropriate for use on some tools, parts of vehicles, and heavy equipment. Use of steam may increase the concentration of the contaminant in air, thus increasing the inhalation hazard to decon station operators. This must be considered when selecting protective equipment for decon personnel.

Freezing

Freezing or otherwise lowering the temperature will sometimes enhance the removal of certain highly viscous contaminants. When the temperature approaches the freezing point of the chemical, the latter will become solid rather than sticky and may let go of the surface. Care should be taken in using this method, as certain chemical-resistant fabrics may become brittle, crack, and be irreparably damaged at low temperatures. Literature provided with clothing should allow evaluation of each fabric relative to freezing.

Chemical Solutions

Advantage

Chemical solutions can be formulated which will change one chemical into another chemical or change its form; in some cases, the use of solutions makes contaminant removal much more efficient. When physical methods are found to be inadequate, chemical solutions should be considered.

Disadvantage

The major disadvantage of the use of chemical solutions is the possibility that they will introduce additional hazard to the decon procedure. Only two types of solutions will be discussed here: detergents and neutralizers.

Solutions

Detergent Solutions

Detergent solutions work by chemical means to remove a contaminant from a surface. They are most effective against oil-based chemicals, which are the most difficult to remove by purely physical means.

Detergents work by reducing the surface tension which forms between oil and water or oil and any other surface. These tensions hold the two together, sometimes quite strongly.

Detergents, including soaps, emulsifiers, and surfactants, are made up of molecules which have two very different ends. One end is a long carbon chain, represented by the straight lines in Figure 10-3; the other end may be made of many different things but is always polar. This means that the short end, represented by the balls in Figure 10-3, has its positive and negative charges permanently separated. The carbon chain end, which looks like the sucker stick in the drawing, is attracted to oil, and the polar end, the round part of the sucker, is attracted to water. Many, many molecules of detergent surround droplets of oil, separating them and floating them off the surface of whatever they are stuck to, whether it be greasy dishes or a moon suit. Flushing with water will then wash away the droplets of oil with their surrounding detergent molecules.

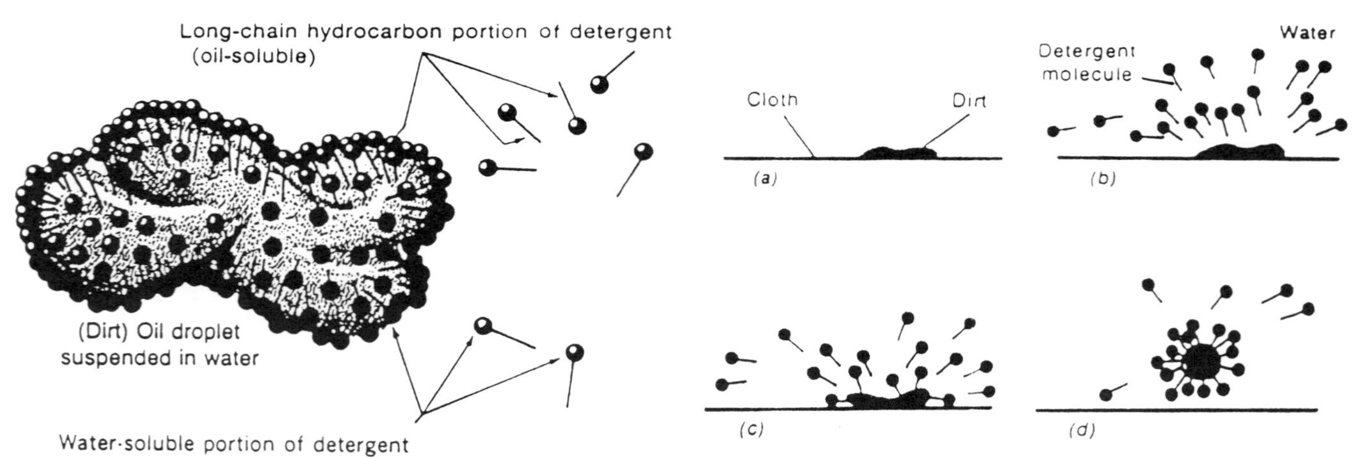

FIGURE 10-3. Detergent removes oily and greasy contaminants from surfaces.

Neutralizing Solutions

Neutralizing solutions can be used to alter a contaminant chemically so that the resulting chemical is harmless. The recipes for these solutions are found in a number of "how-to" manuals; we hesitate to recommend them, since most contain hazardous chemicals, and they are not likely to be needed in emergency response. They are most useful in ensuring the total cleanliness of sample bottles, when laboratory analysis of samples is desired, or for decontamination of sampling equipment to prevent cross-contamination of these devices. If further information regarding other solutions is desired, a chemist should be consulted. It should also be remembered that each neutralizing solution must be tailored to the specific chemical contaminant one wishes to alter.

Level of Decontamination Required

How much decontamination is enough? It is often difficult to be certain of the level of decontamination required to remove all chemicals adequately. Many contaminants are not visible, and those that are give only an indication of surface contamination, with no information about permeation or penetration of clothing.

"Swipe tests" can be conducted in which a suit or tool thought to be clean is wiped with a bit of material known to be clean and the swipe sent to a lab to be analyzed. No immediate information will be available, but when the information does come back, it will indicate whether the procedures outlined in the decon plan have been effective or need to be altered for future use.

Two criteria are generally used in writing a decontamination plan: The type and amount of contamination offer useful guidelines.

Type of Contamination

Consider the chemical itself and the likelihood that it has gotten on clothing and equipment surfaces, and evaluate the health hazard it presents. To estimate the amount which may have gotten on surfaces, we probably want to know the physical state of the contaminant, and of any environmental substances, like mud or water, with which it has mixed.

The type of contamination also determines the degree of hazard it presents; as you have already learned, this information can be gained only after the chemical has been identified. Extremely hazardous chemicals require extremely vigorous decontamination methods and a higher level of protection for decon station operators than might otherwise be needed. More stations may be required, wash/rinse steps repeated, and more distance left between stations to protect operators at adjacent stations.

Solids

Solids are not likely to get on the surfaces of suits or respirators unless they are powdery and drifting in air, or have formed from liquids adsorbed and dried on dust particles. Particulates released in a fire or in a plume from a spill may also cover almost every surface. Solids in a surface release, unless carried by wind or dissolved in mud or water, will contaminate only boot soles, gloves if they are touched, and any tools or wheels that encounter them.

Liquids

Liquids may splash anywhere, especially on the clothing of responders who approach the point of the release. Obviously, avoidance of splashing or walking through puddles will limit contamination. Liquids which splash on the suit will run down into creases and boots, if these are not eliminated or protected by taping, and make decontamination more difficult.

Gases

Unless they dissolve in rain or highly humid air, or on the naturally moist surfaces of skin, gases will result in little surface contamination. In cold weather, condensation on any surfaces which are cooler than ambient temperature may occur.

Amount of Contamination

The amount of contamination will depend partly on the physical state of the chemical and partly on the task performed by the responder or his equipment. It has already been mentioned that liquid or sludge materials are most likely to cause heavy contamination.

Responders who walk into puddles, pick up contaminated objects, or perform patching or plugging operations on spewing liquids must be assumed to be heavily contaminated. The use of disposable protective outer gloves and boots should be considered in these situations, as well as for any entry which must be made into an area of contaminated mud or oil.

SETTING UP A DECONTAMINATION AREA

Location

The location of the decontamination area will depend on certain parameters of the incident location. These considerations should be made as rapidly as possible, since it is an axiom of emergency response that no entry into a potentially contaminated area should be made until the decon area is ready.

Incident Parameters

The conditions under which the incident occurs are important considerations. Notable factors are the nature and degree of the hazard presented by the chemicals, the wind conditions present, and the topography of the incident scene.

The Nature and Degree of the Hazard

If the chemical is a solid or nonvolatile liquid, the hazard zone may be relatively small and the decon area close to the release. A vapor or other gas creates a larger hazard zone, as does the threat of fire or explosion. The shorter the distance responders must walk to be decontaminated, the longer their time on task can be. Once the chemical has been identified, the degree of hazard can be assessed further.

Wind

Wind direction is the most critical weather parameter for determining the location of the decon area. Since wind direction is subject to change, the decon area should be located in the place most likely to remain safe if the wind changes.

Topography

The topography of the land on which the incident has occurred should also be considered. Since decontamination always includes water which must be contained, the area should be set up so that this task is made easier. Water runs downhill, and contaminated water cannot be allowed to flow into clean areas or onto soil or other permeable surfaces.

Access to Water and Equipment

Access to water and other necessary equipment is vital. If a mobile tanker is the water source, it must be able to be moved into the hose range of the area. If water is provided by a fixed hydrant, hose length will limit access. Other equipment which may be needed in close proximity to the decon area includes the air tank refill system, emergency medical equipment, and salvage drums.

Personnel Decontamination Line

Access Control Points

The line where people and their suits are cleaned should be set up separately from the area where tools, equipment, and vehicles are cleaned.

The area should be clearly marked and control points set up to restrict personnel to those wearing the appropriate protective clothing. Vehicle access should be completely separate from personnel access for obvious safety reasons. The personnel entry point should be separate from the exit point to avoid cross-contamination.

The decontamination line should consist of a number of distinct stations set up in such a way as to reduce contamination as the responder moves down the line. Stations should be clearly separated from each other by enough distance to eliminate any splashing from one station to another. No flow of liquid from one station to another should be permitted.

Stations in the Line

The number of stations required for adequate decontamination varies, depending on the nature and degree of hazard presented by the chemical(s) involved in the incident. Hazmat teams may also find themselves limited by the number of people available to staff the stations; however, it is wiser to find ways to gain additional personnel than to limit the number of stations.

EPA has made recommendations for minimum decontamination station numbers. These range from 19-station decon procedures to a minimum of 7-station procedures for responders leaving the contaminated zone, and the steps at each station depend on the level of personal protection worn by the entry teams (see Appendix 1). Seven stations will suffice for incidents in which responders have not been heavily contaminated with a relatively hazardous toxin.

The overall goal at each station is to remove contamination from the garment or other item in such a way that the contaminant will not touch the worker as he doffs the garment. Specific objectives at each station in a 7-station line set up to decontaminate Level A protective clothing are listed below. They can be used with appropriate modifications for a 7-station line for Level B-protected responders. Table 10-1 lists suggested equipment for each station.

Station 1: Equipment Drop

This station consists of a tarp or sheet of plastic just inside the hot zone, where tools or monitoring equipment can be left by a team as they exit. The equipment can be picked up and used by the next entry team, and it and the tarp decontaminated or disposed of at the termination of the incident.

Equipment for this station is a plastic sheet.

TABLE 10-1. Suggested Decontamination Equipment at Each Station (For a Seven-Station, Level A Protective Clothing Line)

Station	Equipment
1	Plastic ground sheet
2	Container of detergent solution
	Source of running water
	Containment basin
3	Containers (lined) for gloves, boots, tape
	Bench or stool
4	Full air tanks
	Outer boots
	Outer gloves
	Tape
	Optional: resting bench, drinking water and cups, medical monitoring equipment
5	Containers (lined) for boots and gloves
	Bench or stool
	Container or rack for suits
6	Plastic ground sheet
7	Soap
	Running water
	Container for runoff
	Towels

Station 2: Outer Clothing Wash and Rinse

This is really two stations in one, since wash water will probably include detergent and rinse water will not. Scrubbing or scraping can accompany both processes. The objective is the removal of contaminants from all outer garments, including the suit, outer gloves, and outer boots (Figure 10-4). The equipment needed at this station includes

- A container of detergent solution (e.g., garden sprayer, bucket, drum)
- A source of copious amounts of water for rinsing
- A containment basin (e.g., kiddie pool, shower enclosure, lined pit, leak-protected drum pallet)
- Long-handled brushes

Station 3: Outer Boot and Glove Removal

Outer boots and gloves are removed here and are placed in a plastic-lined container for disposal or later drying. Responders can remove their boots

FIGURE 10-4. Gross decontamination is done at station 2.

more easily if they are able to sit down (Figure 10-5). Equipment includes

- Containers (one for gloves, one for boots, one for tape)
- Plastic liners (garbage bags)
- A bench or stool

Station 4: Tank Change

This station is located off to the side of the decon line (Figure 10-6) and is visited only by responders who are going back on standby to return to the hot zone. They may return after donning clean outer gloves and boots and being retaped. This station is placed after station 3 so that it will remain clean, since suits are unzipped here and responders may be exposed to contaminants in the area. Equipment includes

- Full air tanks
- Outer boots
- Outer gloves

- Tape
- Optional: a bench for resting, drinking water and disposable cups, trash can, medical monitoring equipment

Station 5: Boots, Gloves, and Outer Garment Removal

Boots, the fully encapsulating suit, and inner gloves are removed and deposited in separate containers lined with plastic. Assistants should help the responder with the doffing so that he does not touch the outer surfaces of any of the garments (Figure 10-7). As at station 3, a bench should be provided so that the responder can sit to have his boots removed. Equipment needed at this station includes

- Containers
- Plastic liners
- A bench or stool
- A container or hanging rack for nondisposable suits

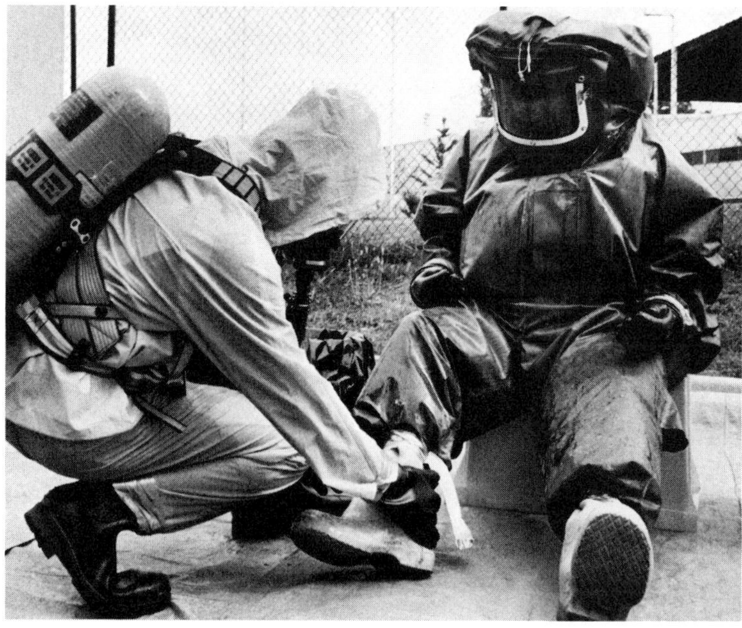

FIGURE 10-5. The doffing assistant removes the boots without help from the responder.

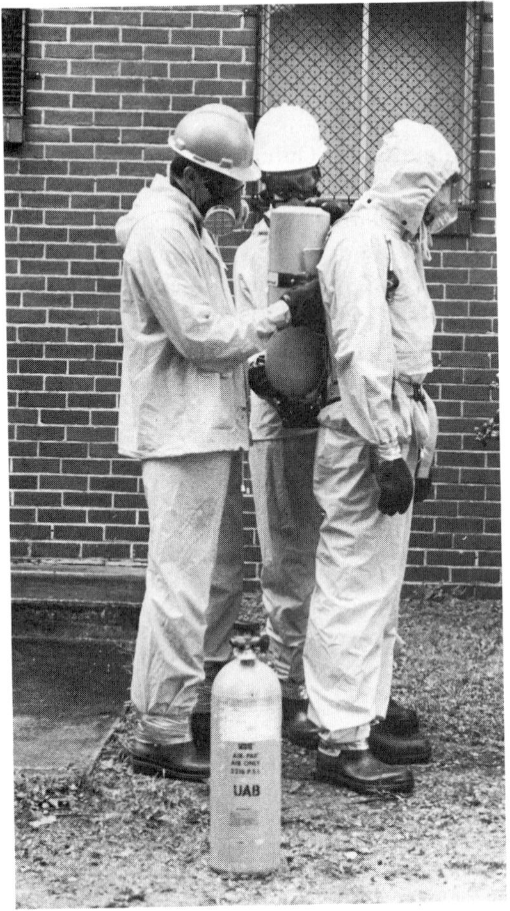

FIGURE 10-6. Tank changes should be accomplished without allowing contamination of any surface inside the suit.

Station 6: SCBA Removal

The SCBA backpack and facepiece are removed by an assistant, and the responder is reminded not to touch his face until he has washed his hands. The unit is placed on plastic, to be cleaned later. The only equipment needed here is a plastic sheet on the ground.

Station 7: Field Wash

Soap and water are provided here for the responder to wash his hands and face, and clean towels should be supplied. Ideally, a private shower should be available; lacking this, responders should take a full shower as soon as possible. Equipment at this station includes

- Soap and running water
- A container to catch runoff water

Decontamination of Tools, Equipment, and Vehicles

The decontamination of tools and equipment can be made much easier by protecting them from contamination. Some air monitoring devices, in fact, are practically impossible to decontaminate if allowed to be splashed with liquid chemical. All sensitive equipment should be protected by clear plastic; a large bag will fit such units as the photoanalyzer, the flame analyzer, and the Geiger counter. A rubber band or long twist tie can be used to close the bag while allowing the probe to extend through the

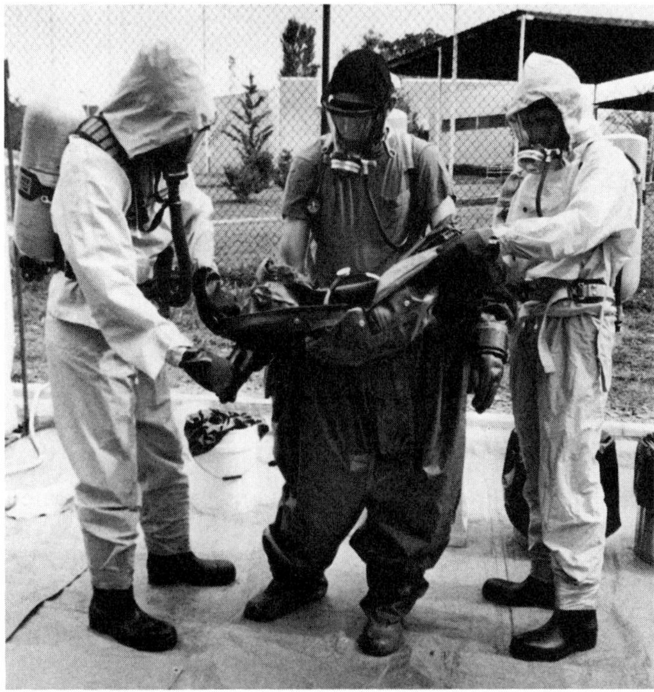

FIGURE 10-7. The responder is warned not to help in the removal of his suit and not to wipe his face until he has washed his hands.

opening of the bag. Care should be taken to collect only air samples; liquids will damage not only the outside of the probe but also the sensor.

The ground rules for avoiding boot contamination by not walking into pooling product or contaminated mud should be extended to vehicles. Even drivers of fire trucks and ambulances should take care to avoid contaminated ground.

Equipment decontamination should be handled at the end of the incident, although gross decontamination, such as removal of contaminated mud, should be done as soon as possible to reduce cross-contamination of personnel. It should be noted that certain tools, for example those with wooden handles, may have soaked up contaminants which cannot be removed. These should be disposed of as hazardous waste.

Metal or plastic tools can be scrubbed with cleaners which are too strong for skin or protective fabrics; the only restrictions are those of incompatibility and limits on the exposure of people using them. Plastics are degraded by some organic solvents; corrosive acids and bases may damage certain metals. If a chemical solution is used, it should be evaluated by the same criteria used in assessing the hazards of any chemicals in the incident, and protection of decon personnel must be appropriate.

Emergency Medicine and Decontamination

Decontamination and emergency medical services (EMS) may have to be performed on a sick or injured worker. When a responder or worker in the hot zone needs medical help, he cannot be treated like an ordinary victim. There is a good possibility that the surface of his protective clothing contains enough hazardous material to contaminate rescuers and first aid providers or to enter the space inside the clothing as it is removed and further injure the victim.

No person should enter the hot zone for any reason unless the proper protective gear is worn, and a person unfamiliar with the gear should not attempt to wear it. It must be emphasized that EMS responders who have not been trained to wear respirators and protective clothing should stay out of the hot zone and the decontamination area. Sophisticated protective gear should be used only by those with proper knowledge and experience. If decontamination is carried out prior to delivery of the victim to the EMS personnel at the perimeter, then no special gear may be needed.

Primary goals for emergency personnel in a situation involving hazardous materials include termination of exposure to the patient, removal of the patient from danger, and patient treatment—while not jeopardizing the safety of rescue personnel.

If EMS responders are trained to do so, they should wear appropriate protective gear while undertaking primary assessment, giving priority to the ABCs: airway, breathing, and circulation. Other personnel may, at the same time, begin to decontaminate the patient so that protective gear may be downgraded as contamination is removed. A call to the local or regional poison control center may provide information useful in determining whether medical personnel are at risk of secondary contamination from the patient; this information is available only if the name of the chemical is known.

During initial patient stabilization, all clothing suspected of being contaminated should be removed or cut away. The decon and medical personnel must avoid contact with potentially hazardous substances. Any clothing that cannot be removed should be wrapped in plastic or whatever is available in order to contain the contaminants and prevent their migration to the patient or others.

Decontamination should be an orderly process. It should begin at the head, with particular attention to the eyes, and move down. Since intact skin is, in most cases, more resistant to contaminant permeation than eyes or damaged skin, eyes and open wounds must be thoroughly flushed. Wounds should be covered with waterproof dressing after washing.

Decontamination should be performed using the least aggressive methods. Mechanical or chemical irritation to the skin must be limited to prevent increased permeability. Contaminated areas should be washed under a gentle spray of warm (never hot) water, wiping with a soft sponge and using a mild soap such as dishwashing liquid. Care should be taken so that contaminants are not introduced into open wounds. The degree of decontamination should be based upon the nature of the contaminant, the form of contaminant, and the patient's condition. It is important that runoff water be contained, if possible, since it may be contaminated and will require treatment before disposal.

All potentially contaminated clothing should be removed from the patient and placed in labeled bags for later decontamination or disposal.

Treatment of the chemically exposed patient will include primary and secondary surveys; the primary survey can be accomplished simultaneously with decontamination, and secondary surveys should be completed as conditions allow. Unless required by life-threatening conditions, invasive procedures such as giving intravenous medication or intubation should be performed only in fully decontaminated areas. These procedures may create a direct route of introduction of the hazardous substances into the patient's body.

In the case of a victim of accident or illness in a contaminated zone, it may not be known immediately whether the patient is suffering from chemical exposure. Whatever the results of later assessment, the names of all potentially hazardous chemicals in the vicinity of the patient must be determined as quickly as possible. Since exposure of EMS personnel must be prevented and decontamination procedures started, references must be consulted. This is possible only if the chemicals known or suspected to be present are identified. It may also be necessary to continue to monitor the patient for latent effects of chemical exposure; symptoms of exposure are described in reference sources.

If it becomes necessary to transport a contaminated patient by ambulance, special care should be exercised in preventing contamination of the ambulance and subsequent patients. Exposed surfaces that the contaminated patient is likely to come into contact with should be covered with plastic sheeting (Figure 10-8). The patient should be as clean as reasonably possible prior to transport, and further contact with contaminants should be avoided. Responders should make every attempt to prevent the

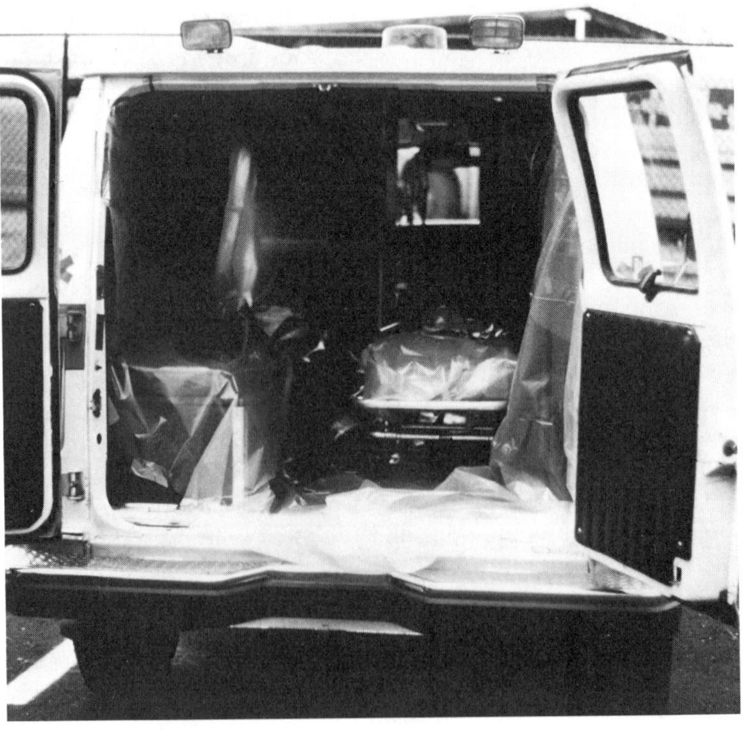

FIGURE 10-8. Protect the inside of the ambulance by lining it with plastic.

spread of contamination and, at the very least, should remove the patient's clothing and wrap the patient in blankets, followed by body bags or plastic or rubber sheets. If the chemical is one that presents the added danger of accelerated skin absorption due to heat, plastic or rubber wrap should not be used.

In an ambulance during transport, personnel must use appropriate respiratory protection if contaminants are present. If weather conditions permit, opening windows in the patient's and driver's compartments will provide maximum fresh air ventilation. The receiving hospital should be contacted as soon as possible and given any information that has been gained about the chemicals involved.

Transportation by helicopter presents special, more serious hazards. Exposure of the flight crew may interfere with their ability to fly safely. It is even possible that downdraft from the helicopter could spread contamination on the site. This means of transport should not be used without careful consideration of all factors.

MANAGEMENT OF THE DECONTAMINATION AREA

In considering criteria for good management of the decon area, it is appropriate to recall that the overall goal of the process is to eliminate the transfer of contaminants from exposed equipment and clothing to any other surfaces which cannot be contained and disposed of. These other surfaces include all areas outside the decon line, all materials which will later leave the decon area, and all personnel who attend or walk through the decon line.

Orderly Cleaning and Doffing

Let us first consider how exposure to the responder can be prevented as he removes contaminated garments. This goal can be accomplished through orderly cleaning and doffing of the items the responder is wearing (we assume that he is not carrying anything, having left all tools and equipment behind at station 1).

Touching Contaminated Surfaces

The responder never touches anything inside the decon area. Doffing assistants unzip, untape, and remove all his protective clothing for him. At station 3, the assistant removes the outer boots and gloves without touching the suit. Station 5 helpers remove all other garments, and station

6 operators remove the SCBA. At none of these stations does the responder help, but rather avoids touching any of his equipment. He touches nothing until he has washed his hands and face at station 7, at which time he exits the decon corridor.

At no time does a doffing assistant touch a part of the clothing not necessary in the removal appropriate to his station. Nor does he touch the skin or inner clothing of the responder.

Physical Safety

The decon stations are set up with the responder's safety in mind. All tripping hazards are eliminated, and no station requires him to hold someone or something to keep his balance. A seat is available wherever outer or inner boots must be removed.

Contaminant Movement

No decon personnel move about between stations. Contamination should decrease throughout the line; assistants can easily spread the chemical by walking down the line. It is easy to understand how an operator at one station would wish to assist others when he is not busy, but this must be prevented.

Protection of Decontamination Personnel

People who operate the decon stations must be protected from exposure to the chemical being removed and, if potentially harmful chemical solutions are used on vehicles or equipment, from these solutions as well. They must therefore be trained in and equipped with the appropriate PPE. As a general rule, decon assistants wear protective clothing one level below that of the responders they are cleaning, but ideally, selection will be based on the type and amount of hazard the chemical presents.

Is the amount of the chemical on the surface being decontaminated large enough to be splashed onto the assistant as it is washed or rinsed? If so, splash protection clothing should be worn by the assistant.

Can the chemical be expected to volatilize under decon conditions? If so, respiratory protection must be worn.

Is the chemical one which is known to allow secondary contamination of assistants? Chemicals which fall into this category include corrosives, phenols, pesticides, PCBs, asbestos, hydrogen cyanide gases and salts, and hydrofluoric acid solutions. Protection from these chemicals, and others judged to be potential causes of secondary contamination, must be afforded decon workers.

Containment of Liquids

All contaminated materials, including all wash and rinse fluids, must be contained for proper disposal. This will necessitate pumping or otherwise removing wash and rinse water from confinement pools as they fill, and providing a large container to hold it for disposal. Since all wash and rinse waters generated in the decontamination process are contaminated with the incident chemicals, they must be contained until they are disposed of as hazardous waste, treated to render them nonhazardous, or analyzed and determined to be safe for release into the environment. Close attention to where these liquids are flowing during decontamination will suggest capture methods.

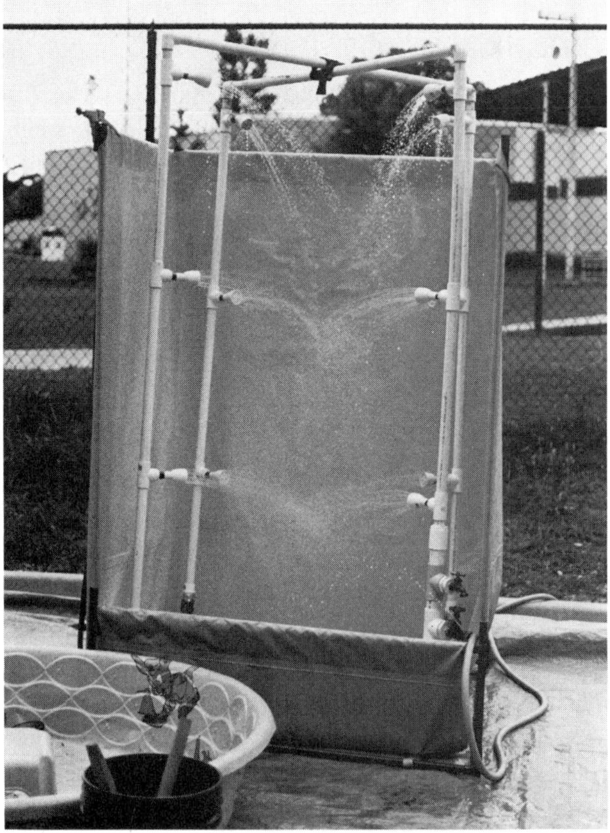

FIGURE 10-9. A portable shower stall prevents splashing of the ground or nearby personnel.

Several containment devices can be purchased, and others can be fairly easily built. Children's swimming pools work well, and are inexpensive and readily available. The inflatable type is convenient if the hazmat brigade carries its equipment on a truck or van, as this pool takes up little space when deflated and can easily be inflated using an air bottle. The rigid type may last longer, since it is not as likely to be punctured, but it is harder to handle and store.

Portable shower stalls are available for purchase through safety products catalogs (Figure 10-9); some hazmat units have made their own from polyvinyl chloride pipe covered with sewn and seam-sealed plastic.

The corridor should be lined with plastic, the edges of which can be turned up and supported to direct overflow into a catchment basin if it occurs.

POST-INCIDENT MANAGEMENT

When the incident is over, decon assistants must go through an orderly process of cleaning and doffing their own protective clothing. Starting with the most highly contaminated station, the operator bags up any disposed-of or cleaned materials, labels them, and cleans all reusable equipment and containers. He then passes through the remainder of the line, where he is decontaminated as he goes. The other operators do the same, from most to least contaminated.

Contaminated water and solutions, disposable clothing, tape, and any other items removed in the decon process should be contained, closed, and clearly labeled for proper disposal.

The definitions to be used in determining whether decon liquids and trash must be disposed of as hazardous waste are the same ones used in designating any hazardous waste. It is not up to the responder to make these decisions.

The last job for the decontamination team is to restore readiness for the next incident. This requires thorough cleaning of all equipment, assessment of breakage, maintenance and oiling of mechanical equipment, and replacement of any damaged or consumed materials. All equipment and materials are made ready and stored in their proper locations. The decon team members can then meet and evaluate their part in the incident to determine how to help make the next response even more efficient.

SUMMARY

Although hazardous materials released in an incident may contaminate the tools and equipment used by emergency responders or get on the outside of their protective clothing, these materials can be prevented from leaving the scene of the release. If responders follow prearranged SOPs for decontamination, and use materials and methods appropriate to the chemicals encountered, all contaminants can be left behind on disposable materials or in wash and rinse solutions to be disposed of. Proper decontamination protects responders, members of their families, and the people who live in the community.

REFERENCES

Andrews, L.P., editor. 1990. *Worker Protection during Hazardous Waste Remediation*. New York: Van Nostrand Reinhold.

DHHS. n.d. *Chemical Emergency: Guidance for the Management of Chemically Contaminated Patients in the Prehospital Setting*. Washington, DC: U.S. Government Printing Office.

DHHS. 1985. *NIOSH/OSHA/USCG/EPA Occupational Safety and Health Guidance Manual for Hazardous Waste Site Activities*. Washington, DC: U.S. Government Printing Office.

Appendix 1

EPA-Suggested Decontamination Lines

All diagrams are from the *Occupational Safety and Health Guidance Manual for Hazardous Waste Site Activities*, 1985, prepared by the National Institute for Occupational Safety and Health, the Occupational Safety and Health Administration, the U.S. Coast Guard, and the U.S. Environmental Protection Agency.

Maximum Decontamination Layout
Level A Protection

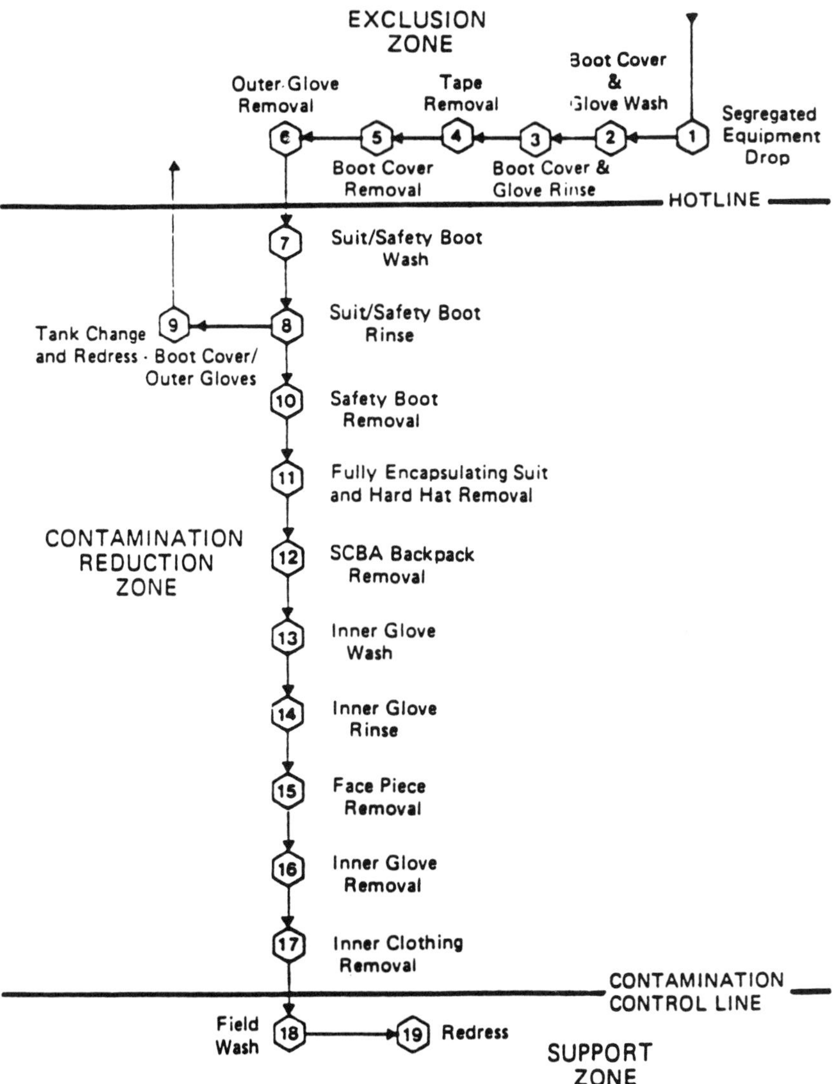

Maximum Decontamination Layout
Level B Protection

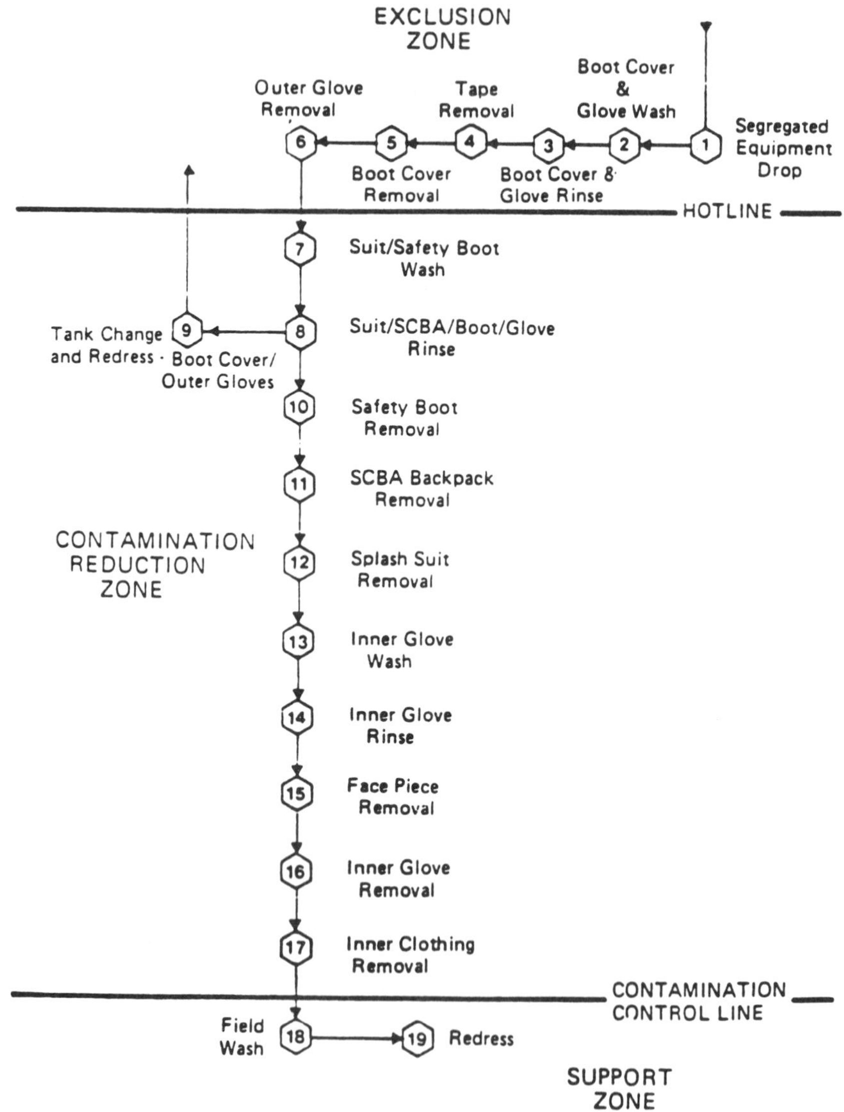

Maximum Decontamination Layout
Level C Protection

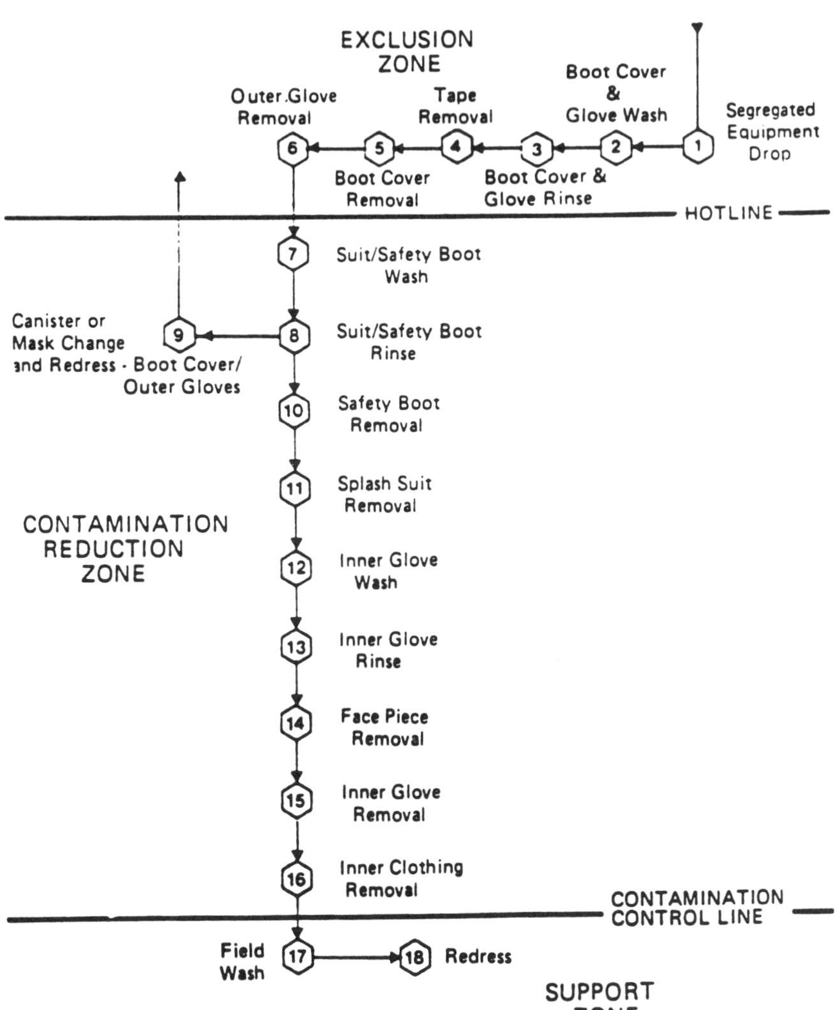

Minimum Decontamination Layout
Levels A & B Protection

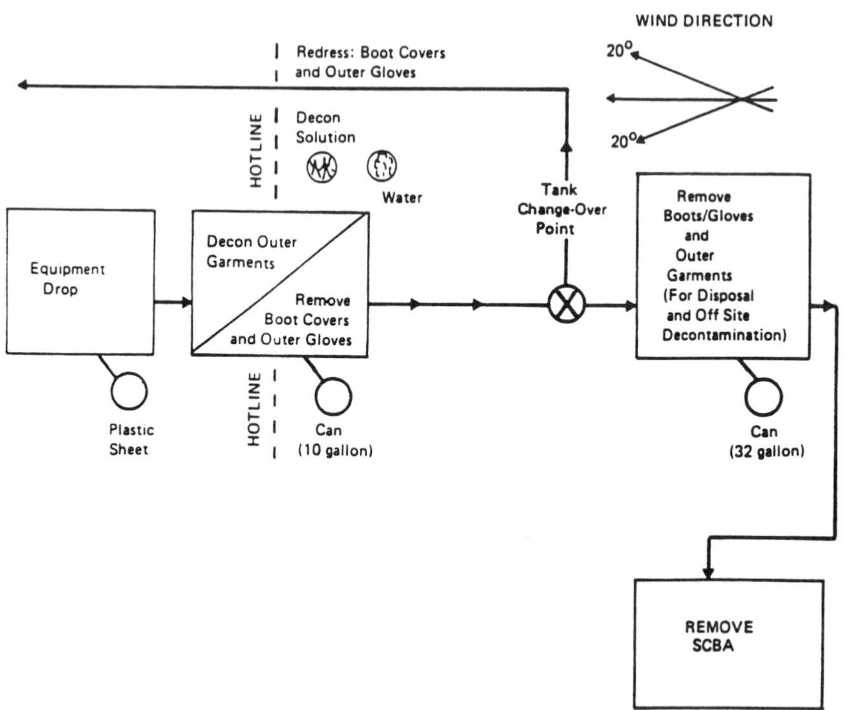

Minimum Decontamination Layout
Levels C Protection

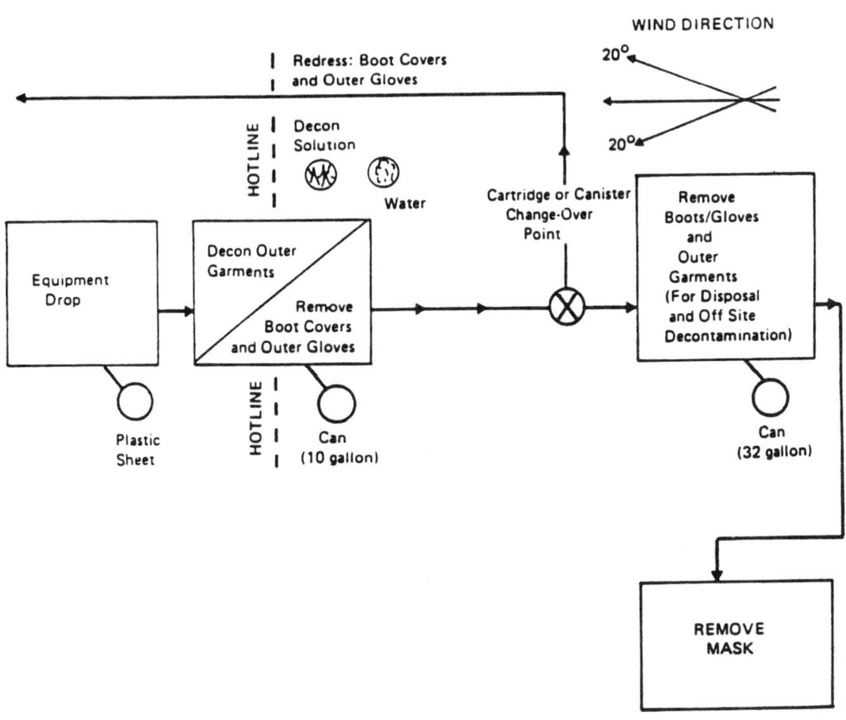

11

Basic Hazardous Materials Control

D. Alan Veasey, M.A. Ed., and
Barbara M. Hilyer, M.S., M.S.P.H.

At an industrial park located in Newark, New Jersey, a ruptured seam on a storage tank released approximately 2,000 gallons of hydrochloric acid. Vapors generated from the surface of the spilled acid produced an acidic cloud which drifted across nearby tracks of the major rail artery linking Boston, New York, Newark, and Washington, D.C. The rail lines were temporarily closed, leaving thousands of commuters stranded. By spraying the surface of the spilled hydrochloric acid with a hazardous materials foam, responders were able to suppress vaporization of the acid so that the rail lines could be reopened.

As commuters stranded by the vapor cloud probably realized, bringing the hazardous materials involved in an incident under control is a critical aspect of incident mitigation. Effective control operations can prevent or minimize undesirable outcomes which may result from the release, or potential release, of hazardous materials. For this reason, hazmat control operations should be initiated as soon as appropriate procedures can be selected and determined to be safe through hazard and risk assessment and the use of decision-making guidelines such as the DECIDE process.

For the purposes of this textbook, hazardous materials control procedures can be generally divided into two categories: confinement procedures and containment procedures. Confinement procedures are utilized in order to limit the migration of a hazardous material within the environment to the smallest area possible following a release. We typically think of confinement procedures as being performed in response to spills, which exist when a substance has escaped from its container. In contrast, containment procedures are performed in order to terminate a release of hazardous materials at the source. We tend to think of containment procedures as being utilized in order to stop container leaks, which exist when a substance is in the process of escaping from its container.

Confinement procedures differ fundamentally from containment procedures. Procedures used in confinement are considered to be defensive in nature, since they can be performed at a location remote from the actual point of release of hazardous materials. While confinement procedures require that a certain amount of ground be sacrificed to the released substance, responders involved in confinement operations are typically exposed to lower hazard levels than responders involved in containment operations. Procedures used in containment are considered to be offensive in nature, involving work in close proximity to the point of release. Prompt containment operations will minimize the amount of released material which must be confined. However, these operations typically involve a higher potential level of exposure of personnel to the hazards involved and may require special tools, equipment, and protective gear, as well as special training.

This chapter is intended to provide knowledge of the basic hazardous materials control procedures utilized in confinement and other defensive operations. This basic knowledge is appropriate for emergency response personnel who will function as first responders at the operations level. Advanced hazardous materials control procedures, utilized by hazardous material technicians in containment operations, will be covered in the following chapter.

OBJECTIVES

- Be aware of the general types of hazardous materials releases and related confinement operations
- Understand the concept of defensive response to hazardous materials releases as related to the role of first responder at the operations level
- Know the various types of equipment and techniques used in confining spilled hazardous materials
- Be familiar with the equipment and techniques used in collecting hazardous materials for placement in containers
- Have a general understanding of how the following procedures or techniques can be applied in basic hazardous materials control:
  Diking
  Diversion
  Inlet blockage
  Catch basin construction
  Overflow damming

Underflow damming
Booming
Chemical treatment
Vapor knockdown and dispersion
Vapor suppression
Solidification
Gelation
Use of dispersants
Dilution
Use of sorbents
Use of pumps

THE ROLE OF HAZARD AND RISK ASSESSMENT AND DECISION MAKING IN HAZARDOUS MATERIALS CONTROL

In order to be effective, hazardous materials control procedures must be selected based on factors such as (1) the properties of the released material, (2) characteristics of the release event and subsequent migration of the material, and (3) site-specific conditions and considerations. Information gathered during the hazard and risk assessment process is vital. This information should be used in conjunction with decision-making guidelines (such as Benner's DECIDE process) in order to select and implement appropriate control procedures while ensuring the safety of personnel involved in control operations.

Types of Releases

In selecting appropriate confinement procedures, the type of release involved is a critical consideration. For the purposes of this chapter, releases will be classified as either land, air, or water releases, and appropriate confinement techniques will be presented for each type of release. As described in Chapter 5, these releases may occur in the ways that will now be discussed.

Land Releases

Land releases occur anytime a container is breached and the contents spill on the ground. In the case of a liquid product, the released material may then migrate freely along the ground, perhaps reaching the surface water system to create a water release. Land spills may also vaporize, forming air releases. Land releases typically lead to soil contamination and may also produce groundwater contamination, both of which must be addressed in postemergency response cleanup operations and are therefore beyond the scope of this text.

Water Releases

Water releases can occur when a hazardous material is released directly into a stream or other body of surface water or migrates to reach a body of water following a land release. Contaminants in groundwater may also pollute surface water bodies.

Air Releases

Air releases typically occur when a hazardous material escapes its container in gaseous form or when a liquid vaporizes after escaping the container. The contaminant is then able to migrate through the atmosphere, as discussed in Chapter 5.

Other Considerations

It is important to note the relationship between the different types of releases, as shown in Figure 5-5. For example, a single large-scale land release may conceivably result in air and water releases, as well as soil and groundwater contamination, as an incident runs its course. To be fully effective in minimizing harm, responders may be required to engage simultaneously in different types of confinement operations at multiple locations. Thus, it is important that responders fully assess all potential outcomes of the release event before selecting the confinement procedures intended to intervene in the course of events. Otherwise, available resources may not be used to the best advantage.

As in all other aspects of hazardous materials emergency response, the safety of response personnel should be given the highest priority in selecting and performing control operations.

PROCEDURES AND CONSIDERATIONS FOR BASIC HAZARDOUS MATERIALS CONTROL

As previously discussed, the ultimate goal of basic hazardous materials control procedures is to utilize low-risk, defensive operations to prevent or minimize harm resulting from a release. Ideally, this involves using appropriate procedures to retain or capture the released material at a given location so that it can be collected, treated, or otherwise dealt with.

In some cases, it may be impossible to capture a released product, as when large quantities are involved and resources for the initial response are in short supply. In such cases, responders may be able to divert the flow of product away from the predicted migration pathway and into an area where it will do less harm. As an example, responders in some incidents have created diversion dikes or ditches (as described below) to shunt flowing product away from streams or storm sewer inlets in order to prevent contamination of a municipal water supply.

In worst case situations, there may be very little that responders can do to prevent exposure of an area to hazardous materials released in an incident. For example, it may be impossible to prevent a vapor cloud from migrating into an occupied area. In such situations, responders may still be able to utilize defensive tactics to minimize harm resulting from the incident. For example, after establishing control of the potentially affected area, responders may act in advance of the vapor cloud to limit property damage by removing mobile equipment and supplies to a safe area, covering stationary equipment to minimize exposure, and isolating buildings by closing doors and windows and shutting off heating, air conditioning, and ventilation systems.

In situations in which adequate planning and preparation for emergency response have been carried out prior to an accidental release, responders may quickly be able to confine and collect the released material so that minimal environmental exposure occurs. In other cases, the available resources for initial response may be overwhelmed. In such cases, tough decision making will be required to determine how available resources can be best utilized so as to limit most effectively the harm resulting from the incident. Figure 11-1 illustrates how some of the procedures described in this chapter might be applied in the situation shown in Figure 5-5.

Controlling Land Releases

When hazardous materials are spilled on the ground, the speed and direction of their migration within the environment will be largely dependent on the physical properties of the material involved and the topography, or shape of the earth's surface, in the vicinity of the release location. Spilled solids may accumulate on the ground immediately adjacent to the point of release, allowing responders to collect the materials with shovels and place them in a suitable container. In contrast, spilled liquid products may flow for great distances, and may be very difficult to confine and place in containers.

Spilled liquid products tend to flow from higher to lower points of elevation under the influence of gravity. In many instances, the liquid

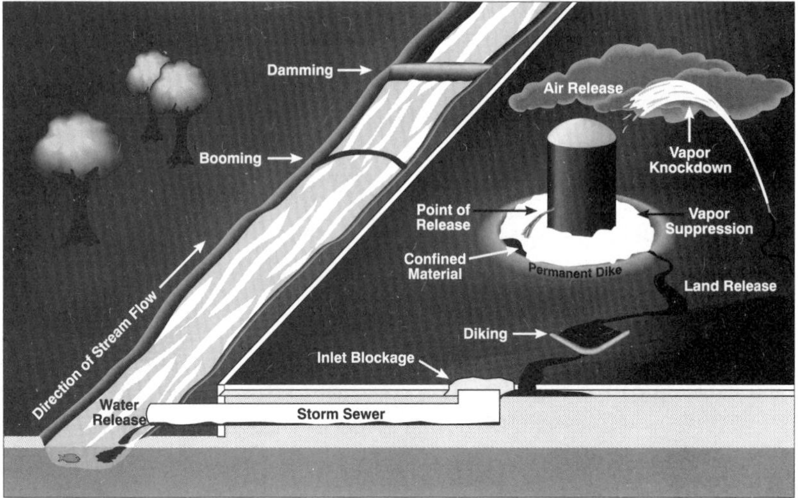

FIGURE 11-1. Overview of basic hazardous materials control procedures.

product will enter and flow along drainageways, such as ditches, gutters, hollows, or dry stream beds. In so doing, the material will tend to migrate as a narrow stream of product down the drainageway. In the absence of a well-defined drainageway, spilled product may tend to flow downslope in a braided pattern or as a sheet.

Effective control of land releases requires that we predict the pathway which the released material will follow, and attempt to intercept and confine the product at some downstream point along that route. This interception point must be far enough in advance of the spill to allow adequate lead time for responders to create confinement structures, such as dikes, before the arrival of the product. This will maximize confinement efficiency and minimize exposure of the responders involved. However, excessive lead time should be avoided, as this sacrifices needless ground to the hazardous material.

Basic Control Procedures

A variety of procedures can be utilized for controlling spilled materials. In many instances, these procedures involve the creation of temporary control structures which are constructed using whatever materials are available or can be rapidly obtained. Ingenuity on the part of response personnel may be required. Equipment and supplies which may be used in performing basic control procedures include the following:

Shovels	Dry granular sorbent
Picks or mattocks	Salvage pumps
Sandbags	Salvage drums
Bagged materials (such as mulch bags)	Plastic sheets, tarps, and/or salvage covers
Sorbent socks, pillows, and/or sheets	Sorbent booms
	Barrier booms

Diking
Diking involves the construction of relatively impermeable barriers located so as to block the flow of a spilled hazardous material across the earth's surface. Dikes can be constructed in straight, V-shaped, or circular configurations, as shown in Figure 11-2. Dikes must be of the appropriate size and configuration given the characteristics of the spill involved. For example, material flowing along a ditch may be readily confined by constructing a simple dike or dam across the ditch. In contrast, for product migrating in sheet flow down a gradual slope or along a broad, shallow depression, a longer V-shaped dike may be called for. For small or slow-moving spills on flat terrain, it may be possible to build a circular dike which completely surrounds the point of release. Dikes should be used in series so that incidental leakage from one structure will be confined by the next. This arrangement also provides for backup in case of dike failure.

Diversion
Diversion involves the placement of control structures so as to divert liquid products away from places where we do not want them to go (such as storm sewer inlets) and into locations where we do want them to go (such as pits excavated as "catch basins" in the ground). For example, ditches or trenches can be created in advance of a spill in order to divert the spill so as to prevent a significant exposure or to direct the spill into a confinement structure, as shown in Figure 11-3. Such a confinement structure may be a diked area or a catch basin. Linear dikes and barrier booms (as described below) may also be employed in diversion operations. Barrier booms can be improvised using sections of firehose which are wrapped in plastic (to minimize contamination), placed in the path of the spill, and charged.

Inlet Blockage
Inlet blockage is used to prevent hazardous materials from entering storm sewer systems or other highly undesirable locations. For example, plastic

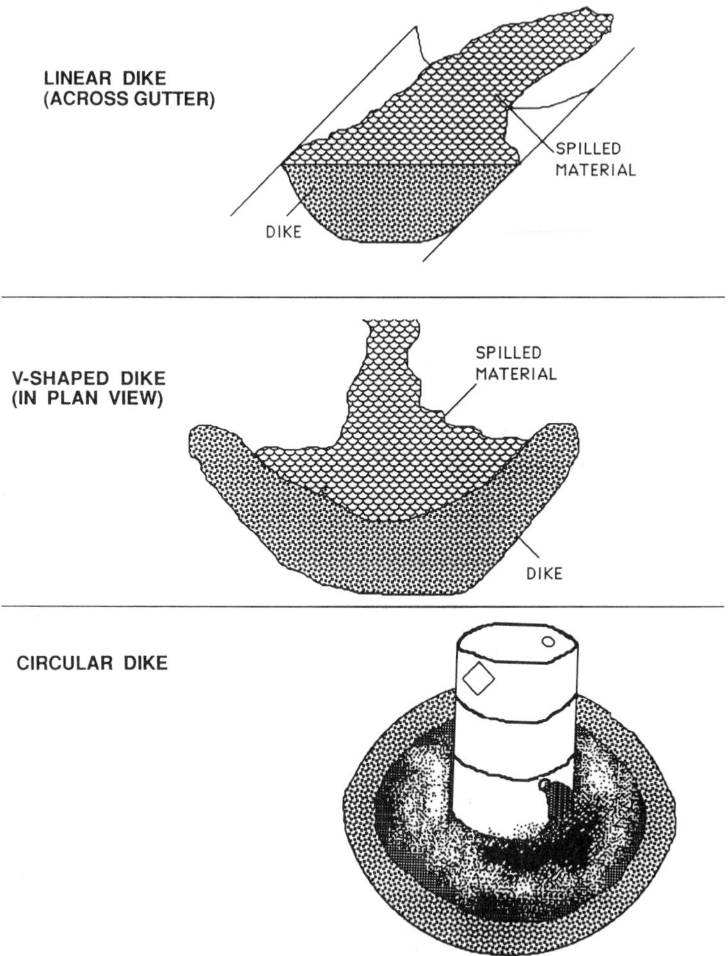

FIGURE 11-2. Basic dike configurations.

sheets, tarps, salvage covers, and/or mats can be used in conjunction with soil, sand, or other suitable materials, as shown in Figure 11-4, to completely block off gutter inlets, drainage grates, and manhole covers which feed into the storm sewer system and thereby the surface water system. Inflatable bags which may be placed in drain and sewer pipes and inflated to form a tight seal are commercially available. In some instances, barrier booms or sorbent booms (as described below) may be placed around entry points to confine or divert the product.

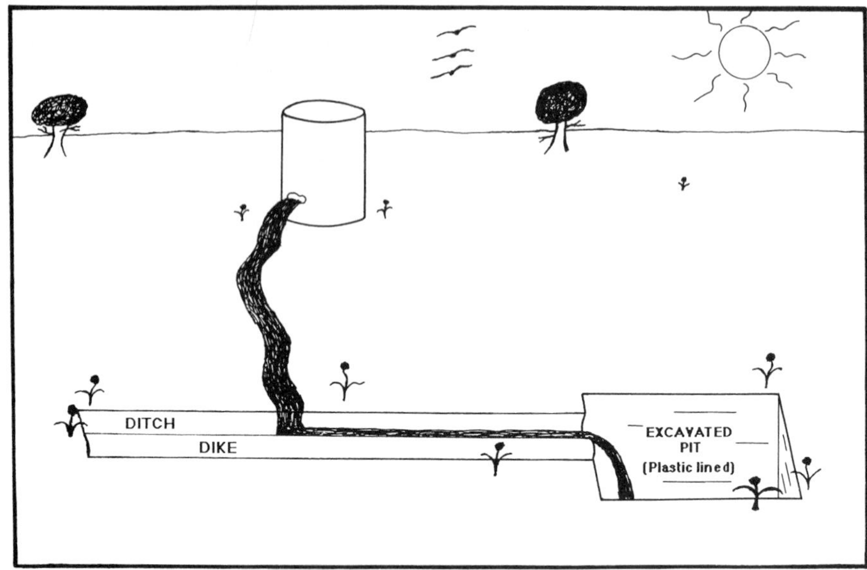

FIGURE 11-3. Use of a ditch and dike to divert spilled liquid into an excavated pit.

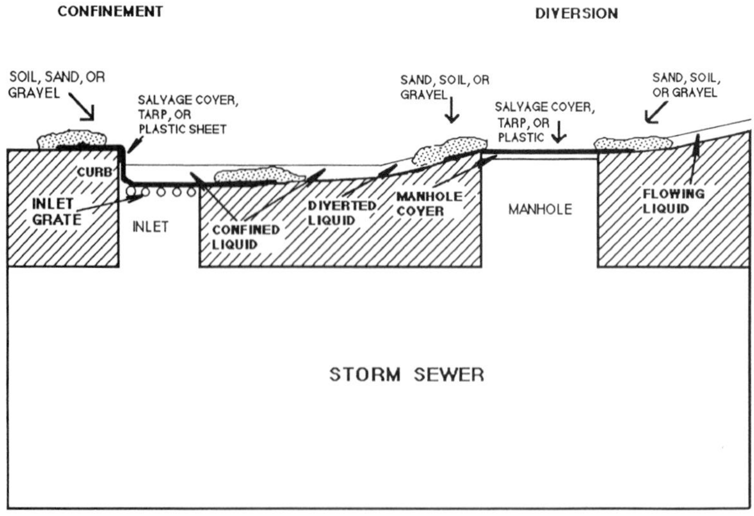

FIGURE 11-4. Use of inlet blockage for diversion and confinement, shown in cross-sectional view.

Implementing Control Procedures

Simple construction operations are usually required for implementing basic control procedures. Earth moving is frequently required so that, as a minimum, shovels and picks or mattocks will be needed. However, response personnel's ability for manual earth moving may be quickly overtaxed if anything beyond a small dike or ditch is required, especially if work must be carried out in PPE. For this reason, it is recommended that earth-moving equipment, such as backhoes, loaders, and dump trucks be located during preemergency planning and placed on standby if it is anticipated during assessment that larger control structures may be required.

If manual construction techniques are to be used, it is advisable to try to utilize bagged materials, such as sandbags or mulch bags, in creating control structures, as shown in Figure 11-5. This is much faster and easier than moving dirt or sand by the shovel load.

Whenever possible, responders should use available items such as plastic sheets, tarps, or salvage covers to enhance the effectiveness of control structures. For example, plastic sheets can be used to prevent liquid product from soaking into and through soil or other materials used to construct a dike. Likewise, these items can be used to line diversion ditches or catch basins so as to minimize soil contamination and related cleanup costs. In all cases, the potential for a reaction between the materials used in creating confinement structures and the material to be confined must be assessed.

It should be noted that commercially available products for dike construction (such as foamed concrete or polyurethane) work very well in some instances. However, use of these products requires equipment which is not commonly available and personnel with special training. Booms (both barrier and sorbent types) and the other commercially available products described below may also be utilized in basic confinement operations on land.

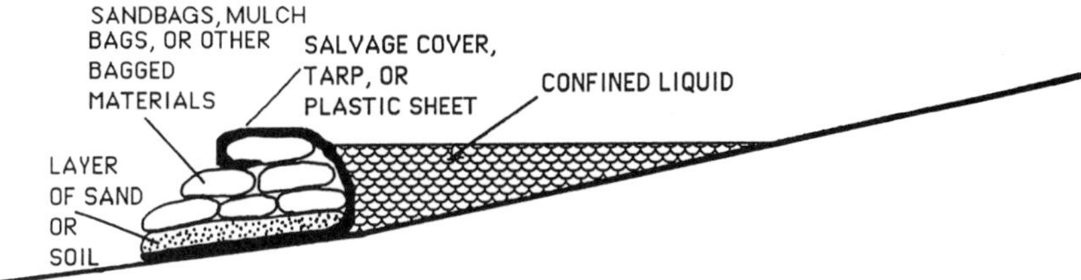

FIGURE 11-5. Improvised dike using bagged materials, shown in cross-sectional view.

Controlling Water Releases

In selecting appropriate techniques for mitigating releases of hazardous materials into water bodies, important assessment considerations include the degree of water solubility and the specific gravity of the product involved. These properties are fully discussed in Chapter 5. Based on its solubility, the hazardous material may tend to intermix with or remain separate from the water body involved. Heavy insoluble products, or those with a specific gravity greater than 1, tend to sink to the bottom of a body of water. In contrast, light insoluble products, or those with a specific gravity less than 1, tend to float on the surface of the water. Based on these properties, a number of options may be available to the responder attempting to control the hazardous material. It must be remembered that agitation of a water body, such as through turbulence or wave action, may have the effect of physically mixing an insoluble contaminant with water.

Controlling Heavy Insoluble Materials

When a heavy insoluble material enters a water body such as a stream, it tends to migrate to the stream bottom and then along the bottom under the influence of gravity. Movement will typically be in the downstream direction for flowing streams. This behavior allows several procedures to be used in confining the product involved so that it can be collected and placed in a suitable container for recycling or disposal.

Overflow Dams

Overflow dams can be used in some instances for confining heavy insolubles. This procedure involves constructing a dam, as shown in Figure 11-6, across the stream bed at some point downstream from the point of release. If properly located and constructed, the overflow dam should

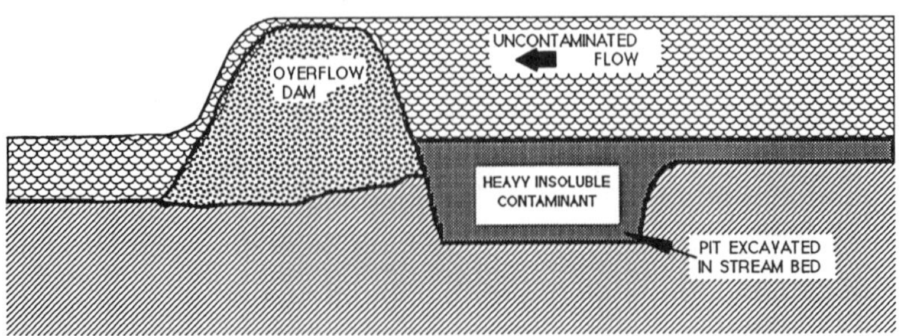

FIGURE 11-6. Use of an overflow dam and pit in a stream bed to confine a heavy insoluble material, shown in cross-sectional view.

confine the released material at the base of the dam while allowing uncontaminated water near the surface of the stream to flow over the dam and continue downstream. This is highly preferable to attempting to capture the entire stream flow.

Catch Basins

Catch basins are another type of structure which may be effective in the confinement of heavy insolubles. This procedure requires the excavation of a pit or catch basin in the floor of the stream in advance of the material (Figure 11-6). Ideally, the product will accumulate due to the influence of gravity and remain at this location.

Both of these procedures for confining heavy insolubles tend to work best with small, slow-moving streams. When larger or fast-flowing streams are involved, these procedures may be ineffective or extremely difficult to implement.

Controlling Light Insoluble Materials

When light insoluble materials enter a stream, river, or lake, they tend to float on the surface of the water. The direction and speed of migration of the material are then influenced by factors such as stream flow and wind. Several procedures may be utilized in confining this type of material.

Booms

Booms can be very efficient devices in some situations for confining light insoluble materials, as shown in Figure 11-7. Simply defined, booms are elongated tubular devices which float on the surface of a water body. They are commercially available or can be improvised, for example, by capping and inflating sections of hose. Barrier booms are designed simply to confine a product. In contrast, sorbent booms are filled with a sorbent material designed to soak up the spilled product (typically, a petroleum-based product) while remaining impervious to water.

Booms can be connected end-to-end and deployed so as to surround a slick of floating product on a water body such as a lake or large, slow-moving river. Booms can also be deployed across a stream at locations downstream from a point of release in order to achieve confinement.

One major weakness of booming as a confinement procedure is that wave action and stream turbulence tend to splash the floating material over the booms. This problem may be addressed by placing booms in series. However, if the water is sufficiently rough, booms may not be effective in confining the released material.

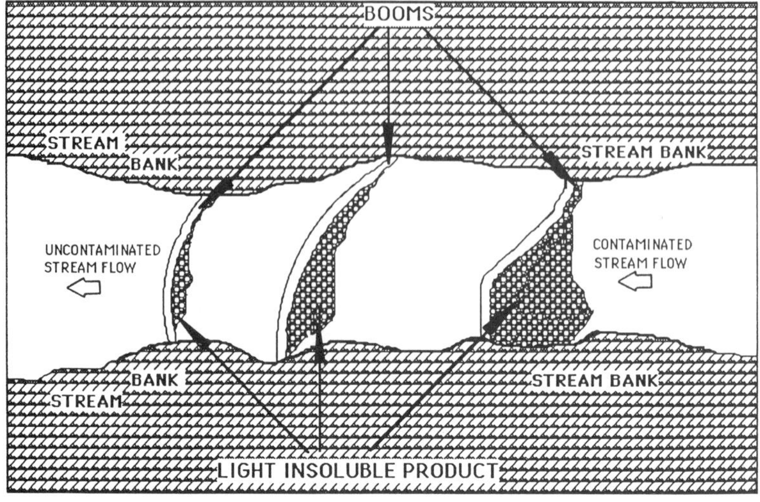

FIGURE 11-7. Use of booms to confine a light, insoluble material, shown in plan view.

Underflow Dams

Underflow dams are quite similar to the dams or dikes previously discussed, except that underflow dams incorporate pipes or tubing sections, as shown in Figure 11-8. The pipes are placed so that the downstream ends are elevated relative to the upstream ends. This configuration causes the floating contaminant to be confined behind the dam while allowing uncontaminated water along the base of the stream to flow through the

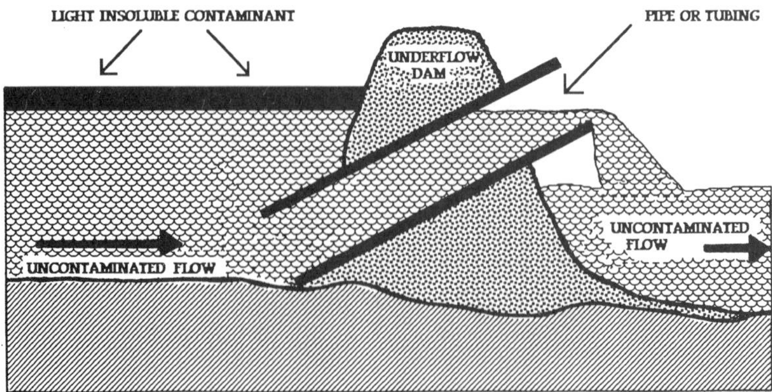

FIGURE 11-8. Use of underflow dam to confine a light, insoluble product, shown in cross-sectional view.

dam and continue downstream. Underflow dams are frequently used in conjunction with booms.

Filter Fences

Filter fences are improvised confinement devices constructed using posts, fencing, and suitable sorbent materials, as shown in Figure 11-9. The fencing holds the sorbent material in place, and the sorbent material confines and soaks up the contaminant as the stream flows through the filter fence. Like dikes or dams, filter fences should be placed in series for greater efficiency and to provide backup in case of structural failure.

Controlling Materials Which Are Water-Soluble

Materials which are soluble in water or have a specific gravity of approximately 1 can be extremely difficult to control when released into a body of water. Since these types of products tend to disperse rapidly throughout a

FIGURE 11-9. Use of a filter fence to confine a light, insoluble material.

water body, confinement techniques typically involve capturing the entire volume of contaminated water so that it can be treated to remove the contaminant. This will most probably involve expertise and equipment, such as mobile treatment plants, which are not readily available to the responder.

Sealed Booms

Sealed booms may be utilized to limit dispersion of soluble contaminants in some instances. These booms support curtains which are anchored to the bottom. This allows the entire depth of water to be confined within the boomed area.

Dams

Dams may be constructed for the purpose of confining the entire volume of contaminated water for treatment. However, this is usually an effective approach only for small streams. Otherwise, the size of the dam required would be beyond the capabilities of the response team.

Chemical Treatment

Chemical treatment processes may be a viable option for dealing with soluble contaminants in water. For example, neutralization may be utilized to render a corrosive contaminant noncorrosive. Likewise, flocculants may be used to cause precipitation of a dissolved contaminant out of the water column. The precipitant can then be pumped or dredged from the bottom for disposal. It is important to note that chemical treatment for hazard control requires specialized expertise and close coordination of activities between responders, personnel with expert knowledge of the chemicals involved, officials of the environmental agencies having jurisdiction, and cleanup personnel.

Controlling Air Releases

In selecting procedures to control airborne gases and vapors, it is important to consider the vapor density of the material involved relative to that of air. As discussed in Chapter 5, lighter-than-air gases and vapors (i.e., those with a vapor density less than 1) tend to migrate up into the atmosphere and disperse. In contrast, heavier-than-air gases and vapors (i.e., those which form a vapor-air mixture which is heavier than air) tend to migrate to the ground and may accumulate in low-lying areas.

In response to air releases of hazardous substances, there may be little that responders can do other than to terminate the release at the source, if possible (see Chapter 12), evacuate the potentially affected area, and allow the released vapors to disperse. However, it must be remembered that heavier-than-air vapors do not disperse readily. Restriction of the airspace above the incident may be required if lighter-than-air vapors are involved. In some instances, specific procedures, such as those discussed below, may be utilized in controlling air releases.

It is important to determine whether or not the substance involved is flammable. If so, it will be necessary to eliminate all potential sources of ignition in areas where a flammable atmosphere may develop. Otherwise, ignition of the vapor-air mixture can occur, with disastrous results.

Vapor Knockdown or Dispersion

In some instances, it may be possible to knock down a vapor cloud by directing fog patterns from fire hoses through the cloud (as shown in Figure 11-10). This is most effective with water-soluble materials. For air releases of substances which are not water-soluble, the air turbulence created by the fog patterns may enhance the dispersion of vapors, thus reducing the atmospheric hazard resulting from the release. In using this technique, large quantities of contaminated water are produced. The contaminated runoff should be confined in order to minimize environmental pollution (Figure 11-10). The Association of American Railroads' *Emer-*

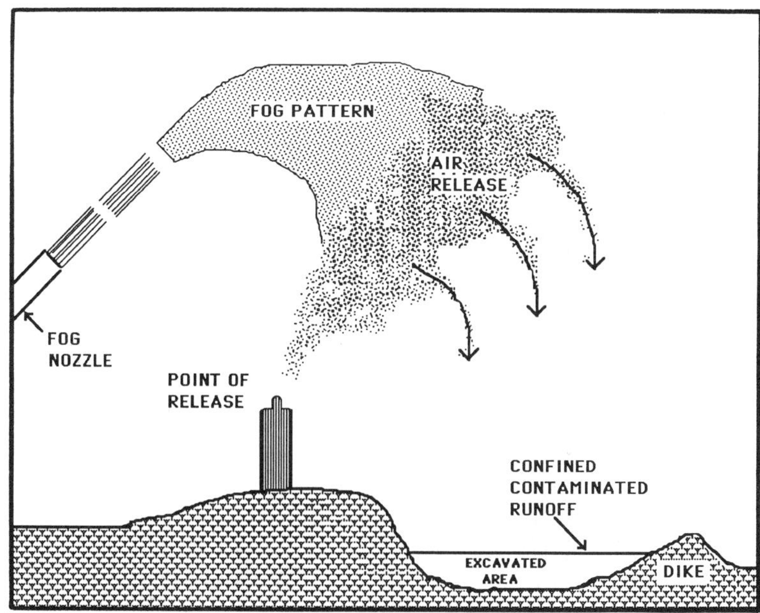

FIGURE 11-10. Use of fog patterns to control an air release, shown in cross-sectional view.

gency Action Guides is a good source to consult in order to determine whether or not this technique is applicable for a given substance. Fans, blowers, and compressed air have also been used to disperse vapors in some instances.

Vapor Suppression

For incidents in which vaporization of a spilled liquid is producing an air release, vapor suppression at the surface of the liquid pool may be a viable control technique. Vapor suppression techniques use some sort of foam blanket to cover a volatile material and prevent the evolution of vapors from its surface (Figure 11-11). Firefighters are familiar with fire-fighting foams; however, hazardous materials foams utilize different agents and require different application techniques. Thus, specific training is required for their safe and effective use.

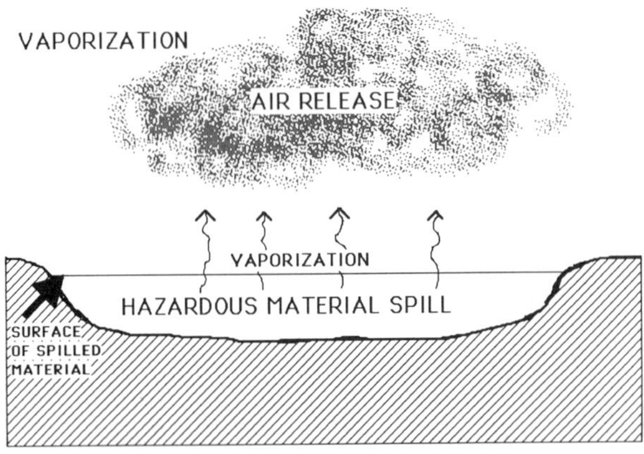

VAPOR SUPPRESSION

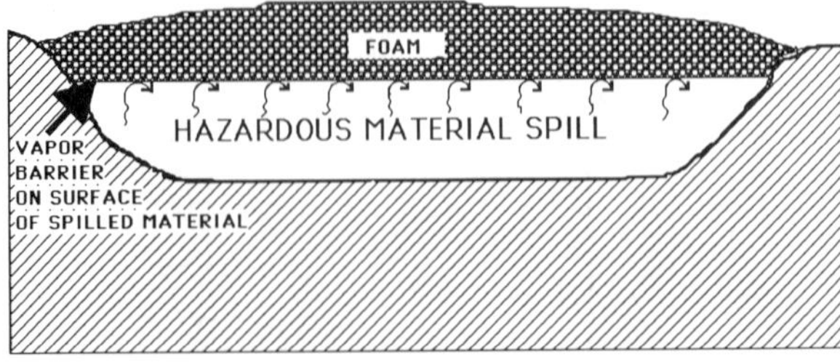

FIGURE 11-11. Vaporization and vapor suppression of a hazardous materials spill, shown in cross-sectional view.

Firefighting Foams

Aqueous film-forming foam (AFFF) is one example of a firefighting foam which can also be used as a vapor suppressant on certain kinds of spills. When AFFF is applied to a spill, draining foam bubbles create a continuous film of aqueous solution on the surface of the spilled liquid. This film acts as a vapor barrier. AFFF is often used on flammable liquid spills to suppress vapors and prevent ignition. However, if the flammable liquid is a polar solvent or has a polar solvent added, the aqueous film will be dissolved by the polar solvent, rendering the foam ineffective. Such liquids are considered to be foam-destructive. This characteristic can cause problems during vapor suppression procedures in response to fuel spills because polar solvents like ethanol and methanol are added to gasoline and some other fuels.

Alcohol-resistant AFFFs are effective on both hydrocarbon and polar solvent fuels. They form a polar solvent–resistant membrane on the surface of the spill and are therefore resistant to the destructive properties of polar solvent fuels.

Hazardous Material Foams

Hazardous material foams are of specialized formulations developed for use against nonflammable chemicals which emit large quantities of vapors. They are not suitable for use in firefighting. These specially formulated foams are designed to handle hazardous materials other than fuels. One type of agent is used for acidic materials and another for alkaline materials. Hazmat foams are resistant to pH extremes which would rapidly destroy fire-fighting foam. They are applied using air-aspirating foam nozzles. In some cases, these new foams convert the spilled material to a less hazardous one.

Considerations for the Use of Foams

Using any foam to mitigate a volatile spill requires expertise in choosing the foaming agent and selecting the application technique, as well as specialized training in using foams for vapor suppression. Training in the use of foams for firefighting is not sufficient. It is important to remember that foams are mostly water, and will add considerable volume to confinement and cleanup liquids.

For effective use of foams in vapor suppression, the following materials must be available at the scene:

- Appropriate application equipment for the foam selected
- Sufficient quantities of foam concentrate to reapply the foam blanket every 30 minutes, or as needed, until the spill stops vaporizing or is cleaned up

- Enough water to maintain the critical ratio of foam concentrate to water

COLLECTION TECHNIQUES

After spills have been confined, the hazardous substances involved must be collected for recycling, treatment, or off-site disposal. In the case of spilled liquids, this is typically done using pumps or vacuum trucks to vacuum up the product or sorbent materials to soak it up. In the public sector, these types of operations are typically performed by cleanup crews who are called in after the emergency is stabilized. However, in the private industrial sector, response personnel may also be involved in cleanup operations.

Pumping and Vacuum Collection

Pumping is the method of choice for collecting large spills after confinement. Liquid materials can be pumped from land, from the bottoms of water bodies, or from the water surface.

A variety of pumps which may be used in collecting liquids are available. General-duty pumps, which are commonly powered by electric motors or gasoline engines, may be suitable for collecting some hazardous materials. Other materials may require specialized equipment, such as manually or pneumatically powered pumps designed for transferring flammable products. In all instances, the potential for reactivity between the product being transferred and the material of which the pump, hoses or tubing, and fittings are constructed must be considered.

Also, when transferring flammables, appropriate grounding and bonding procedures, as described in Appendix 12-1, must be utilized for the equipment and containers used. Figure 11-12 illustrates the use of a manually operated salvage pump.

Usually a residue of product will remain after pump collection. This residue can be removed using sorbent materials or other options, as described below.

Sorbent Collection

When a spill is small, you may be able to use specific materials to soak up the product, as shown in Figure 11-13. Strictly speaking, these materials are classified as either absorbents or adsorbents, depending on the specific process by which they collect and retain the spilled material. For the purposes of this text, the generic term "sorbent" will be used in reference to both types of materials.

FIGURE 11-12. Use of a manually operated salvage pump to collect a spilled liquid material after confinement.

A variety of commonly available and commercially available products can be utilized as sorbent materials. Sorbents which will best handle the materials expected to be involved in incidents on the site should be chosen during preemergency planning. The physical and chemical interactions between the sorbent and the spilled material are important considerations. If sorbents are improperly selected, undesirable chemical reactions can occur, producing toxic by-products and heat, which may result in fire.

Rate of sorbency, convenience of application and collection, and available disposal options are basic criteria for selection. Materials with low pH, low viscosity, and high temperature will sorb more quickly. Bulk materials are hard to carry and clean up and cannot be used in water. Sorbent contained in sheets, rolls, sausages, pillows, socks, or booms are more convenient to use.

Available types of sorbent materials can be generally classified as vegetable by-products, mineral products, or synthetics (Table 11-1). Sorbents

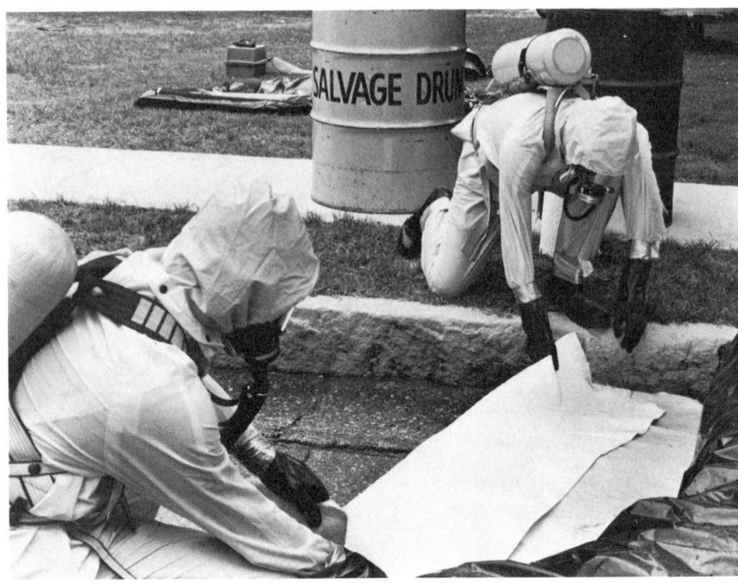

FIGURE 11-13. Use of sorbent sheets to remove the residue remaining after pump collection of material.

in granular form such as vermiculite, fly ash, perlite, granular clays, crushed limestone, and activated carbon are used extensively in spill control operations. These materials have the effect of producing a solid or semisolid mass which can be easily collected and may reduce the hazard associated with the product involved, as when crushed limestone is applied to an acidic spill. However, under existing environmental regula-

TABLE 11-1. Characteristics of Sorbents[a]

Corn Cobs or Other Vegetable By-Products	Minerals (Vermiculite, Crushed Limestone, or Diatomaceous Earth)	Synthetics (Tyvek, Polyolefin, Polypropylene)
− Slow sorbency	− Slow sorbency	+ High rate and capacity of sorption
− Short shelf life	− Limited capacity	
− Nest for bugs and mice	− Heavy	+ Sorbed materials tend not to leach out
+ Can be incinerated	− Create dust	
	− Must be landfilled[b]	+ Can be incinerated
+ Inexpensive	+ Inexpensive	− Expensive

[a] +, advantage; −, disadvantage.
[b] Treatment prior to landfilling may be required under current environmental laws.

tions, the resulting material may require further treatment prior to disposal in a hazardous waste landfill.

Solidification

Certain granular sorbent materials react chemically to completely solidify a spilled liquid in a continuous matrix. This may allow the product to be legally placed in a hazardous waste landfill without further treatment. A number of commercially available solidifiers can be selected specifically for use with spills of acids, bases, fuels, oils, solvents, organic materials, and aqueous products. Selection of a solidification material can be made only by a disposal expert in conjunction with personnel at the facility to which the waste is to be sent.

Gelation

Gelation is a technique in which a gel-forming chemical is introduced into the liquid pool of contaminant or contaminated water. The resulting gel is a semisolid and is therefore easier to remove.

Chemical compatibility is the most important consideration when choosing gelation. The gelling agent must be compatible with the chemical to which it is applied. The gel formed by this method is a hazardous material and must be disposed of as such.

Other Techniques and Considerations

Dispersants

After a floating, oily substance has been removed from the land or water surface by vacuuming and/or sorption, a dispersant can be applied to the thin slick that remains. Dispersants are emulsifiers which break up an oily contaminant. They are generally applied to insoluble liquids spilled in water, like an oil slick on the surface of a water body, or on land, as when fuel spills occur on paved surfaces.

Dispersants are applied with standard firefighting equipment, by using a foam eductor or by pumping from a booster tank into which the dispersant has been directly poured. A fog nozzle is used to apply the stream straight into the spill. Agitation by the hose stream helps the dispersant mix with the spilled material.

As the oily contaminant is broken up, the continuous slick will be broken into tiny droplets which will then be dispersed into the environment. Manufacturers of dispersants state that the contaminant will then be biodegradable (capable of being removed from the environment by bacteria or fungi), and it is true that tiny droplets of chemicals allow better access by microorganisms. Microbes are selective about what they degrade, however, and are unable to metabolize many chemicals. This process should never be used, other than for contaminant residue after collection, without consultation with the appropriate environmental agency, as the contaminant is not removed from the environment, only dispersed.

Dispersant streams can also be directed into manholes or duct openings to emulsify floating layers of volatile fuel oils. The emulsion formed can be recovered by pumping. Chemical dispersants can also be used to remove an oil residue from vegetation, buildings, and vehicles. The emulsified chemical is then rinsed off into the soil. The same environmental considerations discussed earlier are important.

Dilution

Dilution is the process of adding large quantities of water to a spilled material which is hazardous because of its concentration. Diluting the material to a less concentrated solution reduces the hazard.

Dilution with water is effective only when the material is water-soluble. The hazard is reduced, but the volume of material which must then be confined and cleaned up is enlarged.

In some situations, the possibility exists that dilution can render the material nonhazardous if its hazard was due only to its concentration; however, permission of the appropriate environmental agency must be gained before using this method in flushing the spill down a drain or leaving it unconfined in the environment.

Dilution with water should never be considered under the following circumstances:

- The material has not been identified.
- The material is a strong acid.
- The material is water-reactive.

SUMMARY

Basic hazardous materials control operations are performed to confine the hazardous materials released in an incident to the smallest possible area. The considerations and procedures discussed in this chapter are intended to aid emergency response personnel in selecting and implementing appropriate basic control procedures. Although these are relatively simple procedures, the personnel expected to perform them must be provided sufficient equipment and field training in order to respond safely and efficiently. These concerns must be addressed during preemergency planning (see Chapter 8). Even though confinement options may be limited in some instances, responders should identify and perform any defensive actions which can be safely performed at locations remote from the point of release in order to minimize damage resulting from the incident.

REFERENCES

Action Training Systems. 1989. *Diking, Diverting and Retaining Spills* (videocassette). Seattle, WA: Action Training Systems.

Bureau of Explosives, Association of American Railroads. 1990. *Emergency Action Guide*. Washington, DC: Association of American Railroads.

Henry, M.F., ed. 1989. *Hazardous Materials Response Handbook*. Quincy, MA: National Fire Protection Association.

International Fire Service Training Association. 1988. *Hazardous Materials for First Responders*. Stillwater, OK: Fire Protection Publications.

Noll, G.G., M.S. Hildebrand, and J.G. Yvorra. 1988. *Hazardous Materials: Managing the Incident*. Stillwater, OK: Fire Protection Publications.

12

Advanced Hazardous Materials Control

D. Alan Veasey, M.A. Ed., and
Barbara M. Hilyer, M.S., M.S.P.H.

Having addressed the considerations and procedures for basic, or confinement-related, hazardous material control operations in the previous chapter, we will now turn our attention to advanced, or containment-related, control operations. As noted in the introduction to the previous chapter and illustrated by the following scenario, both types of operations are critical to incident mitigation.

As a result of a faulty weld, an oil storage tank located in Jefferson Borough, Pennsylvania, failed. The tank collapsed almost instantaneously, releasing almost 1 million gallons of heavy oil in a 30-foot wave. The initial surge of oil washed over stationary confinement berms, battered nearby tanks, and twisted pipelines before entering the Monongahela River. This release resulted in one of the better-publicized environmental disasters of recent years. Despite a massive effort to confine and collect the spilled oil, approximately 500,000 gallons were not recovered and are presumed to be lost in the environment. As the released oil migrated down the Monongahela River and entered the Ohio River, thousands of people went without water as suppliers along the rivers were forced to shut their intake valves off. However, the incident could have had even more devastating consequences were it not for the actions of response personnel at the scene of the release. The oil-soaked vicinity of the failed tank had a strong odor of gasoline when responders first arrived. This was considered to be very serious, since a nearby gasoline storage tank appeared to have been damaged by the release event. After diligent searching, the responders located the source of the gasoline release: a leak in a pipeline damaged by the sudden release of oil. Response personnel were able to stop the gasoline leak by using a golf tee as a plug. The ingenuity of the responders in stopping the leak removed the threat of impending explosion and fire, which would have made the incident on the Monongahela an even greater disaster.

By terminating a release at the source, containment operations offer the advantage of minimizing environmental contamination. Also, if offensive actions are not taken to terminate a sustained, large-quantity release at the source, the confinement capabilities of response personnel may be overwhelmed. Thus, containment operations should be undertaken in all instances in which they can be safely performed by response personnel. However, since responders must work in close proximity to the actual point of release in order to perform containment operations, the potential for exposure of personnel to hazards related to the incident is high. Safety of personnel must be given the highest priority in these types of operations.

This chapter is intended to provide knowledge and understanding of advanced hazardous materials control procedures utilized in containment and related operations. These operations typically require special tools, equipment, protective gear, and expertise in order to be performed safely and effectively. Only those personnel who have received training to the level of hazardous materials technician should attempt to perform the types of procedures described in this chapter.

OBJECTIVES

- Understand the importance of containment procedures in bringing hazardous materials incidents under control
- Be aware of the high exposure potential of primary responders involved in containment operations
- Know the specific assessment considerations for containment operations
- Be familiar with the types of equipment, supplies, and tools used in containment operations
- Have a general knowledge of the procedures which may be required for containing escaping hazardous materials

ASSESSMENT AND DECISION MAKING FOR CONTAINMENT OPERATIONS

Like all aspects of emergency response, containment operations must evolve directly from the hazard and risk assessment process described in Chapter 5. This process will provide basic information about the incident which can be used, in conjunction with decision-making guidelines (such as Benner's DECIDE process), to select appropriate containment procedures.

Containment-Related Planning and Decision Making

Since emergency response is time-critical, containment procedures must be carried out as efficiently as possible. Thus, preplanning is essential. The site-specific ERP should incorporate general procedural guidelines for containment operations which may reasonably be required during a site emergency. However, specific procedures must be determined based on the information gathered during assessment of the actual incident. Through preplanning and the use of decision-making guidelines, responders should be able to

- Determine the strategic containment objectives.
- Determine the procedural options for containment.
- Choose the best containment procedure and alternate procedures.
- Choose the equipment and tools needed to perform the selected procedure.
- Choose the required PPE.
- Perform the chosen containment procedure.
- Evaluate progress as the containment procedure is performed.
- Change plans or call for additional assistance and equipment as needed to safely terminate the release.

Assessment Considerations For Selection of Containment Procedures

Hazardous materials incidents in the industrial setting may involve uncontrolled releases from a variety of containers, including drums, vats, tanks, cylinders, pipes, and transport vehicles. In order to identify the procedural options for containing these releases, a number of specific factors related to the release event must be considered. These factors can be generally classified as follows:

- Properties of the hazardous material involved, including
  - physical state (solid, liquid, gas).
  - Physical properties (vapor pressure, density, etc.).
  - Toxicity (as related to PPE requirements).
  - Flammability (requiring the use of nonsparking tools and intrinsically safe electrical equipment).

    - Corrosivity (which may destroy materials used in patches or plugs and damage tools used).
    - Reactivity (including reactivity with available patch materials).
    - Composition (aqueous, petroleum-based, etc.).
- Characteristics of the container, including
    - Size of the container
    - Amount of contents remaining
    - Configuration of the container
    - Material of which the container is composed
    - Overall integrity of the container
- Characteristics of the breach and the resulting release, including
    - Size of the opening
    - Shape of the opening
    - Location of the opening on the container
    - Volume and rate of release
    - Expected duration of the release
    - Actual or potential pressure at the point of release

Hazards to Personnel

Since containment operations require personnel to work in areas of high hazard potential on site, safety should be the most important consideration in planning containment operations and choosing containment procedures.

A careful assessment of the nature and degree of stress to containers damaged in an incident is required in order to ensure that a sudden violent release will not occur during the course of operations. This is especially important for operations involving work on pressurized containers. Additional safety considerations are also required for situations in which fire or explosion could occur during the operation.

If the information gathered during assessment indicates that a containment procedure being considered may involve undue risk to responders, then other, less hazardous offensive options should be considered. Response personnel should never attempt to perform containment operations at the point of release before options for stopping the release remotely have been explored. This can sometimes be accomplished in the industrial setting by closing valves at some other location within the facility. However, close coordination with personnel familiar with the design of systems within the facility is required for this type of operation.

Containment procedures should never be attempted unless the equipment, protective gear, and training provided to response personnel are adequate to allow the procedure to be carried out in a safe, effective manner. For situations in which containment cannot be achieved without undue risks to the safety of response personnel, it is permissible to remain in a defensive mode and explore additional options as the incident runs its course.

EQUIPMENT, SUPPLIES, AND TOOLS USED IN CONTAINMENT

A variety of items may be used in containment procedures (Figure 12-1 and Tables 12-1 through 12-4). Some of these are general-purpose items which are readily available, while others are highly specialized and available only from specific sources.

Preassembled kits containing a variety of general-purpose containment-related items are commercially available (see Table 12-5). It is also possible to assemble similar kits in-house using supplies, equipment, and tools which are locally available. "Specialty kits," such as chlorine repair kits (see Table 12-5), are designed for use only in highly specialized applications.

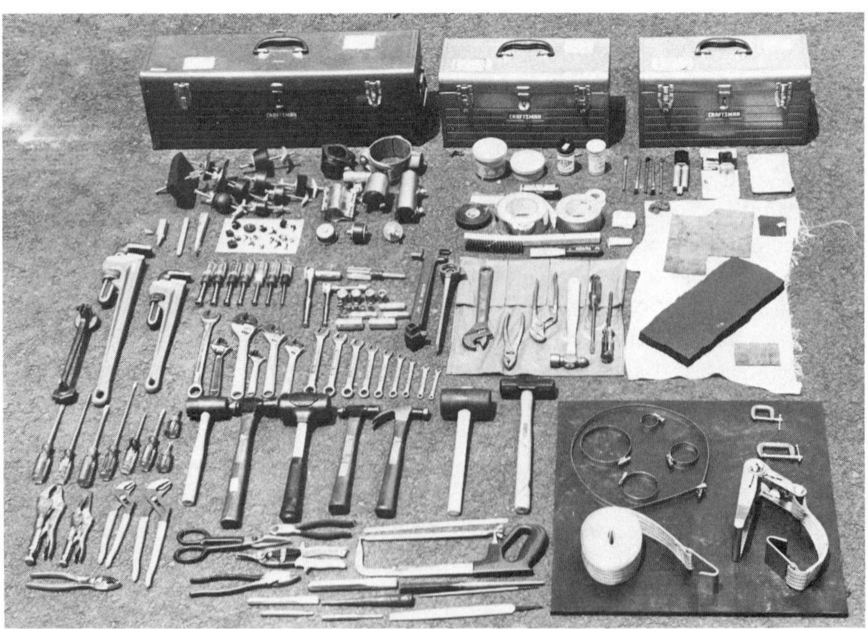

FIGURE 12-1. Overview of useful tools, equipment, and supplies for advanced hazardous materials control.

Equipment which may be used in containment operations can be roughly divided by function into four categories, which will now be discussed.

Plugs, Patches, and Related Items

Plugs are designed to stop leakage by filling holes in containers through which materials are escaping. Patches stop leaks by covering holes in containers. Clamps, fasteners, straps, and similar items are used to hold patches or plugs in place. Plugs, patches, and related items are described in Table 12-1.

TABLE 12-1. Plugs, Patches, and Related Items

Item/Description	Function/Use	Procedural Considerations/Comments
Expandable pipe plugs	Placed inside pipe and tightened; internal expansion of plug blocks pipe.	Various sizes available; must be appropriate size for pipe to be plugged. Some types require a special tool for installation. Some are fitted with vent pipes and/or valves for pressure installation
Pipe repair clamps (e.g., "pipe saver" or "bandaid" clamps)	Placed around leaking pipes, fastened, and tightened to clamp off leaks.	Clamps must be selected for specific outside diameter of leaking pipe. Gasket used must be compatible with leaking material
Leak-sealing bandages (product of Vetter Systems)	Used to stop pipe leaks.	Perform same function as pipe clamps but are pneumatically operated.
Boiler plugs or "doogit" plugs (e.g., Mueller brand)	Screwed into holes in containers, pipes, etc.	Various sizes available.
Sheet metal screws with gasket	Screwed into small, simple punctures in drums.	Can be improvised readily. Gasket materials used must be compatible with material to be contained.
Toggle bolt patches	Toggle is pushed through a small hole in the container wall, and patch is tightened into place to cover the hole.	Can be fitted with wing nut, washer, and neoprene pad, lab stopper, or rubber ball for patching small holes. Avoid overtorquing, as toggle may fail.

TABLE 12-1. (Continued)

Item/Description	Function/Use	Procedural Considerations/Comments
T-bolt patches	T is placed through small, elongated tears in container walls and rotated 90°. Patch is then tightened into place to cover the hole.	Can be fitted with wing nut, washer, and neoprene pad, lab stopper, or rubber ball for patching small holes. Avoid overtorquing, as T may be pulled through thin container walls.
Wooden plugs and wedges	Used individually or in combination to plug holes of various sizes and shapes.	Use with suitable gasket or sealant material to stop leakage. Soft wood should be used. Wooden plugs may react with some corrosives and ignite.
Neoprene or viton plugs, etc.	Used to plug small leaks of various shapes.	Material used must be compatible with chemicals involved. Foam materials used must be closed-cell. Can be preshaped to fit expected configuration of hole (e.g., forklift tine).
Drift pins	Driven into small holes in containers.	Should be nonsparking for flammables.
Metal sheeting	Used as backing in conjunction with gasket materials, sealant, or adhesive to patch leaks of various sizes and shapes.	Must be flexible or preshaped. Can be held in place with toggle bolts, T-bolts, ratchet straps, ropes, etc.
Strap iron "bandaid"	Used to patch large and/or irregularly shaped leaks.	Can be attached by toggle bolts, T-bolts, ratchet straps, ropes, etc. Can be fitted with relief pipe and valve for application under pressure.
Ratchet straps, banding straps, etc.	Used to hold large patches in place on containers.	Can be used with adhesive patches and removed after adhesive cures.
Dome cover clamps	Used to clamp tank truck dome covers into closed position.	Can be improvised or fabricated inhouse.
Leak-sealing bags (product of Vetter Systems)	Used to stop leaks from holes in tank trucks or rail tankers.	Bags are strapped in place over breach and inflated to stop leaks.

Adhesives, Sealants, and Gaskets

Adhesives, sealants, and gaskets are frequently used in conjunction with patches and plugs. For example, adhesives can be used to secure patches in place on a container. Sealants and gaskets are often used to enhance the sealing action of patches and plugs. Adhesives, sealants, and gaskets are described in Table 12-2.

Tools

Tools of various types can be used for purposes such as preparing a surface for placement of an adhesive or sealant, screwing or driving plugs into place, installing patches, and tightening leaking pipe connections. Tools which may be used during containment operations are listed in Table 12-3.

Other Containment Related Items

Additional items which do not fall within the categories above may be utilized during containment operations. For example, drip pans may be used to confine leaking product at the point of release during plug/patch

TABLE 12-2. Adhesives, Sealants, and Gaskets

Item/Description	Function/Use	Procedural Considerations/Comments
Plastic steel sealant (e.g., Devcon brand) Aluminum putty sealant (e.g., Devcon brand) Plastic carbide sealant (e.g., Devcon brand	Used as sealant in conjunction with various plug and patch materials.	Should be selected based on composition of container material and degree of durability required.
5-minute epoxy and epoxy gel (e.g., Devcon brand)	Used as adhesive in conjunction with various plug and patch materials.	Syringes provide a convenient means of premixing resin and hardener.
Petroseal Putty (product of Essex Fire and Safety)	Used to stop leaks of petroleum-based products.	Can be held in place by various patches (e.g., fiberglass fabric and epoxy or tape).
Aquaseal Putty (product of Essex Fire and Safety)	Used to stop leaks of water-based products	Can be held in place by various patches.
Plug-n-dike putty	Used to stop leaks, as well as to construct small dikes.	Can be held in place by various patches.

TABLE 12-2. (Continued)

Item/Description	Function/Use	Procedural Considerations/Comments
Viton sheets, patches, and washers; closed-cell neoprene sheets, patches, and washers.	Used as gaskets to enhance seal of patches, plugs, etc.	Material used must be compatible with chemicals involved. Can be used as whole sheet or cut to the size and configuration needed, depending on the application. Thickness varies with application.
Duct tape	Provides temporary seal for cracks and small leaks.	Can be used over putty or foam packing as a temporary patch. Product may dissolve glue holding tape in place.
Aluminum repair tape	Provides temporary seal for cracks and small leaks.	Backless type preferred due to difficulty in peeling backing from tape while wearing gloves.
Teflon sealant tape	Used to enhance seal of threads on plugs and pipe connections.	
Lead wool	Used as packing to stop leakage along container seams (e.g., drum chimes).	
Lead foil	Used as packing or to wrap wedges or plugs.	
Cloth or felt	Used to wrap wedges and plugs for enhanced seal.	
Tubeless tire plug/patch kit	Use to stop small leaks in a variety of containers.	
Plumber's oakum	Use as filler material and wrapping for plugs and wedges.	Resin-impregnated, fibrous substance; swells when wet.
Fiberglass fabric	Used in conjunction with epoxy to hold patch/plug materials (e.g., putty) in place.	
Assorted O-rings, washers, nuts	Used in conjunction with various plugs and fasteners to enhance seal.	
Bar soap (e.g., Octagon brand)	Used as packing to stop leaks in containers such as fuel tanks.	For temporary use only.

TABLE 12-3. Tools Which May be Required for Containment Operations

Item/Description	Function/Use	Procedural Considerations/Comments
Acid brushes	Application of adhesives to containers and patches.	
Utility shears	Cut and shape gasket materials.	
Cotter key extractor	Placement or removal of gaskets, packing, etc.	
Wire brushes	Surface preparation prior to application of adhesives.	Must be nonsparking for flammability hazards.
Emery cloth	Surface preparation prior to application of adhesives	
Putty knives	Surface preparation or application of sealants and adhesives.	Must be nonsparking for flammability hazards.
Hammers, various sizes and types Sledge Mallet Ball-Pen Claw Deadblow Etc.	Used in installation of wedges and plugs, as well as for various general uses.	Must be nonsparking for use in potentially flammable situation (e.g., brass, rubber). Dead blow hammer is loaded to prevent rebound and possible injury.
C-clamps	Used to hold patches in place while adhesive cures.	
Tubing clamps (e.g., Entex brand)	Used to pinch leaking tubing closed.	Other items, such as vice grip pliers, can sometimes be adapted for this function.
Bung wrench	Tightening of drum bungs to stop leakage. Removal of bungs for transfer of product to sound container.	For flammable product, use nonsparking wrench, bond and ground containers.
Expandable pipe plug wrench	Required for installation of some expandable pipe plugs.	
Pipe wrench Adjustable Wrench (e.g., Crescent brand) Combination wrench set Socket wrench set Slip-joint pliers Needle-nose pliers Vice-grip pliers Lineman's pliers Wire cutters Punch-chisel set Hacksaw Flat bastard file Round bastard file Screwdriver set Pop-rivet tool	Various general applications related to confinement procedures.	Must be nonsparking for use in potentially flammable work areas.

TABLE 12-4. Other Containment-Related Items

Item/Description	Function/Use	Procedural Considerations/Comments
Drums and other containers	Transfer product from leaking container to sound container.	Must be bonded and grounded for transfer of flammables.
Salvage drums	Storage of used sorbents or overpacking of damaged drums for transport.	See Appendix 2. Drums must be DOT certified for materials to be transported.
Drip pans	Confinement of leaking product prior to and during patching procedure.	Must be bonded and grounded for use with flammables.
Sorbent materials and related equipment	Confinement and collection of product at point of release in conjunction with containment procedures.	See Chapter 11.
Grounding and bonding equipment	Prevent ignition through discharge of static electricity during operations involving flammables.	See Appendix 1.

TABLE 12-5. Containment Kits

Item	Function/Use	Procedural Considerations/Comments
Essex quick kit	Plugging/patching pipes, drums, transport vehicles and various large containers.	Contains various tools and supplies.
Vetter containment kit	Plugging/patching railcar leaks.	Contains inflatable patches which can be strapped in place and inflated with compressed air.
Hazardous materials response kits, series A through D (product of Edwards and Cromwell of Baton Rouge, LA)	Used for containing leaks from a variety of sources, including drums, large containers, and pipes, depending on the kit selected.	Series A used for leaks of various sizes from containers of various sizes. Series B used for large leaks in containers. Series C used for pipe leaks. Series D used for small containers.
Chlorine A kit	100 and 150 lb. chlorine cylinders.	Specialty kits, specifically designed for use with chlorine. Available from the Chlorine Institute, Washington, DC.
Chlorine B kit	1 ton chlorine containers.	
Chlorine C kit	Chlorine railcars.	
Chlorine D kit	Chlorine barges.	

procedures. Grounding and bonding equipment is required for safe transfer of a flammable liquid to a sound container (see Appendix 1). Overpack drums may be required for transportation of leaking drums after patching (see Appendix 2). Any number of additional items may be required for safe and effective containment operations. Some of these items are listed in Table 12-4.

CONTAINMENT PROCEDURES

A great variety of procedures and techniques may be utilized for containment during hazardous materials incidents. The procedures used must be tailored to the specifics of each situation. Thus, it is entirely possible that no two containment procedures used in the field will be exactly alike. Ingenuity, resourcefulness, and flexibility are called for on the part of the responder.

It is not possible to describe in detail all the containment operations that responders may be called on to perform in the field. However, general procedures that may be applicable to various types of releases are incorporated in the information which follows.

General Procedures

In all containment operations, general procedures for responder safety and minimization of the spread of contamination should be followed.

Basic safe work practices should be utilized to minimize personal exposure. For example, avoid walking through or standing in puddles of spilled product, if possible. Also, try to minimize direct contamination of protective clothing through splashing or spraying during procedures such as patch installation or valve repair.

The responder should attempt to minimize spillage during the course of all containment operations. This will minimize contamination of the work area and thereby reduce the likelihood of contamination of personnel. If possible, change the position of the container involved in order to place the point of release in the vapor phase. Since a given volume of spilled liquid may produce a much larger volume of vapor (in some cases, several hundred times the original volume of liquid), this will significantly reduce the concentration of hazardous vapors in the work area. For containers which cannot be repositioned, it may be possible to place a suitable container beneath the point of release to confine the escaping material.

In some situations involving pressurized containers or systems, it may be possible to reduce the amount of pressure involved prior to the onset of operations. In some instances, pressure on a system may be lowered by manipulating valves or pump pressures. Obviously, these operations will require expert knowledge of the system involved. Container pressures may be lowered significantly by cooling the containers involved, such as through the use of hose streams. Cooling containers may be vital for situations in which container temperatures are elevated.

For all operations involving flammable materials, extra precautions must be followed. Container grounding and bonding procedures, as described in Appendix 1, should be utilized. All tools used should be constructed of nonsparking materials, and any electrical equipment used (such as direct-read atmospheric instruments, radios, and flashlights) must be intrinsically safe. Furthermore, suitable protective equipment should be utilized, with appropriate fire-extinguishing equipment and personnel on standby in case of flash fire.

In all containment operations, compatibility between the leaking material and the components of plugs, patches, and related items used is a critical consideration. If any evidence of a reaction is noted, cease the procedure immediately and explore other options.

Procedures for Small Containers

For the purpose of this text, small containers will be considered those which can reasonably be repositioned by a two-person entry team. A commonly encountered container of this type is a 55-gallon steel drum.

Small, Simple Punctures

A variety of procedures may be used for plugging or patching small, simple punctures in small containers. Boiler plugs or sheet metal screws fitted with a suitable gasket material may be very effective for this type of leak (Figure 12-2). Likewise, wooden plugs or wedges may be driven into the hole (Figure 12-3). Small toggle bolt patches may be installed if the opening is large enough to allow the toggle to pass through in the closed position (Figure 12-4).

In some cases, a suitable epoxy, such as fuel tank repair epoxy, may be placed over small holes in containers. Another option is to pack the opening with putty or closed cell foam (see Figure 12-7). It may be desirable to precut items such as neoprene foam or neoprene wedges into the shapes anticipated to be required, such as the shape of a forklift tine. Tape may be utilized to hold the packing material temporarily in place, or an epoxy-fiberglass patch (as described below) may be used.

After the container is repaired, it may be desirable to place it in an overpack container. If so, the patch or plug installed must not interfere with the overpacking procedure. For this reason, it may be desirable to

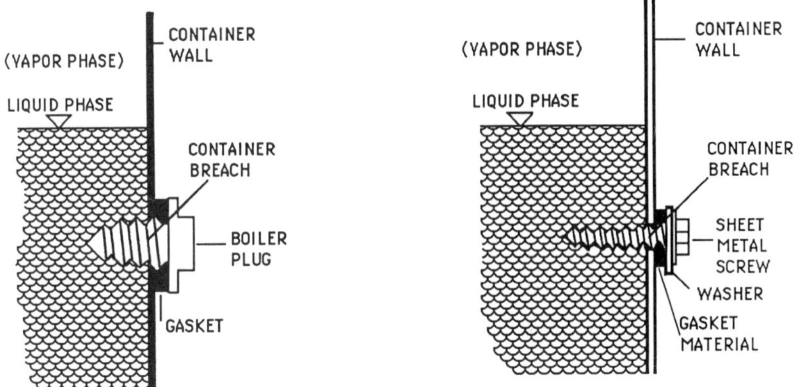

FIGURE 12-2. Screw plugs, shown in cross-sectional view.

trim items such as toggle bolts or wooden wedges flush, using bolt cutters or a saw. Overpacking procedures are described in Appendix 2.

Large or Irregular Holes

In repairing large or irregularly shaped holes in small containers, techniques similar to those described above are applicable, with certain modifications.

Toggle patches or T patches, fitted with a backing plate and gasket of suitable size and shape, may work for some leaks of this type (Figure 12-4). It may also be possible to place a number of wooden wedges, available in a variety of sizes and shapes, in the breach and drive them

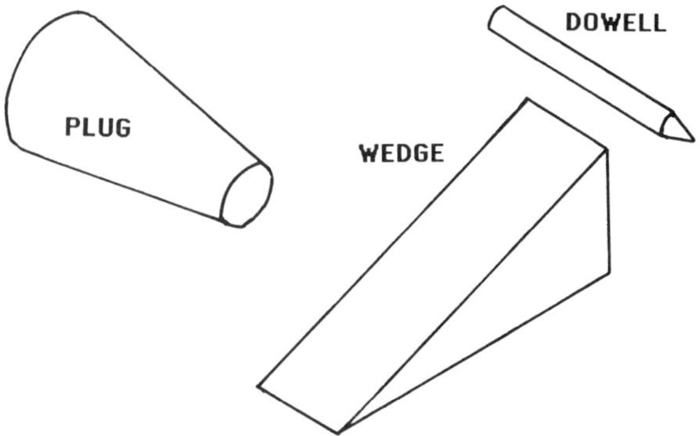

FIGURE 12-3. Basic shapes of wooden plugs and wedges.

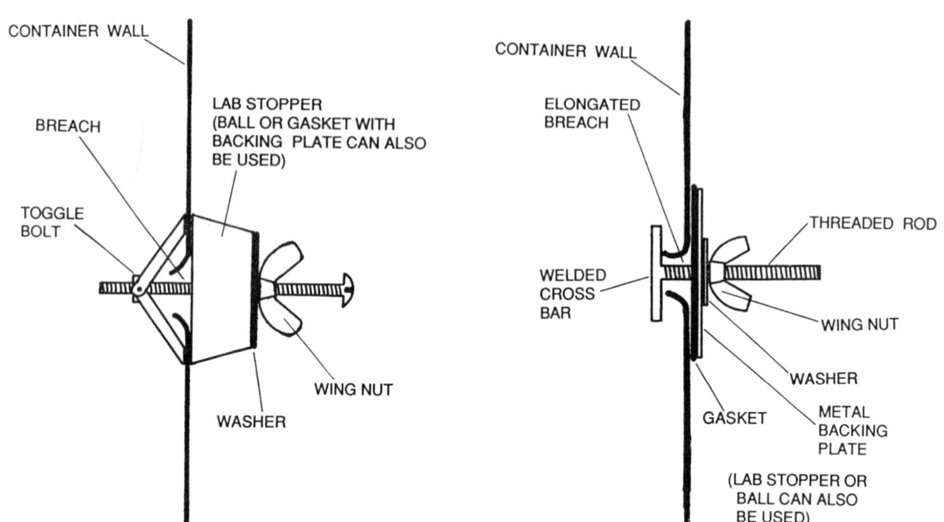

FIGURE 12-4. Toggle and T patches, shown in cross-sectional view.

tight (Figure 12-5). For very large leaks, it may also be possible to place a patch consisting of a suitable gasket with a sheet metal backing plate over the breach, then tighten the patch into position using worm drive clamps or load binding straps (Figure 12-6).

In some cases, it may be desirable to utilize a semipermanent patching procedure, such as a fiberglass-epoxy patch (Figure 12-7). This type of patch may be installed as follows:

- After the container has been grounded (if required) and repositioned, as discussed above, pack the breach with closed cell foam, putty, or other suitable material.
- Prepare the surface of the container for the adhesive by scraping, brushing, or sanding to remove loose paint, rust, scale, or dirt. Only nonsparking items should be used if the material is flammable. The entire area to which adhesive is to be applied should be cleaned.
- Prepare the required amount of epoxy adhesives by mixing resin and hardener. The fast-drying (5-minute-type) epoxy should be used. Small, stiff-bristled brushes, such as acid brushes, are good for mixing and applying adhesive material.

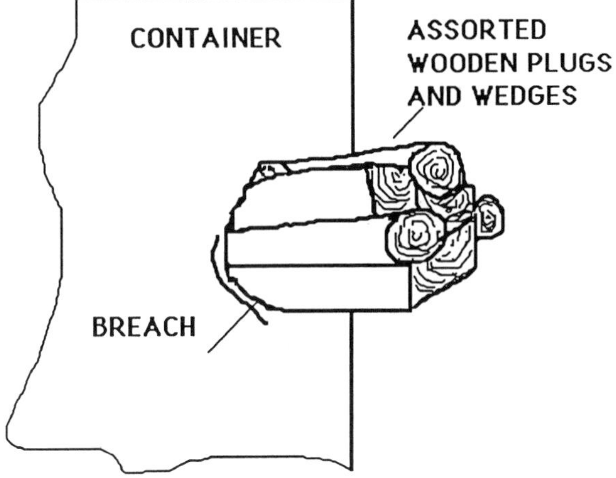

FIGURE 12-5. Use of multiple wedges and plugs to close a large opening.

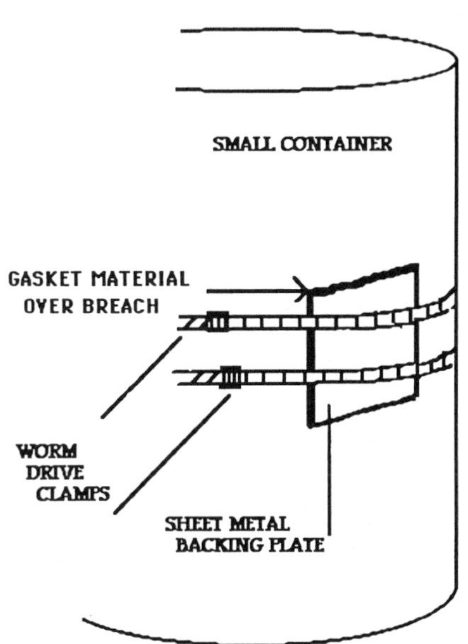

FIGURE 12-6. Small container patch using a gasket and backing plate held in position by worm drive clamps.

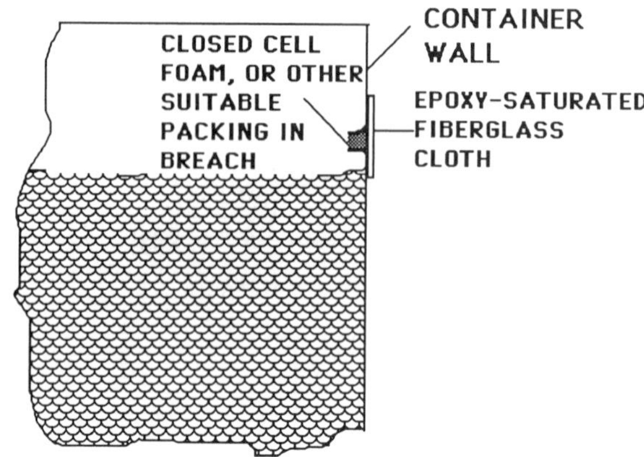

FIGURE 12-7. Fiberglass-epoxy patch, shown in cross-sectional view.

- Apply a thin film of epoxy to the area of the container surrounding the breach.
- Place a single layer of fiberglass cloth (as used in boat and automobile body repair) over the breach and the surrounding area. The epoxy film applied in the previous step will hold the cloth in place.
- Completely saturate the fiberglass cloth and surrounding areas of the container surface with the remaining epoxy.
- Allow the epoxy to dry before moving the container.

Other Small Container Leaks

Many additional types of leaks may be encountered in small containers. For example, leaks may occur along the chime of steel drums. The chime is the crimped area where the top or bottom is joined to the cylindrical side or wall section of the drum. These types of leaks are commonly patched by peening a suitable material, such as lead wool, into the chime along the area of leakage.

Closure failure may also be encountered in small containers. For example, the threads of drum bungs may be stripped or otherwise damaged, especially if the threads are corroded. In such situations, the bung hole in the drum can be closed by driving a plug of wood or some other suitable material into place.

In some cases, drums may be so deteriorated that repair is not feasible. For example, corrosion may produce pinholes in numbers too great to be patched, or may weaken the container wall to the point where it cannot reasonably be patched. In such cases, transfer of the drum's contents into a sound container is probably the best option, since it requires minimal disturbance of the container. Overpacking may also be an option if the container has enough integrity to survive being moved (see Appendix 2 for overpacking procedures). Overpacking, and any other operations attempted, should be performed with extreme care if deteriorated containers are involved, since total container failure may occur.

Procedures For Large Containers

Procedures for controlling releases from large containers are generally analogous to those described for leaks from small containers. However, certain limitations are imposed by larger containers. For example, it is typically not possible to change the position of these containers in order to place the point of release in the vapor phase. Thus, responders may have no choice but to terminate a release while a hazardous material is actively flowing through the breach. Transfer of contents to a sound container may be a viable option to perform in conjunction with, or instead of, containment operations for large container leaks. However, this should be attempted only by personnel with appropriate knowledge of the valves, fittings, and transfer procedures for the containers involved.

Transfer is considered a specialized operation. Unless responders have special training in the transfer operation required, or have knowledge of the operation gained in the course of their routine work activities, personnel with the appropriate experience and expertise should be called in. If transport vehicles such as tank trucks or railcars are involved in incidents on site, coordination with the transport companies and/or the manufacturer of the product involved may be required to effect transfer.

Small, Simple Punctures

In stopping leaks from small, simple punctures in large containers, any of the procedures described for small, simple punctures in small containers may also be applicable. However, these procedures may be more difficult to perform for larger containers, and may involve a higher potential for exposure of personnel if the point of release is located below the liquid-vapor interface of the product in the container. In these instances, as in all containment operations, the general safety procedures discussed above should be carefully followed.

Large, Irregular Holes

Releases from large or irregularly shaped holes in large containers can be especially difficult to control, given the high potential flow rates from such openings. However, a variety of options exist for addressing this problem.

Large or irregular holes may be patched in some instances by placing a metal plate and gasket over the breach and tightening the plate against the container wall, utilizing ratchet-equipped load binding straps, or chains or ropes tensioned with load binders or power pullers (Figure 12-8). A similar alternative to this procedure is the use of a "strap iron bandaid" patch which is held in place by T bolts placed through the breach (Figure 12-9). For either of these options to work well, the backing plate used must either be preformed to the shape of the container or flexible enough to assume that shape under tension.

In either of the procedures described above, it may be possible to utilize a suitable adhesive to seal the backing plate to the container. The backing plate can be fitted with a valved outlet for use in transfer of product and for pressure relief during installation (Figures 12-8 and 12-9). In the latter case, the patch should be installed with the valve in the open position. After installation, it may be possible to close the valve in order to completely terminate the release.

Commercially available patch items suitable for large releases from large containers can also be used. One such item, designed with railcar or

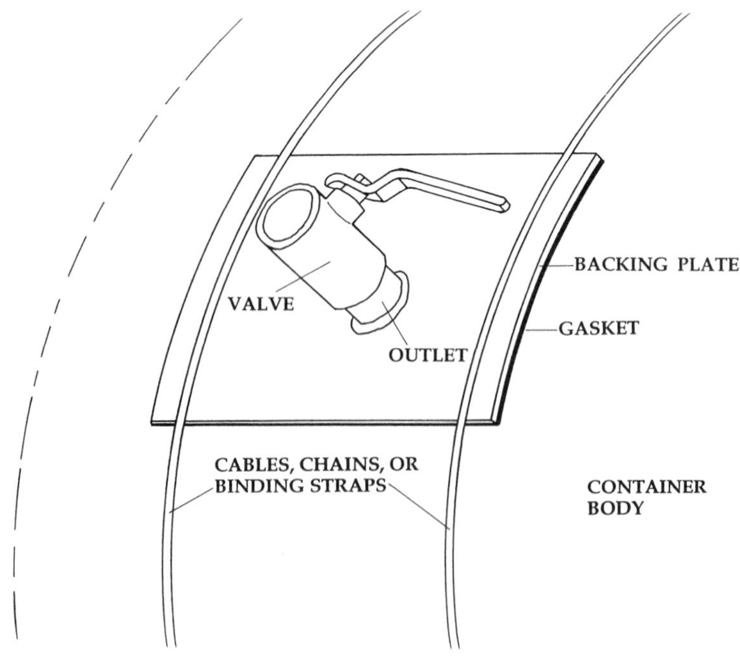

FIGURE 12-8. Large container patch using a gasket and backing plate with outlet valve.

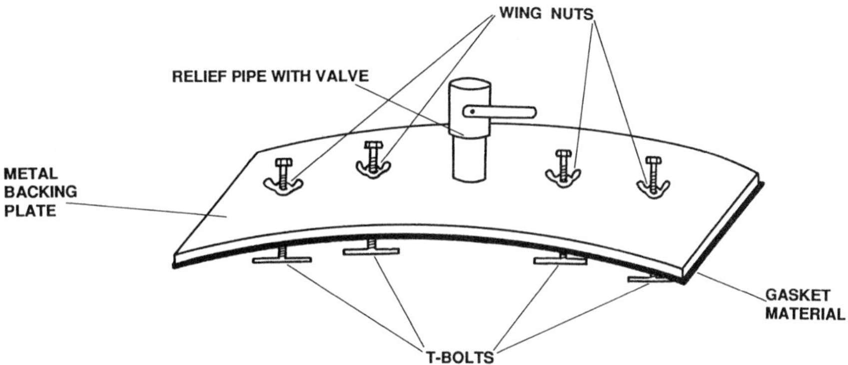

FIGURE 12-9. Strap iron "bandaid" suitable for long tears in containers.

tank truck releases in mind, is an inflatable patch which can be bound to the container wall over the breach and then inflated to seal off the leak. A similar item consists of a pad designed to be held in place by a vacuum created by a special vacuum pump. The pad is fitted with a valved outlet line for offloading the container's contents.

Procedures For Plumbing Leaks

A variety of hazardous materials releases may originate from valves, piping, and other components of plumbing systems in an industrial facility. In all cases, it is highly preferable to shut down the system involved in order to terminate the release and allow repair or replacement of defective or damaged parts. However, in some instances, immediate shutdown may not be a feasible option, and response personnel will be called on to handle an ongoing release from a piping system. In responding to such releases, a working knowledge of the system involved is vital. Response personnel should possess this knowledge firsthand, or coordinate closely with personnel who do, for initiation of control procedures.

Leaking Pipe Connections

Loose pipe connections or fittings in plumbing systems can cause significant leaks, especially if the system involved is under pressure. In many instances, leaks at joints can be stopped by tightening the connections, using pipe wrenches or other suitable tools. Persistent leaks at pipe connections may require shutdown for repair or replacement of defective components.

Holes in Piping

Holes in piping may occur due to accidental physical damage to plumbing systems. Holes may also develop through routine wear and tear or corrosion while a system is in operation.

It may be possible to terminate releases from these holes by installing a pipe repair clamp or "pipe saver" (Figure 12-10). These clamps must be selected based on the outside diameter of the leaking pipe. They are installed by placing the device around the pipe so that the gasket material covers the leak, latching the bolt(s) in place to fasten the clamp around the pipe, and tightening the bolts so as to compress the clamp and stop the

FIGURE 12-10. Pipe repair clamp (bottom) and improvised repair item (top).

leak. It is worth noting that if a pipe repair clamp is not available, a similar item can be improvised, as shown in Figure 12-10, using suitable gasket material and small worm drive clamps (such as automobile radiator hose clamps).

Broken Piping

Pipes may be completely broken as a result of accidental damage to plumbing systems. A handy device for temporarily stopping the flow of hazardous materials from broken pipes is the expandable pipe plug (Figure 12-11).

These plugs must be selected based on the inside diameter of the broken pipe. Pipe plugs are installed by inserting the plug several inches into the pipe through the broken end. The plug is then tightened, shortening the distance along its length and thereby forcing it to swell or expand around its circumference. As a result of this swelling the plug should tighten against the inside of the pipe, thus sealing off the leak. Some plugs of this type can be installed using a socket wrench with a drive extension, while others (such as those used by public utility companies) require a special tool for installation. They should be installed in pairs, with one plug directly behind the other.

Standard pipe plugs are not suitable for applications involving a significant amount of pressure. However, for pressure applications, a vented pipe plug may be used. This plug is fitted with a vent tube, as shown in Figure 12-11. Flexible tubing can be fitted to the threaded nipple on the end of the vent pipe so that the escaping liquid product can be diverted into a secondary container, or a gaseous product can be vented at a safe location or neutralized. It is also possible to fit the nipple with a valve. The plug can then be installed with the valve in the open position to vent the pressure during installation. After installation, it may be possible to stop the release by closing the valve. However, in this procedure, as in all operations involving pressure, extra caution is required. It must be assumed that the plug may be blown out at high velocity at any time during the operation. This can result in serious injury to responders.

Leaking Valves

Leakage of hazardous materials from valves on containers or in piping systems may occur due to a variety of causes. Efficient response to releases involving damaged or malfunctioning valves may require specific knowledge of the valves involved, as well as special tools, equipment, and parts. Coordination with knowledgeable personnel (such as plant maintenance personnel) should be initiated during the preemergency planning phase.

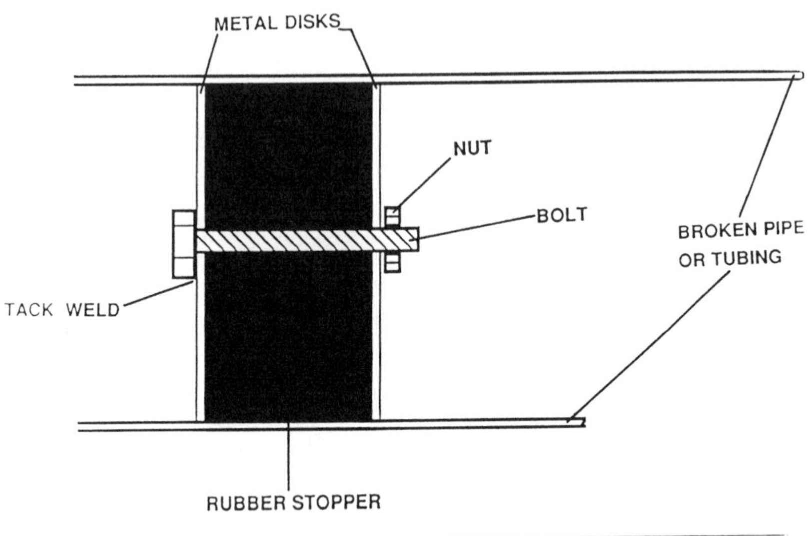

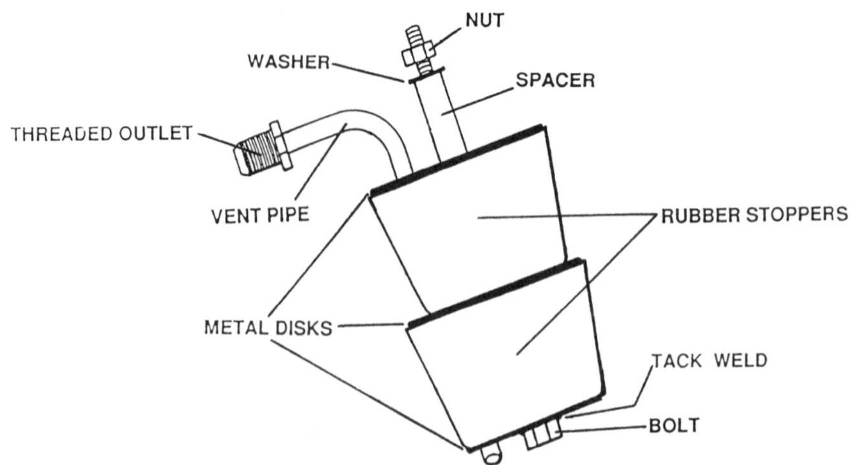

FIGURE 12-11. Expandable pipe plugs, shown in cross-sectional view.

Leakage past valve seats (Figure 12-12) sometimes occurs when valves are in the closed position. In addressing valve leaks, it should initially be determined that the valve is fully closed, since failure of personnel to close valves fully is a common cause of accidental releases. In some instances, a piece of scale or debris within the system may have lodged in the valve seat, preventing full closure. In such a case, the debris may be dislodged by slightly opening and then fully closing the valve.

Leakage around valve packing glands is also a common cause of release at valve locations (Figure 12-12). In some cases, this may be remedied by tightening down on the packing nut to increase the pressure on the packing.

Valve leakage due to problems such as worn or damaged valve seats or stripped valve stem threads will require extensive repair or replacement of the valve involved. This will typically require shutdown of the system involved. However, simple procedures, such as capping outlets on leaky valves, may work as temporary measures.

Procedures For Pressurized Cylinders

A variety of hazardous materials may be stored and used in pressurized containers in the industrial setting. Pressurized cylinders come in a variety of sizes and configurations. Their design and level of pressurization vary according to the material the cylinder is designed to contain.

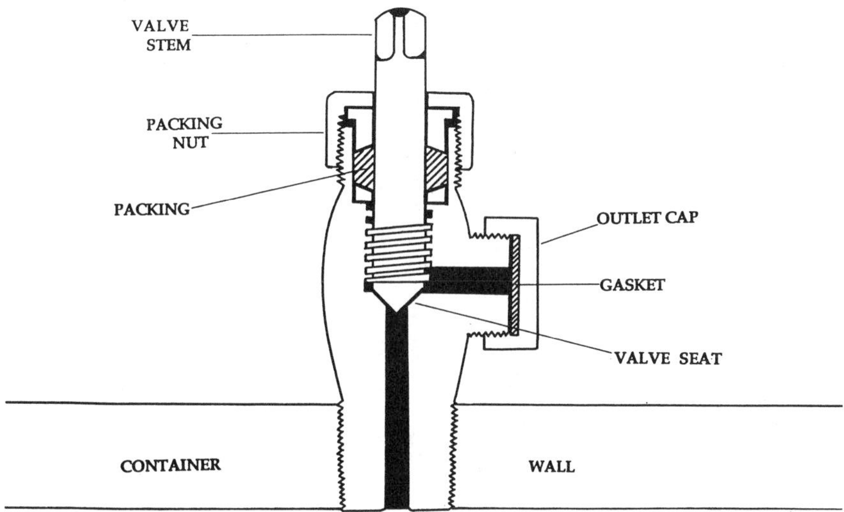

FIGURE 12-12. Basic valve components, shown in cross-sectional view.

Pressurized cylinders normally contain a material with a high vapor pressure. Such materials exist as gases at atmospheric pressure. However, within the pressurized cylinders, these substances may exist as liquids due to the pressure under which they are stored. This pressure can be in excess of 2,000 psi for some pressure cylinders. A breach in one of these cylinders can result in a sudden release of this pressure, with disastrous results. Damaged pressure cylinders have been known to act as projectiles, flying through the air for great distances and smashing through solid objects. In addition to the hazardous properties (such as reactivity, toxicity, flammability, corrosivity, etc.) of the materials stored in pressurized cylinders, the level of pressurization involved can be extremely hazardous to response personnel. Response to incidents involving pressurized cylinders requires great caution.

Releases from pressurized cylinders can occur from leaks in cylinder walls or in attachments such as fittings, valves, and pressure relief de-

FIGURE 12-13. Installation of a specialty kit to contain material leaking from a pressurized cylinder.

vices. If liquid material is escaping from a pressurized cylinder, the cylinder should be repositioned, if possible, to place the point of release in the vapor phase so that only gas escapes from the cylinder. This is required because of the potentially huge liquid-to-gas expansion ratios involved. Other measures, such as cooling a cylinder, may work to reduce pressure, thus reducing the rate of release. Also, it may be possible to move leaking cylinders to a safer area for repair or to allow contents to vent.

For the most part, response operations involving pressurized cylinders will require specialized knowledge and may require special equipment (Figure 12-13). In some instances minor problems, such as leaks due to loose connections or valves which are not fully closed, may be remedied using general skills and commonly available hand tools. Other procedures may require special expertise and special equipment (such as the chlorine repair kits described in Table 12-5).

Materials contained in pressurized cylinders may require special handling and storage procedures. Acetylene, for example, is unstable in pure form and may become shock sensitive if improperly stored. For response to incidents involving pressurized cylinders, hazmat team members should have a thorough knowledge of the specific materials, containers, and procedures involved. Unless hazmat team members have applicable knowledge gained through their normal work activities, specific training must be provided.

SUMMARY

Advanced hazardous materials control operations are performed to contain an uncontrolled release at the source. These offensive operations can be very effective in minimizing the damage resulting from an incident but typically involve a high exposure potential for personnel who must work at the point of release.

Assessment considerations, procedures, equipment, and tools which may be required for containment operations are described in this chapter. However, this information is somewhat generalized and should be supplemented as appropriate for the specific containment operations which may be required at a given facility. In all instances, field training will be required prior to the performance of containment operations in emergency situations. These concerns must be addressed during preemergency planning if response personnel are to be adequately trained and equipped.

In planning and performing advanced control operations, the safety of response personnel must be given the highest priority. No operations which place responders at undue risk should be attempted. Responders should bear in mind that, in some situations, containment will not be a safe option and responders will have to rely on defensive tactics alone in controlling hazardous materials releases.

REFERENCES

Action Training Systems. 1988. *Containing Leaks in Pressurized Cylinders* (videocassette). Seattle, WA: Action Training Systems.

Action Training Systems. 1989. *Incidents Involving Fuel Tank Trucks* (videocassette). Seattle, WA: Action Training Systems.

Action Training Systems. 1989. *Plugging and Patching Drums* (videocassette). Seattle, WA: Action Training Systems.

American Petroleum Institute. 1986. *Precautions Against Electrostatic Ignition during Loading of Tank Motor Vehicles*. Washington, DC: American Petroleum Institute.

Bureau of Explosives, Association of American Railroads. 1990. *Emergency Action Guide*. Washington, DC: Association of American Railroads.

Coastal Video Communication Corporation. 1990. *Static Electricity: Solving the Mystery*. Virginia Beach, VA: Coastal Video Communications Corp.

Henry, M.F., ed. 1989. *Hazardous Materials Response Handbook*. Quincy, MA: National Fire Protection Association.

International Fire Service Training Association. 1984. *HazMat Response Team Leak and Spill Guide*. Stillwater, OK: Fire Protection Publications.

International Fire Servicing Training Association. 1988. *Hazardous Materials for First Responders*. Stillwater, OK: Fire Protection Publications.

Noll, G.G., M.S. Hildebrand, and J.G. Yvorra. 1988. *Hazardous Materials: Managing the Incident*. Stillwater, OK: Fire Protection Publications.

U.S. Department of Transportation. Title 49 CFR Part 178. Washington, DC: U.S. Government Printing Office.

Appendix 1

Grounding and Bonding Flammable Liquid Containers

Whenever liquid products are in motion, such as when they are pumped, flow through pipes or hoses, or simply fall freely through air, it is possible for the liquid to assume a static electrical charge. Even the oscillation of product in a container as it is repositioned to place a leak in the vapor phase can produce a static charge. If an electrostatically charged product contacts an object which is grounded or has a lower electrical potential, static discharge, accompanied by a spark, can occur. If the liquid product involved is flammable, the spark resulting from the static discharge may serve as an ignition source for a flammable vapor-air mixture produced by vaporization of the product, resulting in a flash fire. This is a serious concern for the safety of responders involved in operations such as container handling and product transfer when flammables are involved.

In order to handle flammable liquids safely, static electrical charges must be suppressed or conducted safely to ground as soon as they are formed. Static suppression is achieved through the utilization of grounding and bonding equipment and procedures.

Grounding and Bonding Equipment

Various equipment items will be needed for grounding and bonding procedures (Figure 12-14). These items can be purchased from suppliers and, in some cases, improvised or fabricated in house.

Grounding Clamps and Cables

Several types of clamps may be used for grounding and bonding operations, depending on the types of containers involved. Basic clamp config-

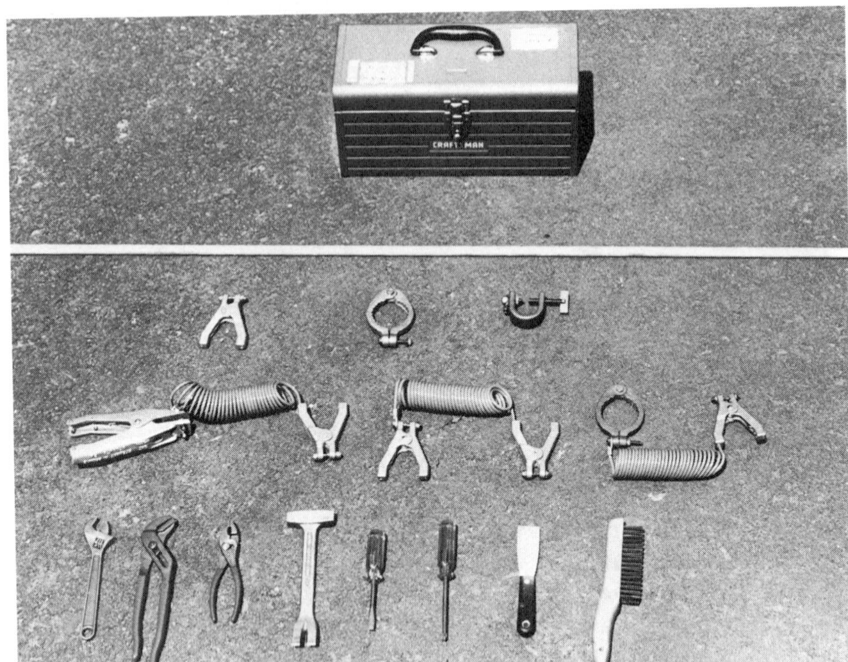

FIGURE 12-14. Grounding and bonding equipment.

urations include plier clamps (which are spring-loaded), c-clamps, and pipe clamps (both of which are screw-tightened). Grounding clamps are equipped with sharp, replaceable points and powerful springs (for plier-type clamps) so that direct metal-to-metal contact can be made through paint, rust, grease, and other foreign material on containers. Grounding clamps must be maintained in good condition or replaced.

Grounding clamps should be connected by cables of suitable length, quality, and durability. Since good conduction by the cables is vital, they should be inspected regularly and checked periodically for electrical resistance. Inspection should incorporate a careful examination of the clamp-to-cable connection.

Grounding Electrodes

Grounding electrodes are the final components of grounding/bonding systems which transfer static charges to earth. In a field operation, the grounding electrode will typically be a ground rod. This is a copper-coated metal rod specifically designed to be driven into the ground in order to conduct electrical charges into the soil. Stationary objects, such as an underground metal water pipe, may also be used in the field setting if good conductance to ground can be ensured. During a response at an industrial facility, responders may be able to utilize permanently installed grounding systems used in routine dispensing of fuels and solvents.

Other Items

A variety of tools or other items may be needed for grounding/bonding operations. For example, scrapers or wire brushes may be needed for removal of paint, rust, or other material from container bodies to ensure good metal-to-metal contact with grounding clamps. It is critical that these items be of nonsparking design.

Procedures For Grounding and Bonding

Static suppression for hazardous materials operations involves two distinct procedures: bonding and grounding.

Bonding Procedures

Bonding is the process of equalizing the electrical potential between two containers. This is done by connecting a cable, using suitable clamps, between the two containers (Figure 12-15). It is important to always make the connection at the hazardous location first. Thus, if static discharge occurs as the circuit is completed, the resulting spark will be at the "safe" container. When attaching clamps, it is important to ensure direct metal-to-metal contact between the container and clamps. For plier-type clamps, use a back-and-forth motion to work the points through paint, rust, and dirt on containers. For some installations, such as when pipe clamps are used, surface preparation may be required to expose bare metal prior to clamp installation.

Grounding Procedures

Grounding is the process of connecting a container or other object to ground so as to suppress any static charges which may develop. This is accomplished by connecting a grounding cable fitted with suitable clamps between the container and a suitable grounding electrode or grounding system (Figure 12-15). Concerns for ensuring metal-to-metal contact at clamp locations, as described for bonding procedures, apply in grounding as well.

In attaching grounding cables, always make the connection to the grounding electrode last in case a spark occurs as the circuit is completed. For transferring operations, both containers should be grounded. Thus, if

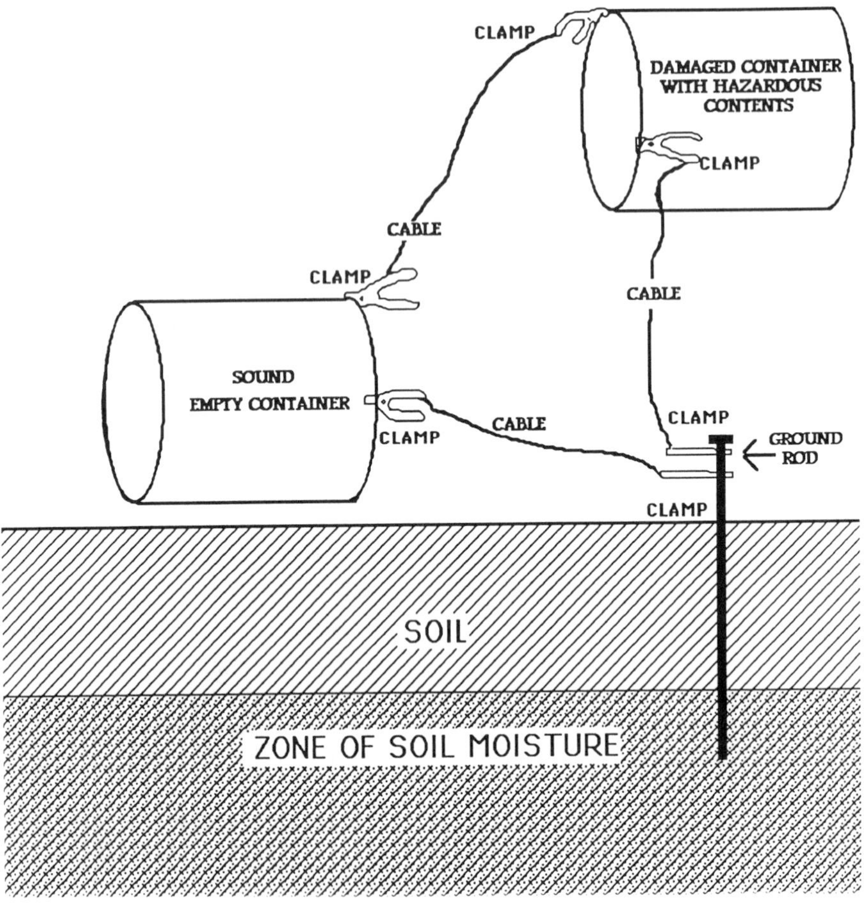

FIGURE 12-15. Grounding and bonding prior to product transfer.

one of the lines involved in grounding and bonding the containers is accidentally disconnected during the operation, charges can be safely conducted to ground as long as the other two lines remain in place (Figure 12-15). Pumps used in product transfer should also be grounded.

When utilizing a ground rod as the grounding electrode, it is important that the rod be driven deeply enough to ensure contact with moist soil for adequate grounding. For field operations, ground rods should be driven to the hilt, or as deeply as possible into the ground, with 3 feet as the absolute minimum depth. This is especially important during extended periods of dry weather.

Other Safety Precautions

During transfer operations involving flammable liquids, personnel can minimize the amount of electrostatic charge generated by transferring the product slowly and minimizing splashing, spraying, and swirling of the product during the operations. All pipes or hoses, pumps, and containers used should be clean. Other safety precautions which have been used include the wearing of special clothes designed to suppress electrostatic buildup and the use of inert gas blanketing during transfer operations.

Appendix 2

Overpacking Damaged Containers

Purpose and Basic Concept of Overpacking

In handling hazardous materials in small containers which have been damaged, it may be desirable to overpack the damaged containers. Basically, this involves placing the damaged container in a slightly oversized, open-topped salvage container. For example, 55-gallon drums are overpacked using 85-gallon salvage drums. The space between the inner and outer drums is packed with a suitable sorbent material, and a tight-fitting lid is clamped to the salvage drum. The salvage container may then be safely handled and transported. For purposes of transportation, it should be noted that DOT regulations related to certification of salvage containers for the product shipped, labeling of the containers, and shipping papers must be strictly adhered to. If the product is considered a hazardous waste, EPA requires that a hazardous waste manifest accompany it during transportation, as in shipment to a disposal site.

Safety Considerations

In overpacking operations, as in all aspects of hazmat response, safety should be the foremost consideration. Chemical threats to personnel involved in these operations are posed by the materials involved, which may be toxic, flammable, corrosive, or otherwise hazardous. If feasible, damaged containers should be temporarily repaired prior to overpacking to minimize spillage and resulting chemical hazards to personnel during the procedure. Safety considerations for the use of PPE and safe handling of hazardous materials, as discussed previously in this text, must be addressed.

Personnel are also at risk due to physical safety hazards related to the mechanics of the overpacking procedure. A 55-gallon drum of liquid product typically weighs several hundred pounds, so that crushing-type injuries, especially to hands and feet, are definitely possible. The use of safety gear, such as hard hats and safety shoes, is important for physical protection. Responders must also bear in mind that improper lifting can produce back strain, or similar injuries, when drums are handled manually.

Procedures for Placing Damaged Containers in Overpack Containers

In describing the following procedures, it will be assumed that the damaged containers involved are 55-gallon drums. In handling containers of

FIGURE 12-16. Overpacking, using the inverted overpack technique.

this size, some type of lifting equipment should be used, if possible, so that personnel will not have to handle the drums manually.

Procedures Using Lifting Equipment

If feasible, equipment such as an overhead crane, fork truck, backhoe, or front-end loader should be employed for lifting drums and placing them in the salvage container. A device suitable for placement around the container to be lifted, such as a nylon sling, should be attached to the lifting equipment. The damaged container should be lifted and lowered gently into the salvage container. The operation can then be completed by adding the sorbent packing and placing the lid on the salvage container. It is vital that basic safety procedures be followed during these operations. Personnel should never be located beneath a suspended container. Also,

FIGURE 12-17. Use of a drum upender bar to rotate a salvage drum into the upright position.

general safety rules related to work around heavy equipment should be followed closely.

Procedures for Manual Overpacking

When overpacking must be performed manually, the weight involved can result in back strain or other types of strain injuries. Also, the threat of crushed hands, feet, or digits is very real. Extreme care and the use of appropriate protective gear is called for. Personnel should never place parts of their bodies where they may be injured. For example, placing hands in possible pinch points, such as between the salvage drum and the damaged drum, should be avoided during overpacking operations.

In performing manual overpack operations, personnel should lift in teams of two, acting in unison, to reduce individual strain. The use of commercially available drum upender bars (Figure 12-17), or similar improvised items (such as boards), to provide leverage is highly recommended. Manhandling of drums should be avoided in all cases.

The Inverted Overpack Technique

This technique involves placement of the overpack container, in the inverted position, over the damaged drum (Figure 12-16). The overpack

FIGURE 12-18. Overpacking, using the roller technique.

container is then rotated manually into the upright position (Figure 12-17), packed with sorbent material, and closed.

The Roller Technique

This technique involves the use of 1- to 2-inch-diameter tubing, such as metal water pipe or other suitable tubing, as rollers. The damaged drum is placed on its side, with the rollers underneath it (Figure 12-18). The salvage drum is then positioned so that the damaged drum can be rolled most of the way into it. The salvage drum can then be turned upright using a suitable lever, and the remaining steps of the process can be performed. If rollers are not available, a 2 × 4-inch board or similar item can be utilized to elevate one end of the damaged drum to start moving it into the salvage drum.

Angle Roll Technique

In using this technique, the damaged drum and the salvage drum are placed end to end at an obtuse angle, as shown in Figure 12-19. The drums can then be simultaneously rolled, so that the damaged drum tends to

FIGURE 12-19. Overpacking, using the angle roll technique.

Richard R. Paul

work its way into the salvage drum. At some point, the inner drum will "foul out" inside the salvage drum and inward progress will stop. At this point, the drums can be repositioned to reverse the original angle and rolled in the opposite direction. By continuing this process, the damaged drum can be worked most of the way into the salvage drum. The salvage drum can then be lifted into the upright position, using a suitable lever, so that the remaining steps of the process can be performed.

13

The Behavior of Chemicals

Barbara M. Hilyer, M.S., M.S.P.H.

In Batavia, New York, a maintenance man was hospitalized and 35 people were evacuated from a building when a worker mixed two chemicals, hypochlorite solution and muriatic acid. Mixing these two products permits a reaction to take place which generates potentially deadly chlorine gas. Clearly, a knowledge of the chemical incompatibility of these products could have prevented this incident from happening.

Emergency responders must know what to expect from the chemicals involved in a hazardous materials incident before they can evaluate the hazards present. The likelihood of reaction, fire, explosion, toxicity, and mobility of the material on land, in air, and in water are all determined by the chemical makeup of the product. In order to control an incident safely and efficiently, a responder must understand enough chemistry to be able to predict the behavior of hazardous materials.

This chapter presents a basic discussion of chemistry. There are exceptions to almost everything stated, but they will not be mentioned.

OBJECTIVES

- Know the basic structure of atoms
- Be able to explain how atomic structure determines the behavior of elements and compounds
- Understand chemical and physical properties and how they influence chemical behavior
- Correlate chemical structure with chemical names, thereby predicting the behavior of identified hazardous chemicals

- Be able to recognize radioactivity and deal with it safely
- Know how to locate additional sources of help and information

ATOMS

Everything in the world is made of atoms of chemicals. Everything—houses, cars, dogs, food, the earth we walk on, the air we breathe, and each of us—is made of atoms. We cannot see atoms or the particles they are made of, but scientists accept their existence because it explains the results that can be seen and measured when observing the way matter behaves.

Structure of the Atom

An atom probably looks like the drawing in Figure 13-1. At the center is the nucleus, made of two kinds of particles, neutrons and protons. The neutron is heavy but has no electrical charge. The proton is as heavy as the neutron and has a positive electrical charge.

Outside the nucleus, spinning around it in orbit, are the electrons. These are light; in fact they have no mass or weight at all, but they do have a negative electrical charge. The electrical charge of the atom as a whole is zero, so we know that atoms contain the same number of electrons as they do protons; the negative and positive charges cancel each other out.

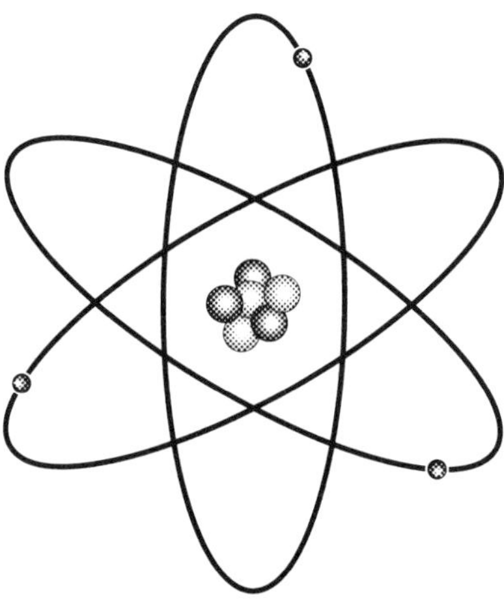

FIGURE 13-1. The structure of the atom.

Sometimes forces outside the atom cause it to gain an extra electron, giving it an overall negative charge. It now contains more negative electrons than positive protons, so the sum of the electrical charge is negative. The atom has become a negative ion. If other forces cause an atom to lose an electron, the overall charge is positive due to this loss of a negatively charged electron. This atom has become a positive ion. Ions are electrically charged atoms; they may be negative or positive. This charge greatly influences an atom's behavior; although it is still the same element (the sodium ion is still sodium; the iron ion is still iron), it will behave very differently from the uncharged form.

Electron orbits are arranged in layers, or shells. The inner shell is full with two electrons, and if it has two it does not seek more. Further layers, shells two and three, are full and satisfied with eight electrons. If an atom is large enough, it may have even more electron layers, which have room for larger numbers of electrons.

The number of electrons in the outer shell of an atom is important because a shell which is not full seeks other electrons, and this leads the atom to react with other atoms.

Combinations of Atoms

There are 92 atoms which occur naturally, and more which can be produced artificially. All matter, living and nonliving, is made from these 92 basic building blocks; the great variety of things around us is due to the different ways in which atoms combine.

For example, different combinations of only the three elements carbon, oxygen, and hydrogen can form very different substances (Table 13-1). The atoms of these three elements are different from each other, so that if we could examine them, we could always recognize an atom of hydrogen, an atom of oxygen, or an atom of carbon (Figure 13-2).

TABLE 13-1. Substances Formed From Carbon, Oxygen, and Hydrogen

Substance	Atomic Structure
Water	2 hydrogens, 1 oxygen
Vinegar	4 hydrogens, 2 oxygens, 2 carbons
Grain alcohol	6 hydrogens, 1 oxygen, 3 carbons
Yellow color in carrots and butter	28 hydrogens, 20 carbons
Propane	8 hydrogens, 3 carbons
Benzene	6 hydrogens, 6 carbons

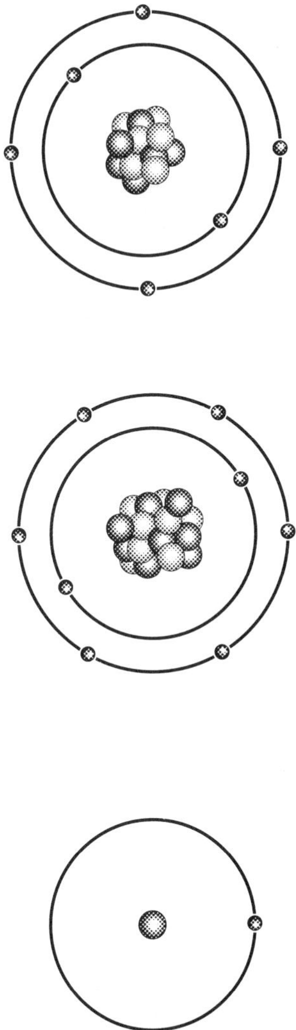

FIGURE 13-2. The structure of carbon (top), oxygen (center), and hydrogen atoms.

Atoms combine with other atoms in many different ways to form molecules. A molecule is made up of two or more atoms. The atoms may be of the same element, as when two atoms of oxygen combine to form O_2, the form of oxygen gas which occurs naturally in air, or the molecule may consist of atoms of different elements. If the atoms in a molecule are different, the substance formed is called a "compound." These may be

H· + ·H ⇨ H:H

FIGURE 13-3. Atomic hydrogens combine to form molecular hydrogen. The symbol denotes the nucleus, and dots denote electrons.

•He• •He• •He•

•He• •He•

FIGURE 13-4. Atomic hellium does not combine with other helium atoms, even though they are plentiful in air. The helium atom's only electron shell is full.

simple compounds, as when water is formed from atoms of hydrogen and oxygen, or very complex compounds made up of hundreds or thousands of atoms.

Atoms combine with other atoms in order to satisfy a need to fill their outer electron shells. Looking at the smallest, most simple atom, the hydrogen atom, we see that it is made up of one proton in the nucleus and one electron in orbit around the nucleus. The positively charged proton balances the negatively charged electron, so the overall charge of the atom is zero, but all is not well with the hydrogen atom. Expressed in human terms, this hydrogen is lonely and unfulfilled because its electron shell has only one electron. It seeks fulfillment by another electron, so when it meets another hydrogen they bond, share electrons, and both are satisfied. Each now has two outer shell electrons; even though they are shared, each atom can count both. Hydrogen gas is normally found in the form of H_2, made of the two combined atoms (Figure 13-3).

Helium gas, by contrast, exists as separate atoms of helium. Each atom contains two electrons, filling the shell, so helium is satisfied and does not combine with other helium atoms (Figure 13-4). Helium is described as being "inert" because it does not easily combine with other atoms. Other inert gases, such as nitrogen, are used in purging flammable gases from containers; they can be trusted not to react with the gases.

Many large atoms have numerous electrons, as many as 100 or more, but their behavior is determined primarily by the number of electrons in the outer shell, as we have seen in small atoms.

FAMILIES OF CHEMICALS

Elements can be grouped into families by their similar properties. All members of a family have similar configurations of outer shell electrons.

We can predict, then, that all members of a family will behave in much the same way. Just as the James Brothers were all outlaws, and the Perry brothers, William ("the Refrigerator") and Michael Dean, look alike and are All-American and NFL football players, all chemical family members show similar chemical behavior.

We can use as an example the halogen family (Table 13-2). Halogen atoms have incompletely filled outer electron shells. Halogens aggressively seek one more electron; they are quite greedy and will grab strongly for electrons on other atoms. When they do, the reaction with the other atom is fast and often produces energy. The halogens are described as being very reactive for this reason.

Atoms of the halogen elements may be attached to various other atoms to form different compounds, often called "halogenated" compounds. Chlorine is probably the most widely used halogen; a compound with chlorine atoms attached is called halogenated or "chlorinated." Often the halogen can be recognized in the chemical name, as in tri*chloro*ethane or hydro*chloric* acid. A compound containing a halogen is often dangerous due to the reactive nature of the halogen atom.

Since chlorine, like all halogens, is reactive, we expect it to form compounds with other atoms even if it has to steal them. In some cases, the compound formed is not hazardous, as when chlorine combines with sodium to form sodium chloride, which we use as table salt. The chlorine steals an electron from sodium, giving chlorine an extra electron and a negative charge. The sodium, having lost an electron, has a positive charge. The negative chlorine and the positive sodium are attracted to each other (opposites attract) and remain together in crystals (Figure 13-5).

In some halogen reactions, a more toxic compound is formed (Figure 13-6). Trichloroethane is widely used as a degreaser and solvent cleaning agent. It is moderately toxic to humans if the vapors are inhaled or the

TABLE 13-2. Members of the Halogen Family

Name	Symbol
Fluorine	Fl
Chlorine	Cl
Bromine	Br
Iodine	I
Astatine	At

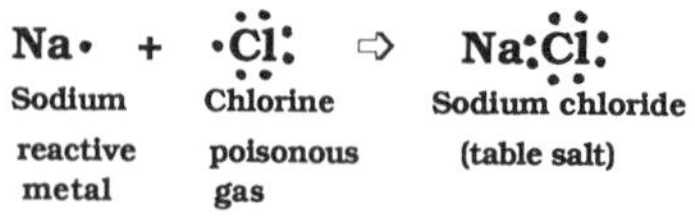

Na• + •Cl: ⇨ Na:Cl:
Sodium reactive metal | Chlorine poisonous gas | Sodium chloride (table salt)

FIGURE 13-5. Sodium and chlorine atoms combine to form table salt.

liquid splashes on their skin. When mixed with water, it may form hydrochloric acid, a strong irritant whose vapors are dangerous if inhaled. If subjected to high temperatures, trichloroethane may combine with oxygen in the air to form deadly phosgene gas.

This same degreaser grabs the sodium atoms if mixed with sodium hydroxide, and the leftover atoms combine into harmless carbon dioxide and dangerous hydrogen gas, which is highly flammable and may be explosive.

All the reactions shown in Figure 13-6 occur because the chlorine atoms are hungry for one more electron to fill their outer shells. Chlorine, like other halogen family atoms, has seven outer shell electrons and aggressively seeks one more.

Names of Chemicals

Each chemical element has a name which refers to one element and only that one element. Some of these are familiar and are mentioned in everyday conversation, such as oxygen, iron, silver, and neon, while many, for example niobium, xenon, and tellurium, are not known to most people.

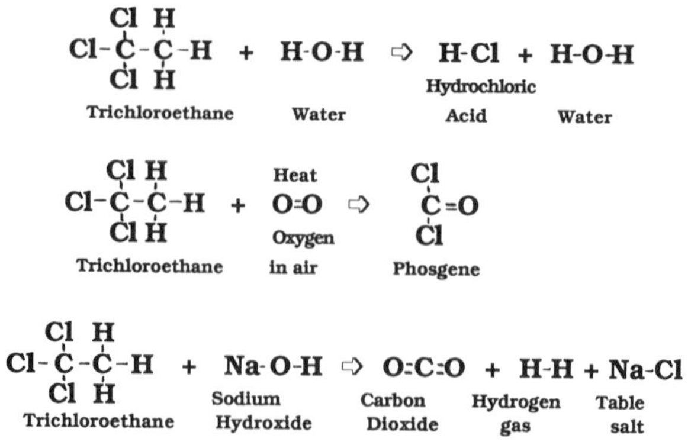

Cl-C(Cl)(Cl)-C(H)(H)-H + H-O-H ⇨ H-Cl + H-O-H
Trichloroethane | Water | Hydrochloric Acid | Water

Cl-C(Cl)(Cl)-C(H)(H)-H + O=O (Heat) ⇨ Cl-C(=O)-Cl
Trichloroethane | Oxygen in air | Phosgene

Cl-C(Cl)(Cl)-C(H)(H)-H + Na-O-H ⇨ O:C:O + H-H + Na-Cl
Trichloroethane | Sodium Hydroxide | Carbon Dioxide | Hydrogen gas | Table salt

FIGURE 13-6. Reactions involving 1,1,1-trichloroethane; lines represent shared electrons.

Technical names, proper shipping names from the DOT 49 CFR Section 172.101 Table of Hazardous Materials, and names complying with International Union of Pure and Applied Chemists guidelines are the names responders should use when seeking information about a product. Trade names, like Dowclean, and common names, like perc, will not be found in reference sources and are not helpful when identifying chemicals. Every chemical has a unique Chemical Abstract Service (CAS) number which can also be used to get information. Several of the widely available reference sources use these numbers, as do the telephone emergency service hotlines.

Symbols

Symbols have been adopted to enable us to write chemical formulas more easily. Each chemical has been assigned a one- or two-letter abbreviation by which it is known. As can be seen in Table 13-3, some chemical symbols seem to make sense and others do not. Those whose reason for being is not obvious have to be memorized.

The names of compounds are sometimes more difficult to interpret. The rules for naming compounds are usually understood only by chemists, and even among chemists more than one set of rules exists.

Prefixes and Suffixes

Prefixes (syllables added to the beginning of a word) and suffixes (added to the end of a word) describe atoms added to a basic molecule. One familiar use of prefixes shows the number of atoms added, for example the "di" in carbon dioxide, denoting the two oxygen atoms attached to each carbon molecule in this gas. Table 13-4 shows some prefixes denoting numbers.

Suffixes often denote groups of atoms which are added, and behave, as a group and not as individual atoms. Since they act as a group, they are

TABLE 13-3. Chemical Symbols

Name	Symbol	Name	Symbol
Sulfur	S	Helium	He
Oxygen	O	Neon	Ne
Fluorine	F	Chlorine	Cl
Potassium	K	Mercury	Hg
Phosphorus	P	Radon	Rn

TABLE 13-4. Prefixes Denoting Numbers

Prefix	Number	Example	Formula
mono	1	Carbon monoxide	CO
di, bi	2	Carbon disulfide	CS_2
tri	3	Trichloroethane	$C_2H_3Cl_3$
tetra	4	Carbon tetrachloride	CCl_4
penta	5	Pentachloroethane	C_2HCl_5
poly	many	Polyvinyl chloride	String of many vinyl chloride molecules

called "functional groups." Functional groups are sometimes used as suffixes, or last names, as they are in many alcohols (Figure 13-7).

Organic Chemicals

The words "organic" and "inorganic," while not part of the names of chemicals, are often used to describe the makeup of compounds.

Organic compounds are those which are based on carbon; if two or more carbons are attached to each other, the structure is called a "carbon chain." The carbon chain forms the skeleton to which other atoms are attached. Inorganic compounds lack carbon chains, although they may contain a carbon atom; for example, carbon dioxide is an inorganic, carbon-containing compound.

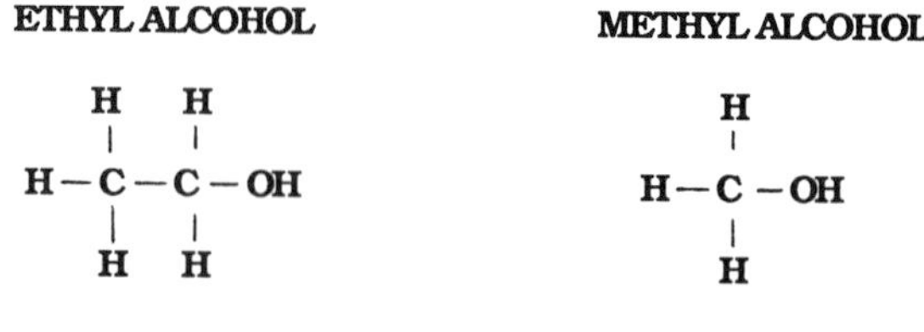

FIGURE 13-7. Alcohols: an example of a functional group of atoms. The OH group is characteristic of alcohols.

In some cases, it may be important to be able to determine if a product is organic. In the presence of certain oxidizers, which are reactive with organics, or when choosing a monitoring instrument, a responder may need this information. Two clues may help: If the product is made from something living or formerly living, like wood or petroleum, or if the formula in the reference documents shows carbons and hydrogens in the molecule, the chemical is organic. Another name frequently used to describe some organics, those made only of hydrogens and carbons, is "hydrocarbon." If a hydrocarbon has halogen atoms attached, it may be referred to as a "halogenated hydrocarbon."

Organic compounds may have one or many carbons in their chains. The simplest organic molecules are named for the number of carbon atoms they contain, as shown in Table 13-5.

Carbon has six electrons. You remember that the first two of these will fill the inner shell, leaving the remaining four in the outer shell. Eight electrons fill an outer shell and satisfy an atom, so carbon seeks four more. It can combine with a variety of atoms to fill its outer shell, and will do so with the result that carbon may form thousands of different compounds.

Since World War II, the chemical industry has taken advantage of newly developed technology and of carbon's ability to combine with other carbon atoms and the atoms of many other elements, and has dramatically increased the production of organic chemicals. Part of the industry is called the petrochemical industry, since naturally occurring petroleum is subjected to a process which separates carbons in the chain (called "cracking") and sometimes puts them back together in new ways to form new chemicals. These man-made organic chemicals are often more toxic than naturally occurring ones. Many of them have halogens added for specific uses, as trichloroethane does, and some of the halogenated hydrocarbons are extremely toxic.

Carbon chains may be straight, branched, or formed into rings (Figure 13-8). Rings with double bonds are said to be "aromatic rings," even

TABLE 13-5. Organic Compounds

Name	No. of Carbons	Formula
Methane	1	CH_4
Ethane	2	C_2H_6
Propane	3	C_3H_8
Butane	4	C_4H_{10}
Pentane	5	C_5H_{12}

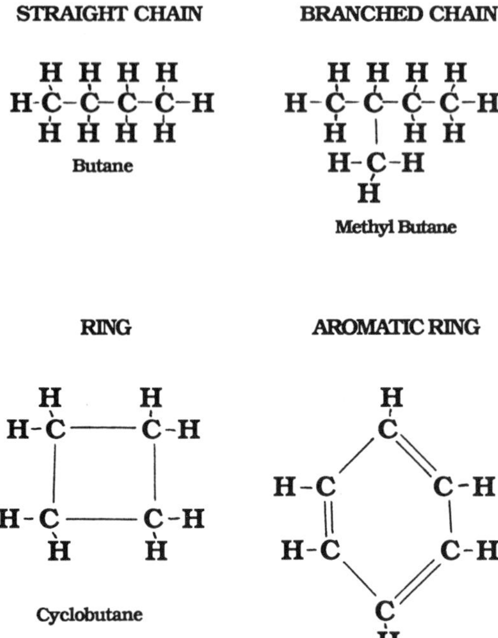

FIGURE 13-8. Carbon chains of organic molecules.

though some have no odor. An aromatic ring always contains six carbon atoms; the double lines represent double bonds, where four electrons are shared by a pair of carbons instead of the usual two.

Aromatic rings occur naturally in plant oils and in many man-made products. Benzene, manufactured by coking coal, is a plain six-carbon aromatic ring with a hydrogen attached to each carbon. Substitution of an OH (alcohol) functional group on one carbon changes benzene into phenol.

The process of "building" chemical compounds is sometimes difficult to do in reality but is easy on paper. In Figure 13-9, carbon and hydrogen are turned into a series of toxic chemicals.

All the compounds in Figure 13-9 are toxic to humans. A number of different phenolic compounds are used in many industries; they are found on many sites and are all toxic. Polychlorinated biphenyls (PCBs), polybrominated biphenyls (PBBs), creosote, and trinitrophenol (picric acid) contain phenol as part of the molecule.

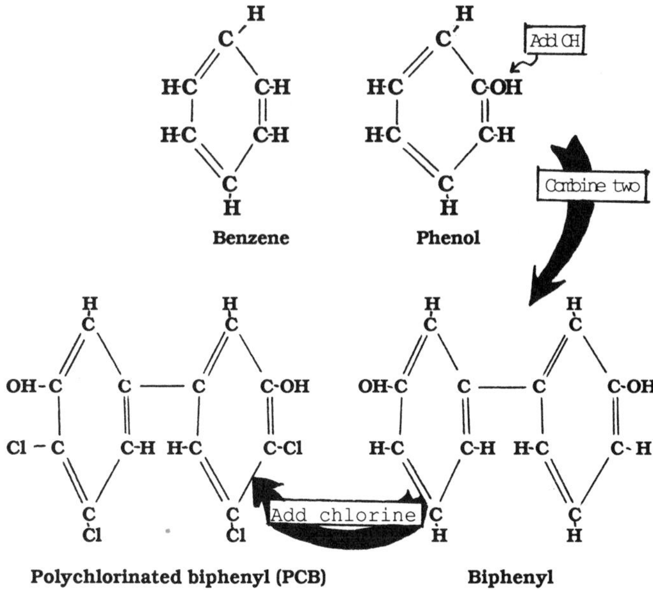

FIGURE 13-9. Building PCBs from an aromatic carbon ring.

PROPERTIES OF CHEMICALS

Since all atoms of a certain element have the same structure, they also have the same properties. The structures and properties of atoms make their behavior predictable. Atoms combine in predictable ways to form compounds, based on atomic structure and on how atoms share or exchange electrons. The combination is always the same if the elements and conditions are the same, so a particular compound always contains the same ratio of elements (Figure 13-10).

A molecule of the compound water, for example, is always made up of one oxygen atom and two hydrogen atoms. Carbon tetrachloride always consists of one carbon atom with four chlorines attached (we can tell this from the name; "tetra" means four).

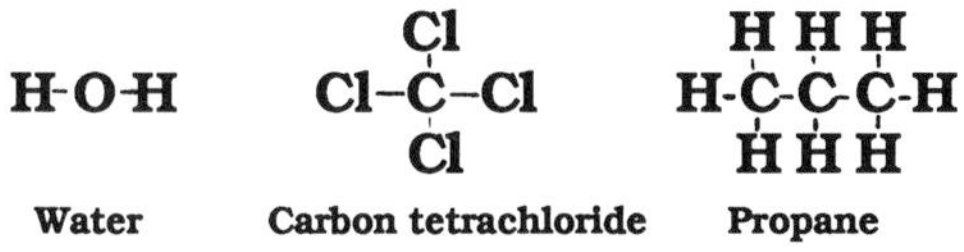

FIGURE 13-10. Compounds showing constant formation.

Since a compound is always made of the same atoms in the same ratio, that compound always has the same properties whenever we find it. We refer to these properties as "characteristic properties," since no two compounds have exactly the same properties.

Physical and chemical properties determine the behavior of elements and compounds. A hazmat responder should always find out the properties of a chemical as soon as possible, so that it can be determined how the chemical will behave with regard to fire, explosion, reactions, dispersal, and human health hazards. Several references can be used to get this information; these will be discussed in a later section.

When assessing properties, it is important to note the temperature scale used in the test which was done to determine the property. Some properties are very temperature dependent, and some indicate dangers which will occur at the ambient temperature in the area where the incident occurs. The Fahrenheit scale is used in describing weather and ambient temperature, but most scientific references use the Centigrade scale. Responders making hazard assessments must be able to convert temperatures from one scale to the other in order to make good judgments about potential hazards (Table 13-6).

Boiling Point

The boiling point (BP) of a liquid chemical is the temperature at which a liquid begins to bubble and rapidly change to a vapor. If the boiling point is lower than the temperature of the air around the chemical, the chemical will boil and vaporize upon release. This does not imply that vapors are

TABLE 13-6. Approximate Temperature Conversions

°F	°C	°F	°C
32	0	80	27
36	2	85	29
40	4	90	32
45	7	95	35
50	10	100	38
55	13	125	52
60	16	150	66
65	18	175	79
70	21	200	93
75	24	212	100

produced only at or above the boiling point temperature; they are generated at lower temperatures, only more slowly.

Examples: Water BP = 100°C
Lead BP = 1,751°C
Hydrogen chloride BP = −85°C
(it is a gas at ambient temperature)
Trichloroethane BP = 74°C

Vapor Pressure

Vapor pressure (VP) is the pressure placed on the inside of a closed container by the vapor in the space above the liquid the container holds. Chemicals with high vapor pressures are more likely to breach their containers when heated, since pressure increases as the temperature rises. Chemicals with high vapor pressures are volatile; more vapors will be present around a spill. The units of vapor pressure in most references are millimeters of mercury, since measuring devices allow pressure to push mercury up a small calibrated tube much like a thermometer. The larger the number, the higher the vapor pressure.

Examples: Water VP = 21 mm mercury
(at 75°F) Alcohol VP = 45 mm mercury
Chlorine VP = 4,800 mm mercury
(it is a gas unless kept under pressure)

Vapor Density

Vapor density (VD) is the mass of the vapor divided by the volume it fills. Vapors with a vapor density greater than 1.0 will sink to the ground or floor, and may accumulate and displace breathing air. Vapors with a vapor density less than 1.0 will rise and float away. If a heavy vapor is flammable, the leaking vapors can travel along the ground to an ignition source and ignite, creating a situation in which flames can travel back to the leaking container.

Examples: Chlorine VD = 2.49
Gasoline VD = 3.0–4.0, depending on the type
Toluene VD = 3.14

Solubility

Solubility indicates the tendency of a chemical to dissolve evenly in a liquid. The liquid may be water or an organic solvent such as benzene or alcohol. Sometimes solubility is described in words and sometimes a number is given, indicating the percentage of the material which will dissolve. The more solute dissolves in the solvent, the more concentrated, or stronger, the solution becomes.

Solubility in water indicates the amount of a material that will dissolve in water. Insoluble or slightly soluble materials will form a separate layer and will either float or sink, depending on their specific gravity (Table 13-7).

Specific Gravity

Specific gravity is the density of the product divided by the density of water, with water density defined as 1.0 at a certain temperature. An insoluble material with a specific gravity less than 1.0 will float; one with a specific gravity greater than 1.0 will sink (Table 13-7). Specific gravity is sometimes given as density, even though the terms do not have the same meaning.

Flash Point

A chemical's flash point (FP) is the minimum liquid temperature at which enough vapors are present above the liquid to ignite. At that temperature a flame, spark, or other source of ignition will cause the vapors to flash or burn. The lower the flash point, the greater the fire hazard of a material. If the flash point is lower than the air temperature, the material is an immediate fire hazard and all ignition sources must be removed from the area.

Examples: Benzene FP = 12°F
Gasoline FP = −45°F
Jet fuel FP = −10° to 30°F
(depending on the constituents)

TABLE 13-7. Chemical Behavior Due to Solubility and Specific Gravity

Chemical	Soluble in Water	Specific Gravity	Behavior in Water
Gasoline	No	0.7	Floats
Trichloroethane	No	1.3	Sinks
Sulfuric acid	Yes	1.8	Dissolves

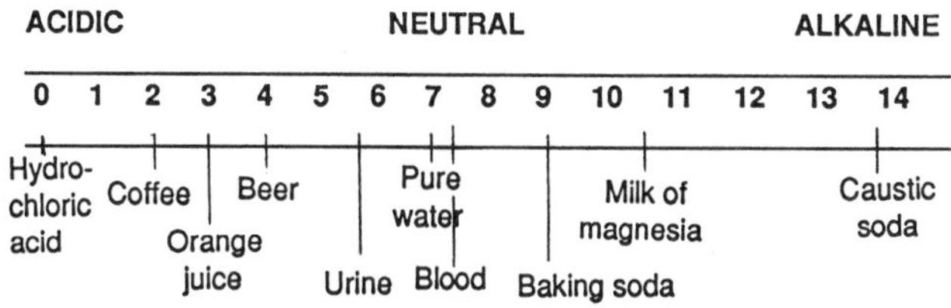

FIGURE 13-11. The pH scale.

Explosive Limits and Range

The explosive range is the range of concentrations of the material in the air which will permit the material to burn. The lowest ignitable concentration of a substance in air is called the "lower explosive limit (LEL)." The highest percentage of a substance in air that will ignite is the "upper explosive limit (UEL)." If the vapor concentration in air is less than the LEL or greater than the UEL, the material will not burn. For example, the LEL for acetylene gas is 3 percent and the UEL is 82 percent; therefore, below 3 percent and above 82 percent acetylene concentrations in air, acetylene will not burn. Acetylene has a wide explosive range (3 to 82 percent) and therefore is a serious fire hazard.

Hydrogen Ion Concentration (pH)

The pH of a chemical is a measure of acidity or alkalinity, where 7 is defined as neutral. A product with a pH lower than 7 is acidic; if the pH is higher than 7, the material is basic or alkaline. The lower the pH, the stronger the acid. The higher the pH, the more corrosive the base (Figure 13-11). Materials with a pH of 2 or lower, or of 12.5 or higher, are classed as corrosive by DOT, and are labeled and placarded as such.

Appearance and Odor

These properties describe the material, including its color, its smell, and its physical state at normal temperature and pressure.

If all properties are considered, it is not possible to find two chemicals with identical properties. The set of properties of a chemical, therefore, is characteristic of that chemical alone.

PHYSICAL STATE

Hazardous materials can be liquids, solids, gases, or sludges that may be part solid and part liquid. Part of the nature of the hazard in an incident depends upon the form of the material; for example, a leaking container of a gas will disperse gas into the surrounding environment and either fill up the area with the gas or create a situation in which the gas is blown by wind into other areas. A spilled or leaked liquid will spread out and seek out the nearest low-lying areas, while a spilled solid generally will form a pile more or less in the vicinity of the spill. The form of the spilled material greatly influences the extent of the spill.

Solids

The primary concerns with spilled solids on land are their flammability and air reactivity. If solids spill into water, the characteristics of solubility and reactivity must be considered. A few solids will form vapors without first becoming liquid; this change from solid to vapor is called "sublimation." The odors of naphthalene (moth balls) and paradichlorobenzene (used as a solid bathroom deodorizer) are due to sublimation. Dry ice is another common product which sublimes. The toxicity and flammability of the vapors must be assessed if subliming solids are spilled.

Liquids

Liquids are able to flow away from a leak, thereby extending the hazard area. Almost all the properties mentioned above should be considered before dealing with a spilled liquid. A volatile liquid creates additional problems as it evaporates and its vapors disperse in air, expanding the hazard zone even further.

The viscosity of a liquid may be a factor in incident response, since the more viscous (thick and sludgy) liquids tend to flow more slowly and to stick to clothing and equipment more readily. Sludges which have thickened due to evaporation may be quite concentrated and are possibly more dangerous than the original liquid.

Gases

There are some special considerations for emergency response to incidents that involve gases. Today many gases are routinely shipped and stored in a variety of forms and containers, so it is likely that they will be encountered in incidents.

Compressed Gases

A compressed gas is any material that, when enclosed in a container, has an absolute pressure of more than 40 pounds per square inch (psi) at 70°F, or an absolute pressure exceeding 104 psi at 130°F, or both. Compressed gases can be either pressurized gases or liquids, depending on the amount of pressure inside the container. A chemical which can exist in both the liquid and gas states has as one of its properties a critical pressure, above which the gas will condense into a liquid as its molecules are pushed closer together by increasing pressure. A gas will become liquid when the pressure is high enough, allowing the container to hold a greater quantity of the material.

Pressurized gases are those compressed gases that are still gases when compressed. Examples of pressurized gases are air, hydrogen, methane, nitrogen, helium, and oxygen.

Other gases are liquefied when compressed and exist in a liquid-vapor relationship inside the containers. Examples of typical liquefied gases are ammonia, butane, chlorine, propane, propylene, and freon. A breached container of liquefied gas may spew liquid which, in changing to the gas form as the pressure is released, expands considerably and may create a large hazard zone.

Acetylene is a gas that does not readily fit into either category. Commercial acetylene cylinders contain gas dissolved in liquid acetone to help make the acetylene more stable. Since acetylene is very reactive chemically and is shock-sensitive, dissolving the gas in acetone decreases its sensitivity and reactivity, thus making it possible to handle the gas safely.

Cryogenic Gases

The cryogenic gases are those with boiling points of less than −150°F that are transported, stored, and used as liquids. A large volume of gas can be stored as a liquid in a much smaller volume at low temperatures, decreasing storage space. The hazards of the cryogenic gases relate to the nature of the particular gas, the tremendous volume ratio of vapor from liquid, and extreme cold. Cryogenic gases include fluorine, oxygen, methane, helium, hydrogen, and nitrogen. Note that several of these gases may also be shipped under high pressure rather than at low temperature. Responders must observe containers to recognize the hazards of shipping or storage methods used for gases.

Changes in Physical State

Some chemical reactions that occur during a leak or a fire cause a change in physical state that can lead to problems. A leaking liquid may change to the gas form, with a tremendous increase in volume. Chlorine, for example, has an expansion ratio of 460 to 1, indicating that the contents of a cylinder of pressurized, liquefied chlorine will expand upon release to fill a space 460 times the size of the capacity of the cylinder. Containers heated in a fire may leak or explode due to an increase in the volume of contained gas or liquid. The expansion ratio of a chemical (the volume of its liquid form compared to the volume of its gas form) is known for all chemicals and is an important consideration in an incident involving the release of gases liquefied by pressure.

CHEMICAL REACTIONS

Stability is the result of strong bonds between atoms. Bonds are formed when atoms exchange or share electrons. More energy is needed to break some bonds than others; generally harder to separate from each other are small atoms, like hydrogen and water, and bonds between atoms with relatively large differences in electrical charge, like hydrogen and fluoride. Heat is the form of energy most often available to break bonds; conversely, when new bonds are formed, heat is released during the reaction.

Release of heat is the most serious hazard in many potentially dangerous reactions. Some reactions produce new products which are hazardous, such as flammable hydrogen gas, irritating hydrogen chloride vapors, fire-supporting oxygen, or deadly phosgene gas. Some reactions occur so rapidly that fire or explosions can occur. Some reactions may cause rupture of containers.

Some of the reactive materials which can cause problems for hazmat responders are water-reactive materials, air-reactive materials, oxidizers, unstable materials, incompatible materials, and materials which polymerize. Polymerization under controlled conditions is a useful technology in the manufacture of plastics such as polystyrene, but if it occurs too rapidly in a spill or fire, it can result in an explosion.

Water-Reactive Materials

Water-reactive materials react with water, often violently, to release heat, a flammable or toxic gas, or a combination of these (Table 13-8). If there is a fire in which water is being used for extinguishment, the presence of water-reactive materials can make the situation much more dangerous.

Air-Reactive Materials

Some of the water-reactive materials mentioned in Table 13-8 are also air reactive and will ignite in air; potassium metal is one example. Diborane

TABLE 13-8. Water-Reactive Materials

Material	Hazard of Reaction
Sulfuric acid	Heat
Potassium, sodium	Flammable hydrogen gas
Calcium carbide	Corrosive and flammable products
Aluminum chloride	Hydrochloric acid burns
Sodium peroxide	Oxygen and heat

and some organic metal compounds, for example, trimethylaluminum, also ignite in air. Materials that ignite in air are termed "pyrophoric."

Another example is white phosphorus, which must be stored underwater to prevent it from igniting. Very finely ground metals, powders, or dusts can be fire and explosion hazards in air; typical examples include metal powders and finely powdered plastics.

Oxidizers

These materials, also called "oxidizing agents," present special hazards because they react chemically with a large number of combustible organic materials, such as oils, greases, solvents, paper, cloth, and wood. Fire is often the result. Also, their reactions generate heat that may be absorbed by other materials, causing ignition. The oxidizers that contain oxygen may release that oxygen as they decompose, helping to sustain a fire.

The halogens are powerful oxidizers. You have learned that fluorine, chlorine, and bromine are strongly reactive. Another group of reactive oxidizing halogens includes the hypochlorites, chlorites, and perchlorates.

Organic peroxides deserve special mention because they are very hazardous. In addition to being strong oxidizing agents, they are inherently unstable chemically. Most of them are sensitive to shock, friction, and heat and can decompose exothermally, releasing a great deal of heat. Examples are acetyl peroxide, benzoyl peroxide, cumene hydroperoxide, and peracetic acid. Since organic peroxides slowly decompose in storage, chemical inhibitors are added to prevent their decomposition.

Similarly, the inorganic peroxides, sodium peroxide and potassium peroxide, are sensitive to shock and are very reactive. They decompose in the same way that organic peroxides decompose.

Other oxidizers include ammonium nitrate, potassium dichromate, chromic acid, potassium permanganate, ammonium persulfate, and sodium nitrate. All of these decompose with heat, sometimes explosively, and release oxygen that can support combustion.

Unstable Materials

Materials designated as unstable are those which have a tendency to decompose by themselves; they do not need to mix with other chemicals to react. They may generate heat or toxic gases, or may burst into flame or explode as they generate flammable vapors. Unstable materials are often stored and shipped with inhibitors in the mixture to prevent decomposition. The organic and inorganic peroxides mentioned earlier are chemically unstable. There is another group of chemicals called "ethers," some of which react with oxygen in the air to form organic peroxides, thus becoming sensitive to shock, friction, and heat. Examples of these peroxide formers include ethyl ether, dioxane, and tetrahydrofuran.

Another group of materials that is chemically unstable contains the monomers, or building blocks that form many types of polymers (resins, plastics, and synthetic rubber materials). Many plastics are made in this way: polyvinyl chloride (PVC) is formed of polymerized vinyl chloride monomers. The polymers polystyrene, polyethylene, and polypropylene are widely used plastics.

Some of these materials can polymerize spontaneously, causing rupture of the container, often explosively. Their presence can compound a fire problem because some of them are sensitive to heat, light, and air. Some monomers, like ethylene, are fairly stable; others, such as butadiene and methyl methacrylate, are much more unstable. Chemical inhibitors are added to some of these materials prior to shipment to prevent unwanted polymerization and ensure safety in handling them. Most monomers are flammable, and many are toxic or irritating materials. These materials are not ordinary flammable materials but present multiple hazards to responders.

Incompatible Materials

No discussion of basic chemistry for first responders would be complete without some mention of incompatibility of chemicals. If two or more chemicals remain in contact with each other without an adverse reaction, they are considered to be compatible. Some materials, when mixed with other materials, can adversely affect human health and the environment in a variety of ways:

- Generation of heat
- Violent reaction

- Formation of toxic fumes or gases
- Formation of flammable gas
- Fire or explosion
- Release of toxic substances if they burn or explode

Some reactions proceed more readily in the presence of a catalyst, a chemical which is added to enhance a desired reaction.

Table 13-9 summarizes some potentially incompatible materials and the potentially adverse consequences of mixing materials.

Toxic Combustion Products

Toxic materials may form from reactions between chemicals during a fire and pose another potential hazard in addition to the numerous ones that

TABLE 13-9. Incompatible Materials
Group A must be separated from Group B or the indicated consequence may occur.

I Group A	I Group B
Acetylene sludge	Acidic sludge
Alkaline/caustic liquids	Acidic water
Alkaline cleaner	Battery acid
Alkaline corrosive liquids	Chemical cleaners ("Chromerge")
Alkaline battery fluid	Electrolyte, acid
Caustic waste waters	Etching acid, liquid, or solvent
Lime sludge	Liquid cleaning compounds (muriatic acid)
Lime waste water	Pickling liquor and corrosive
Spent caustic	Acids
	Spent acid
	Spent acids, mixed
	Spent sulfuric acid

Imminent consequences upon mixing: generation of heat, violent reaction.

II Group A	II Group B
Asbestos waste	Cleaning solvents
Beryllium wastes	Data processing liquid
Unrinsed pesticide containers	Obsolete explosives
Waste pesticides	Petroleum waste
	Refinery waste
	Retrograde explosives
	Solvents
	Waste oil and other flammable explosive waste

Imminent consequence upon mixing: release of toxic substances during fire or explosion.

TABLE 13-9. (Continued)

III Group A	III Group B
Aluminum Beryllium Calcium Lithium Magnesium Potassium Sodium Zinc powder and other reactive materials	Any wastes in I Group A or B
Imminent consequence: fire or explosion due to release of hydrogen gas or intense exothermic reaction.	

IV Group A	IV Group B
Alcohols Water	Any concentrated waste in I Group A or B Calcium Lithium Metal hydrides Potassium Sodium Thionyl chloride, $SOCl_2$ Sulfuryl chloride, SO_2Cl_2 Phosphorus trichloride Tricholor(methyl)silane and other water-reactive wastes
Imminent consequence: contact may generate toxic or flammable gases or cause fire or explosion.	

V Group A	V Group B
Alcohols Aldehydes Halogenated hydrocarbons Nitrated hydrocarbons and reactive organic compounds and solvents Unsaturated hydrocarbons	Concentrated wastes from I Group A or B and III Group A wastes
Imminent consequence: violent reaction, fire or explosion.	

VI Group A	VI Group B
Spent cyanide and sulfide solutions	I Group B wastes
Imminent consequences: release of toxic cyanide or hydrogen sulfide gases.	

TABLE 13-9. (Continued)

VII Group A	VII Group B
Chlorates and other oxidizers	Acetic acids and other organic acids
Chlorine	Concentrated mineral acids
Chlorites	II Group B wastes
Chromic acid	III Group A wastes
Hypochlorites	V Group A wastes and other flammable and combustible wastes
Nitrates	
Nitric acid, fuming	
Perchlorates	
Permanganates	
Peroxides	
Imminent consequence: reaction resulting in fire or explosion (EPA Title 40 CFR 300).	

TABLE 13-10. Toxic Combustion Products

Materials	Toxic Product	Hazard
Organic materials	Carbon monoxide	Asphyxiation at low levels
		Brain damage
		Delayed brain toxicity
	Carbon dioxide	Oxygen deficiency
	Acrolein	Strong irritant to eyes and respiratory system
		Heart and liver toxicity
Materials containing chlorine	Hydrogen chloride	Damage to eyes, skin, and respiratory system
	Phosgene	Fatal by inhalation due to delayed pulmonary edema
Materials containing nitrogen	Ammonia	Pulmonary irritation, possibly fatal
		Flammable if confined
	Hydrogen cyanide	Respiratory arrest due to brain damage
	Nitrogen oxides	Acute respiratory illness
		Decreased resistance to infection
Materials containing phosphorus	Phosphine	Pulmonary edema, possibly fatal
Materials containing sulfur	Sulfur dioxide	Constriction of breathing
	Hydrogen sulfide	Lack of oxygen to cells

firefighters already face. Formation of these toxic products depends upon the nature of the burning material and the amount of oxygen present. Table 13-10 presents information about some dangerous products of combustion that may be encountered in a fire situation, according to the type of material involved in the fire.

RADIOACTIVITY

Radioactivity is a special consideration for some chemicals. Radioactive materials are those which undergo spontaneous transformation and release radiant energy or atomic particles. The energy of this natural transformation process is emitted in the form of alpha or beta particles or gamma radiation. Radioactivity is not affected by the physical state or chemical combination of the element.

Of the three forms of radioactivity, alpha radiation is the least dangerous as long as it remains outside the body. There is enough energy to cause severe damage if an alpha emitter such as radon gas is inhaled, but alpha rays outside the body are stopped by a sheet of paper or aluminum foil, and some low-energy alpha rays are stopped by the dead outer layer of normal skin.

Beta radiation can have higher energy levels than alpha radiation. The particles emitted are smaller, and can penetrate soft body tissues, but can be stopped by a thick sheet of cardboard.

The most dangerous form of ionizing radiation is gamma radiation. It consists of electromagnetic rays like light rays or X-rays, and the rays easily penetrate the human body. Shielding from gamma radiation involves a specially designed thick mass of a dense material such as lead.

Recognition

Tools and equipment containing radioactive sources are used in many industries for checking pipe welds, determining the thickness of metals, and as indicators of the level of liquid product in large vessels. They are also used in portable equipment employed in geology and geotechnical assessment and in well drilling. Other types of radioactive materials are used in medical treatment clinics, many kinds of biological and chemical research laboratories, and in some advanced teaching facilities. They are even found among the chemicals used in illicit drug production labs. Radioactivity can be detected only by using specialized detection equipment, since it cannot be seen, heard, felt, or smelled.

If radioactive materials are properly labeled, the label gives information about the hazard level. Radioactive White I (with 1 vertical bar on the label) is the lowest hazard category, Yellow II (with 2 bars) is the next, and Yellow III (with 3 bars) the highest, based on the radiation activity level of the material. The activity level is required by law to be written on the label.

Response

If radioactivity is detected in an area, responders should not approach its source. Protection from radiation hazards is based on time, distance, and shielding: Less time, greater distance, and more shielding decrease the hazard. At the first sign of detected radioactivity, all responders should back off from the suspected source to an area where monitoring instruments no longer detect radioactivity higher than background level and call the local, state, or federal agency which has jurisdiction over radiation hazards. If no local or state agency has jurisdiction, the U.S. Nuclear Regulatory Agency should be contacted for advice.

SOURCES OF CHEMICAL INFORMATION

There are a number of good written documents which can be used to determine the properties of identified chemicals. A brief list of these follows. Whatever documents are selected should become thoroughly familiar to responders who will use them, since familiarity allows much quicker collection of information when an emergency occurs.

Reference Documents for Use in an Emergency

- *Emergency Handling of Hazardous Materials*. Washington, DC: Bureau of Explosives, Association of American Railroads.
- *Hazardous Chemicals Data*. Quincy, MA: National Fire Protection Association.
- *Hazardous Chemicals Data* (U.S. Coast Guard CHRIS System). Washington, DC: Superintendent of Documents, U.S. Government Printing Office.
- *Manual of Hazardous Chemical Reactions*. Quincy, MA: National Fire Protection Association.

Reference Documents for Desk Use

- *Dangerous Properties of Industrial Materials* by N. Irving Sax. New York: Reinhold Publishing Corp.

- *Hawley's Condensed Chemical Dictionary* by N. Irving Sax and R.J. Lewis. New York: Van Nostrand Reinhold Co.
- *The Merck Index, an Encyclopedia of Chemicals and Drugs*. Rahway, NJ: Merck Co.

In any case where there is confusion about a chemical property, multiple references should be consulted. Remember to convert temperatures and air concentration units so that they agree before attempting to compare numbers. Never make response decisions until the properties of the chemicals have been clearly delineated; every hazmat brigade should have a chemical specialist on call to clarify chemical properties and use good judgment in predicting chemical behavior.

REFERENCES

Andrews, L.P., editor. 1990. *Worker Protection during Hazardous Waste Remediation*. New York: Van Nostrand Reinhold Co.

Cashman, J.F. December 1988. *Hazardous Materials Newsletter*, 6.

DHHS. 1985. *NIOSH/OSHA/USCG/EPA Occupational Safety and Health Guidance Manual for Hazardous Waste Site Activities*. Washington, DC: U.S. Government Printing Office.

DHHS. 1990. *NIOSH Pocket Guide to Chemical Hazards*. Washington, DC: U.S. Government Printing Office.

DOT. Title 49 Part 172.101. Washington, DC: U.S. Government Printing Office.

Gollnick, D.A. 1983. *Basic Radiation Protection Technology*. Altadena, CA: Pacific Radiation Corporation.

Sax, N.I., and R.J. Lewis, Sr. 1987. *Hawley's Condensed Chemical Dictionary*, 11th ed. New York: Van Nostrand Reinhold Co.

Whitten, K.W., and K.D. Gailey. 1984. *General Chemistry with Qualitative Analysis*, 2nd ed. Philadelphia: W.B. Saunders Co.

14

Human Health Effects of Hazardous Materials

Barbara M. Hilyer, M.S., M.S.P.H., and
Kenneth W. Oldfield, M.S.P.H.

Emergency responders are asked to take a great deal of risk when they go to the scene of a hazardous materials incident. This risk can be reduced by employing methods to limit the exposure of responders to toxic chemicals. Decisions regarding protective measures must be made in every incident; they can be made with more assurance if responders understand how toxic chemicals enter the body and cause harm.

In Portland, Connecticut, 47 employees of a manufacturing plant were taken to hospitals after being exposed to a plume of toxic vapors which were released from a tank of degreasing solution. On the same day, nine workers were taken to the hospital after inhaling a floor sealer being used in part of the building. A disaster drill scheduled for an area hospital was called off; it was felt that the staff had had enough practice with these simultaneous incidents.

In this chapter we will present information about toxic chemicals exposure and health effects. The goal of the chapter is to provide emergency responders with enough information to enable them to make good decisions for their own protection and the protection of the community.

OBJECTIVES

- Be familiar with safe exposure terms used in reference documents
- Know the kinds of studies from which safe exposure limits are derived

- Understand the limitations of toxicological studies and the safety factors used in extrapolating study data to human exposure limits
- Be able to determine the health hazards of an identified chemical to responders and members of the community
- Know how toxic materials may enter the human body so that these routes of entry can be protected
- Know how to recognize the symptoms of chemical exposure
- Be aware of the differences between usual workplace and emergency response exposures
- Learn the toxic effects of specific chemicals of concern
- Know when and how to look for immediate and long-term effects of chemical exposure

TOXICITY

A toxic material is one which can make people sick. It may do this by burning the skin; irritating the throat and lungs; damaging the stomach lining; injuring cells in the brain, liver, ovaries, or other organs; or by a number of other means. Some toxins are produced by bacteria and other living organisms such as snakes and insects, and some toxic effects result from the actions of viruses or radioactivity. The toxic materials most likely to be encountered by emergency responders are chemicals, and our focus in this chapter will be on the health effects of toxic chemicals.

In order for a chemical to have an adverse effect on the human body, it must get onto or into the body. The effects are proportional to the dose the body receives. Doses administered experimentally in food or by injection are usually measured in milligrams per kilogram (mg/kg) of body weight of the test subject. In studies where the dose is inhaled with air or is present in drinking water, the environmental concentration may be expressed as parts per million (ppm), meaning the number of parts chemical in every million parts air or water, or as milligrams per cubic meter (mg/m^3), meaning the number of milligrams of chemical in a space of air which measures 1 meter on each side.

The relationship between the amount of chemical exposure and the body's biological response is called the "dose-response relationship." In most cases, the greater the exposure, the greater the response. It follows, then, that reducing or eliminating exposure can reduce or eliminate the body's illness in response to the chemical. A great deal of study has gone into providing people with information they can use to reduce or eliminate exposure to toxic chemicals.

SETTING SAFE HUMAN EXPOSURE LIMITS

Where Do the Numbers Come From?

One of the important things you learn in any job where there may be exposure to toxic chemicals is that there is information available which can help you protect yourself from exposure. The key piece of information to use in choosing protective measures is the number indicating the concentration to which you can safely be exposed. These numbers are determined after extensive consideration by knowledgeable people based on three kinds of studies.

Laboratory Studies Using Animals

Animal studies are the most common studies done with toxic chemicals. They were originally done to provide information about chemicals which had been found to have beneficial, or medicinal, effects and were done in attempts to judge the side effects of prescription drugs. They are still used for this purpose and have been expanded to include studies of workplace chemicals.

The animals used in laboratory studies may be mice, rats, rabbits, guinea pigs, dogs, monkeys, or others, but mice and rats are usually the study animals of choice. Rodents are small, easy to keep, relatively inexpensive to feed, and have a life span of around 3 years. This last characteristic allows researchers to see a lifetime of exposure effects within a reasonably short period. Also, rodents of the same genetic strain can be purchased, allowing scientists to determine that variation in illness was produced by the variations in the dose of the chemical and not by genetic differences in the study animals.

Permission to use experimental animals must be granted by the National Institutes of Health. It must be shown that the research is necessary for human health and that the animals will be treated humanely.

Studies of LD_{50}

Experiments with an untested chemical usually start with a study to determine how large a dose is fatal to the animals. Test animals are grouped, and each group is given one exposure to the chemical, with the measured amount of the dose based on each animal's body weight. The animals are cared for in the normal way for 14 days, with notes kept on how many deaths occur in each dose group. The goal of the study is to determine the dose which is lethal to half of the exposed animals in 14 days.

TABLE 14-1. Results of a Hypothetical Lethal Dose Study[a]

Group	No. Tested	Oral Dose (mg/kg)	No. of Deaths
A	100	12	83
B	100	8	50
C	100	4	12
D	100	0	0

[a] LD_{50} = 8 mg/kg.

The dose which kills half of the animals in 14 days is termed the "LD_{50}," the lethal dose to 50 percent of the animals. Table 14-1 shows how the results of an LD_{50} study might look. The LD_{50} in the study illustrated was found to be 8 mg/kg, since this dose was fatal to 50 percent of the animals that received it.

LD_{50} studies give us only a relative idea of the dangers of the chemical tested. Since the objective of toxicological studies is not to find out how much exposure emergency responders can handle and suffer the loss of only half of their team, the need for further studies is obvious. Table 14-2 shows the LD_{50} values of some chemicals which may be encountered in emergency response.

Once the LD_{50} for a chemical is known, further studies are done with new groups of animals to determine other effects of the same chemical.

Acute Effects Studies

Tests for acute effects are done next, using doses smaller than the known LD_{50}. The objective of these experiments is to determine the dose required to produce immediate effects, but not deaths, in the test animals.

TABLE 14-2. LD_{50} for Several Chemicals

Agent	Oral Rat (mg/kg)	Skin Rat (mg/kg)	Inhalation Rat (ppm)
Ammonia	350		2,000
Aniline	440	1,400	250
Benzene	3,800		10,000
Carbon tetrachloride	2,800	5,070	4,000
Chlorine			293
Creosote	725		
Ethyl alcohol	14,000		
Formaldehyde	800		250

Acute effects are observable within a few weeks or months, and may range from mild hair loss to severe organ damage. They include such biological responses as skin irritation, liver damage, neurological malfunction, and damage to reproductive organs or offspring.

If the chemical being tested is one which is a workplace hazard, the animals are exposed in the same way as workers are expected to be exposed, by breathing or swallowing or skin contact.

The acute effects are noted and are correlated to the dose received by each test group. At the end of the study, the range of doses expected to produce acute effects is known.

Chronic Effects Studies

Chronic effects tests are done following the studies of acute effects. Doses given to study animals are lower than those which produced acute effects, and observations for chronic effects are made for a longer period of time. In some cases, chronic effects last throughout the lifetime of the animal. Table 14-3 shows some chemicals and the acute and chronic effects produced by exposure to them.

Problems with Animal Studies

Problems are encountered when we try to use information gained from animal experiments in setting human safe exposure limits. Obviously, humans are not laboratory rats. Since we do not look like rats, we can immediately assume that we are genetically different from rats. Our genes determine our bodies' ability to handle toxic chemicals, so we may handle some of them in different ways than rats do.

TABLE 14-3. Acute and Chronic Effects of Selected Chemicals

Agent	Acute Effects	Chronic Effects
Chlorine gas	Irritated eyes, nose, mouth; coughing; pulmonary edema	Lung scarring; shortness of breath
Chlordane	Cough; nausea and vomiting; convulsions	Damage to lungs, liver, and kidneys
Methyl isocyanate	Eye and skin injury; cough	Asthma
Benzene	Irritated eyes and respiratory system; nausea; staggering gait	Fatigue; decreased production of blood cells; cancer
Portland cement	Cough; dermatitis	Bronchitis
Turpentine	Irritated eyes and throat; skin damage	Kidney damage; allergic sensitivity

We are much bigger than rats and other laboratory animals. Doses are measured in such a way as to enable researchers to account for this size difference; oral doses are given to animals in milligrams or micrograms per kilogram of their body weight. The dose is adjusted to milligrams per kilograms and can be extrapolated to the weight of humans, which is measured in kilograms. A safety factor is included in this extrapolation, as we will see later.

The conditions under which laboratory animals are tested are carefully controlled in a way that human living conditions are not. The rats do not smoke, drink, eat pesticide-laden fish, or drive to work in polluted air with no seat belts. They are protected from outside disease germs and other factors which might influence a toxic response.

Effects must be measurable. All effects must be observable to be noted, whether visible (hair loss), countable (number of offspring), able to be weighed (liver shrinkage), or other clearly obvious symptoms of disease. It is impossible to get a verifiable answer from a rat to the questions "How do you feel?" and "Does anything hurt today?" Laboratory animals, therefore, are given larger doses than are probably necessary to make them only mildly sick.

Almost all of these studies are done using exposure to only one chemical at a time. It is very difficult to design a study to test the effects of exposure to more than one chemical at a time. In the real world, emergency responders may be exposed to a combination of chemicals in the same incident, or in separate incidents which occur within a short time span. As we will see later, multiple exposures may lead to unexpected results.

Epidemiology Studies

Epidemiologists study epidemics of diseases in humans. They define disease broadly to include accidents and other nontransmissable conditions. Sometimes those diseases are caused by chemical exposure.

Experimental Information

In studying human victims of exposure to a chemical, epidemiologists use a number of different sources of information such as hospital records, insurance company records, and occupational safety and health records. They may also interview people who have worked with the chemical. When enough people have been identified who have been exposed to the chemical, their health records are compared to the health records of an unexposed group. This unexposed group should be as much like the study group as possible in sex, age, and other factors. If all other parameters are similar and the only difference is exposure to a chemical, differences in health history may be credited to the chemical.

Studies of exposure to some chemicals have shown that they cause or contribute to diseases in humans. Figure 14-1 shows the results of one study of the relationship between chemical exposure and human disease. "Expected cases" are those cases of a certain cancer which would be expected to occur in this number of people, based on the percentage of the general population who get the disease. Only 1 person per 100,000 in the general population has the type of cancer studied. Of the group of workers who were definitely known to be exposed to the chemical in question, there were 6.48 cases of the cancer per 100,000 workers. This number of cases is statistically significant; it was found by statistical analysis to be much higher than can be explained by chance. The conclusion was drawn that the chemical exposure was responsible for the increased number of cases of cancer.

Several criteria must be met if a study of this kind is to be valid. Sometimes, when dealing with chemical exposures, these criteria are difficult to reach.

The study must include large numbers of people. If the study group is not large, it cannot be shown by statistical analysis that the effects were caused by the chemical and not by chance.

The exposures must be quantifiable. This is very difficult to do, as it requires that a numerical value be placed on the air concentration, skin

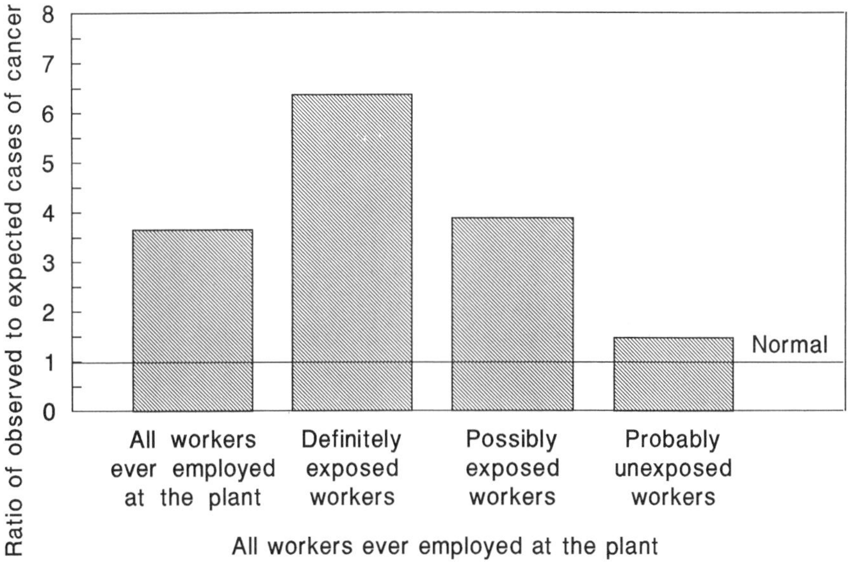

FIGURE 14-1. Incidence of cancer in a group of workers exposed to workplace chemicals.

contact level, or amount swallowed of the chemical. If no one has been doing this kind of monitoring, then estimates of these numbers are hard to figure out. You remember that the dose-response relationship forces us to estimate the amount of the exposure in order to know that the response was due to the dose.

Groups of exposed subjects have to be compared to groups of unexposed subjects who match them as closely as possible in terms of all other parameters. These parameters include sex, age, size, and lifestyle. Lifestyle includes many factors which are controlled by the person, not the researcher, and these almost always affect health and resistance to disease. Some of these factors are smoking, drinking, medications used, and diet.

Problems with Epidemiology Studies

All the criteria mentioned above must be met for epidemiology studies to produce valid results. Two other concerns exist which cannot be controlled.

Every person has a unique genetic makeup. Even identical twins have slightly different genes. These differences may result in different responses to the same chemicals.

Human experimentation is illegal and immoral in modern American society. We cannot put people in controlled, potentially harmful situations where researchers can control exposures and all other variables.

Experiments of Nature

Sometimes groups of people are exposed to chemicals accidentally. These exposures are called "experiments of nature" by toxicologists and epidemiologists who study them. They share some of the characteristics of laboratory experiments: The chemical is known; many individuals are exposed; sometimes dose levels can be determined; and measurable results occur. Some of the most widely studied experiments of nature are mercury-induced disease in Minnemata, Japan; a deadly toluene diisocyanate release in Bhopal, India; and widespread dioxin contamination following a plant explosion in Sevaso, Italy. Surviving victims of these and other accidents have been extensively studied, some for as long as 30 years after exposure.

Let us look at an example of data that result from combining laboratory studies of rat exposures, epidemiology studies of human exposures, and follow-up studies on victims of an experiment of nature.

Hexachlorobenzene is a chlorinated aromatic compound formerly used to prevent stored seed and grain from becoming moldy. Table 14-4 shows exposure data for this chemical. We can conclude that hexachloroben-

TABLE 14-4. Results of Studies of Hexachlorobenzene

Test Subject	Dose	Response
Animal Studies		
Rat	5 mg/kg	Half of the subjects died
Rat	0.08 mg/kg/day	Kidney damage in offspring No observable liver effects
Rat	0.4 mg/kg/day	Chronic liver damage
Rat	2 mg/kg/day	Acute liver damage Increased death in offspring
Rat	6 mg/kg/day	Liver cell tumors
Epidemiology Studies		
Industrial workers	Not measured	Increased blood levels
Vegetable sprayers	Not measured	Blood levels 4–287 ppb Kidney damage
Hazwaste workers	Not measured	Increased blood level
General population	0.2 μg/day	Fat retention 18–35 ng/g fat Retention for 15 years
Experiments of Nature Humans in Turkey	Variable; ingested in bread over 4 years	100% death in children who drank milk of exposed mothers; 10% mortality in adults; liver disease; skin ulcers; neurological effects; short stature and small hands in adults exposed as children

zene is toxic to both rats and people, that the effects are dose related, and that the chemical causes acute and chronic health effects in several organ systems.

Extrapolating Study Data to Human Exposure Limits

Following laboratory and epidemiological studies, scientists from several organizations set legal or suggested safe exposure limits for humans. The aim of setting these limits is to give potentially exposed people guidelines for protecting themselves from hazardous dose levels. Considerations in setting safe limits include the relative degree of danger presented by the chemical and the sort of study data (animal or human) on which the limits will be based.

Safety Factors

Safety factors are calculated into the final numbers. These safety factors take into account the severity of the hazard, the differences between lab animals and humans, and the differences between human individuals. The dose which produced no observable effects in animals is divided by 100 or

1,000 for suggested safe exposure in humans; the magnitude of the safety factor depends on the seriousness of the potential response. A dose which was found to produce no observable effect in human studies is divided by a safety factor of 10. These safety factors are only general guidelines; researchers who are thoroughly familiar with the chemical use their judgment in setting limits.

Exposure limits are designed to protect most people, but they may not protect the most sensitive individual. It is known that certain individuals are more likely to have a negative response to chemical exposure than others, and these people may need more protection than is provided by exposure limits. It is easy to understand differences in sensitivity to a chemical when we consider the differences we are aware of in allergic responses to agents such as plant pollens or poison ivy.

Different Limits Set by Different Agencies

Several agencies set safe exposure limits; their criteria and suggestions for use of the limits they set may be different from those of other agencies. Safe exposure limits are useful only when understood and used in the context in which they are set.

OSHA sets legal workplace limits to protect workers who are exposed to a chemical on a regular basis in their daily workplaces. It must be understood that even though OSHA limits cannot be legally exceeded in the workplace, they are not precise cutoff numbers dividing safe from unsafe levels of exposure. OSHA encourages employers to provide protection to workers in atmospheres where levels are below published exposure limits.

NIOSH sets recommended limits, some of which are more conservative than the OSHA limits. They are often different interpretations of the same experimental results.

The American Conference of Governmental Industrial Hygienists (ACGIH) also sets suggested limits. They are designed to be used in the practice of industrial hygiene. In some cases, these limits have been adopted by OSHA as legal standards.

Although safe exposure limits are listed for all common airborne chemical hazards, their use is not the most accurate way to quantify human exposure. They only inform us about levels in the air; they cannot tell us how these levels translate into the dose received inside the body at the place where dose leads to response. Monitoring would be most useful if it could detect the internal dose before an illness begins; usually this is not possible. Comparing listed safe exposure limits with measured concentrations in air is the most readily available means of judging chemical hazard exposure.

Making Words Out of Alphabet Soup

Each agency which sets exposure limits uses different terminology. The terms are frequently shortened to abbreviations or acronyms, which must be learned in order to understand what is being said about a chemical.

Exposure Limit Terminology

The following are acronyms (new words formed by using the first letters of words) and abbreviations (initials which do not make new words) which may be encountered in seeking toxicological information about a chemical.

- PEL (permissible exposure level; in newer listings, it may refer to the published exposure level). This is the OSHA legal limit for workplace exposure.
- TLV (threshold limit value): The ACGIH suggested limit for workplace exposures. This is defined as the air concentration to which almost every worker can be exposed for five 8-hour days each week throughout his working life without suffering adverse effect.
- TWA (time-weighted average): The 8-hour day, 40-hour work week limit. If not otherwise stated, PELs and TLVs are for an 8-hour exposure period.
- STEL (short-term exposure limit): This concentration is safe for a 15-minute exposure period, as long as there are only four such periods a day separated by 1-hour intervals, and the overall TWA exposure is not exceeded for the day. It is intended to guard against acute effects from brief, high exposures.
- C (ceiling): This concentration should never be exceeded.
- IDLH (immediately dangerous to life and health): The definition of this term states that most people could be exposed to this concentration of chemical for 30 minutes without the likelihood of harm that would be fatal or severe enough to prevent self-rescue. The practical use of the term suggests that a responder avoid all exposure to IDLH levels of chemicals.
- LD_{Lo}: The lowest concentration which is expected to be lethal. The usual route of the dose is oral.
- LC_{Lo}: The lowest concentration expected to be lethal by inhalation.
- TD_{Lo}: The lowest ingested dose expected to be toxic.
- TC_{Lo}: The lowest inhaled concentration expected to cause toxic effects.
- ALARA (as low as reasonably achievable): This new designation avoids setting numerical limits for carcinogenic or other highly hazardous chemicals.

Workplace versus Emergency Response Exposure Limits

Since PELs and TLVs are set to provide safe working conditions for employees who encounter exposure for 8 hours each day, 5 days a week, throughout their working lives, it has been questioned whether these levels are appropriate for hazmat responders whose exposures will be rare and relatively short. In addition to determining safe exposure limits for themselves, responders may be required to make decisions which affect the exposures of members of the community living within a potential gas or vapor cloud area.

Several expert groups have addressed this problem, and have published exposure limits for emergency responders and the general public. They are the National Research Council of the National Academy of Sciences, EPA, and the American Industrial Hygiene Association.

The National Research Council's Committee on Toxicology has published two lists: emergency exposure guidance limits (EEGLs) for emergency response personnel and short-term public emergency guidance levels (SPEGLs) for the protection of people in the community. The EEGLs are set for defined occupational groups such as firefighter hazmat responders, a population which typically is younger and healthier than the general public. For this reason, EEGLs may be higher than SPEGLs. SPEGLs are set to protect the general public from effects which might be incapacitating or irreversible. They are considered acceptable for public exposure during an emergency. Table 14-5 shows EEGLs and SPEGLs for selected chemicals.

The research necessary to set safe emergency limits has been completed for only a limited number of chemicals. For protecting general populations (the public, not emergency responders), the Federal Emergency Management Agency (FEMA) lists the following options, in order of decreasing preference:

- Consult a toxicologist or another similarly qualified individual for advice based on a formal review of the toxicity of the material of concern.
- Use the highest value among the following:
  - IDLH value divided by 10
  - TLV-STEL
  - TLV-TWA
  - TLV-C

FEMA suggests that these values be used only for exposures lasting for less than 1 hour and for noncarcinogenic chemicals.

TABLE 14-5. Summary of Emergency Exposure Guidance Levels

Chemical	EEGL (ppm)	Chemical	EEGL (ppm)
Acetone	8,500	Hydrogen chloride	20
Acrolein	0.05	Hydrogen chloride	1[b]
Aluminum oxide	15 mg/m^3	Hydrogen sulfide	10 (24 hr)
Ammonia	100	Isopropyl alcohol	400
Arsine	1	Lithium bromide	15 mg/m^3
Benzene	1,000[a]	Mercury vapor	0.2 mg/m^3
Bromotrifluoromethane	25,000 (24 hr)	Methane	5,000 (24 hr)
Carbon disulfide	50	Methanol	200
Carbon monoxide	400	Monomethyl hydrazine	0.24[b]
Chlorine	3	Nitrogen dioxide	1[b]
Chlorine trifluoride	1	Nitrous oxide	10,000
Chloroform	100	Ozone	1
Dichlorodifluoromethane (Freon-12)	10,000	Phosgene	0.2
		Sodium hydroxide	2 mg/m^3
Dichlorofluoromethane (Freon-21)	100	Sulfur dioxide	10
		Sulfuric acid	1 mg/m^3
Dichlorotetrafluoromethane (Freon-114)	100	Toluene	200
1,1-Dimethylhydrazine	0.24[b]	Trichlorofluoromethane (Freon-11)	1,500
Ethanolamine	50	Trichlorotrifluoroethane (Freon-113)	1,500
Ethylene oxide	20		
Ethylene glycol	40	Vinylidene chloride	10 (24 hr)
Fluorine	7.5	Xylene	200
Hydrazine	0.12		

[a] Proposed.
[b] SPEGLs.
Source: FEMA, DOT, EPA. 1988. *Handbook of Chemical Hazard Analysis Procedures*. Washington, DC: U.S. Government Printing Office.

EPA suggests levels of concern (LOCs) for protection of the public from the 390 chemicals listed by the agency as extremely hazardous substances (EHSs). LOCs are derived by dividing IDLH concentrations by 10. Some examples of LOCs are shown in Table 14-6.

All the above exposure limits are for relatively pure substances. Mixtures of harmful gases or vapors require a more complicated scheme for setting safe exposure limits, which is beyond the scope of this book.

Special Terminology

Some words describing toxic effects may need explanation, as they are often encountered in reference documents.

TABLE 14-6. Levels of Concern for Selected Chemicals

Chemical	LOC (ppm)	Chemical	LOC (ppm)
Acrolein	0.5	LP gas	1,900
Ammonia	50	Malathion	500
Butyl alcohol	800	Naphtha	1,000
Chlorine	3	Nitric acid	10
Furfuryl alcohol	25	Ozone	1
Hydrogen chloride	10	Phenol	25
Hydrogen cyanide	5	Sulfuric acid	8 mg/m^3

Mutagens

A mutagen is a chemical whose toxic effect is damage to DNA, the genetic material. This may be repaired by the body or, if it is not, may lead to a wide variety of illnesses, including cancer.

Genes are made up of a series of proteins arranged much like beads on a string. The string relays information to newly forming cells, telling them what shape, form, or function to accomplish. Beads are read three at a time, in a sort of three-letter code. If one bead is damaged, changed, or missing, or if the strands get tangled or crossed, the code no longer makes sense and cannot be used as a pattern. The following example may help the reader to understand the process.

The following three-letter-word sentence makes sense to people who read English. If we remove a letter from one word and the reader still reads the sentence three letters at a time, the sentence is no longer meaningful.

THE CAT ATE THE BIG FAT RAT
THE CAA TET HEB IGF ATR AT

This is not the only way genetic damage occurs; other kinds of damage also result in misinterpretation of the code. Caffeine and ultraviolet light are known mutagens, as are a few chemicals used in industry, such as hydroxylamine. Other industrial chemicals are under study to determine if they are mutagenic.

Teratogens

A teratogen is a chemical which is known to cause damage in an unborn baby. The word "teratogen" comes from the Greek and literally means "make a monster." Teratogen exposure should obviously be avoided by pregnant women, but some teratogens can influence embryos which are not yet conceived by damaging eggs or sperm within the bodies of future parents. Since a woman already carries, at her birth, all the eggs she will ever produce, her exposure period lasts from birth until her last pregnancy.

Lead is a known teratogen, causing brain damage in offspring. A recent Supreme Court decision (*U.S.* v. *Johnson Controls,* 1991) determined that an employer cannot discriminate against women of childbearing age in assigning workers to higher-paying, lead-exposed jobs. A number of other chemicals are known to be maternal teratogens, operating through exposure to the mother.

Recent research has found evidence of human sperm teratogenicity from exposure to vinyl chloride. Other studies indicate that children of male auto body workers exposed to a combination of hydrocarbons, metals, oils, and paints had a four- to eightfold increase in the incidence of kidney cancer. After the *Johnson Controls* discrimination suit was initiated, lead was also found to be a sperm teratogen.

Neurotoxins

Neurotoxic agents are toxic to the nervous system. They may damage the brain or spinal cord (together called the "central nervous system") or sensory or motor nerve communications in other parts of the body. One well-known neurotoxic effect is tremor of the limbs caused by exposure to several organic solvents; another results in a shuffling walk known as "Ginger Jake paralysis," since it was first identified in persons who consumed contaminated ginger liquor imported from Jamaica in the 1930s.

Hepatotoxins

Hepatotoxic chemicals damage liver tissue. The liver is a large organ, and damage may not be detected until much of it is irreversibly harmed. Hydrogen cyanide, formic acid, fluorine, and many chlorinated solvents are hepatotoxic.

Nephrotoxins

Nephrotoxic chemicals cause kidney damage. Certain workers—particularly auto assembly plant and metal work employees who clean machines—have a high risk of developing serious kidney disease. These workers have more than double the risk of developing kidney problems that require dialysis or transplants. Many cases of kidney damage are not found until the person has lost three-fourths of the function of the organ.

Other Toxins

Other prefixes may precede "toxic" to indicate toxic effects on other organ systems, such as the spleen (splenotoxic), sperm (spermatotoxic), and genetic material (genotoxic).

Carcinogens

Carcinogens are chemicals which are known or suspected to cause cancer in humans. These toxins are grouped according to the strength of the evidence that they cause cancer—whether the evidence is found in lab animals or humans, or whether the chemical is suspected because of its chemical similarity to a known carcinogen.

Cancer in humans is more difficult to predict than some other effects of exposure to toxic materials. There are two reasons for this.

First, the vast majority of chemicals in use have not been tested on laboratory animals to determine their cancer-causing effects. Of those that have been tested, about half have been shown to cause cancer in lab animals; however, this is not direct evidence of human carcinogenicity.

Second, it can take 10 to 40 years to see the results of exposure to a cancer-causing chemical. You may be healthy for 20 years and develop cancer the next year. The time it takes for the cancer to show up after exposure is called the "latency period." Figure 14-2 shows some of the latency periods for different known human carcinogens.

Is there a safe level of exposure to carcinogens? They have listed PELs, which implies that there is. Many scientists agree that for some toxic chemicals there are safe levels of exposure. Below the safe level, called the "threshold," exposure will not cause any bad health effects. However, there is good reason to believe that this is not the case with carcinogens. Some scientists believe that exposure to any amount of a carcinogen is unsafe and may lead to cancer.

The designation of carcinogenesis has always been based on animal studies and workplace exposures. Emergency responders expect a very different type of exposure; theirs will be relatively short-term, infrequent exposures. Some evidence exists to support the statement that certain known carcinogens, like benzene, do not appear to cause cancer in these people. It now appears that a responder who survives the immediate harmful effects of a benzene exposure is able to detoxify it and to have no further effects without further exposure. Because there is always new information on potential carcinogens and other delayed-effect toxins, it is important that responders keep records of their chemical exposures.

Sources of Information

A list of sources of information about specific chemicals, including their health effects, is provided in Chapter 5. For more extensive data concerning toxicology studies, the reader may consult one of the following sources:

- *RTECS* (*Registry of the Toxic Effects of Chemical Substances*). This multivolume reference is published by NIOSH, a department of the U.S. Public Health Service

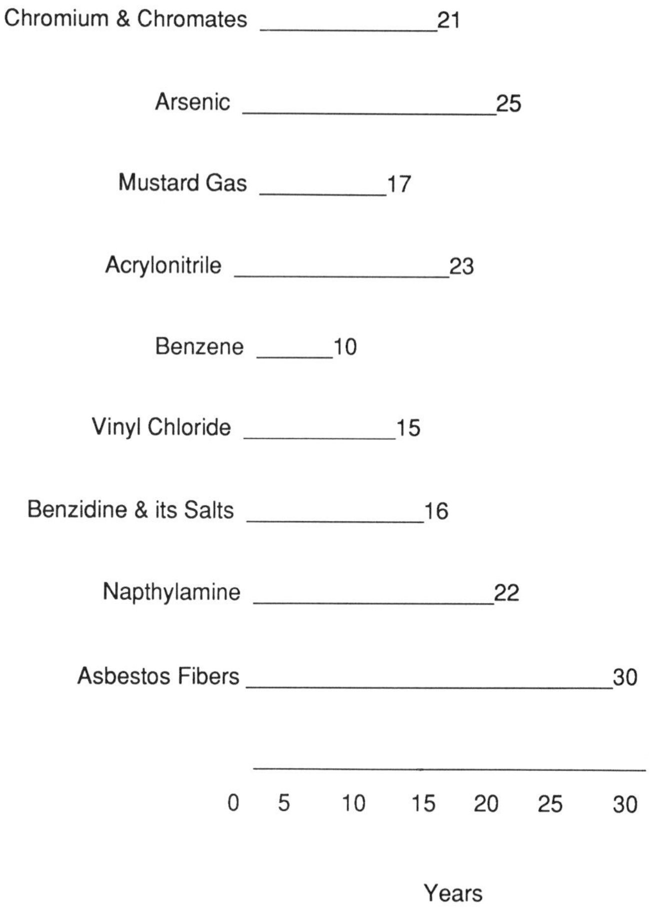

Chromium & Chromates 21

Arsenic 25

Mustard Gas 17

Acrylonitrile 23

Benzene 10

Vinyl Chloride 15

Benzidine & its Salts 16

Napthylamine 22

Asbestos Fibers 30

0 5 10 15 20 25 30

Years

FIGURE 14-2. The average number of years after exposure for the appearance of cancer.

- *Toxicological Profiles* for individual chemicals, published by the Agency for Toxic Substances and Disease Registry of the U.S. Public Health Service
- Computer data bases such as TOXNET and MEDLINE. Arrangements must be made for contact and payment for use

ROUTES OF ENTRY INTO THE BODY

Chemicals enter the human body by three major routes: by inhalation (breathing), by skin contact, or by ingestion (swallowing). Contact with a route of entry may cause harm at the site of contact, or may lead to absorption by skin, respiratory surfaces, or the digestive tract, allowing harmful effects to occur at some distance from the site of entry. Once the route of entry of a chemical is known, protection from the chemical can be established.

The physical state of the chemical is important in determining the probable route of entry. Solids are unlikely to cause harm if they are in large pieces. It is only when they become finely divided by grinding, sanding, welding, or burning that they are small enough to be transported by air to the skin or lungs, or carried to the mouth to be swallowed. Liquids may also get into the mouth or contact skin when they splash. Liquids which form aerosol mists or vapors can be inhaled. Gases are likely to be inhaled and may, under some conditions, condense on skin or other surfaces as liquids.

Contact with the Body Surface

Skin is an excellent protector against harmful agents in the natural environment. It covers the complete surface of the body and prevents entry by naturally occurring germs and toxins. Most of the chemicals used in industry today are man-made, and humans have not had time to evolve protection from them. Two mechanisms may account for their harmful effects.

Damage to the skin itself results from physical injury to skin cells by the chemical or by an allergic-type response in the skin. Strong corrosives such as hydrochloric and sulfuric acids, and bases like sodium hydroxide, may burn the skin severely. Mild to severe irritation can be caused by many halogenated solvents. Some metal compounds, particularly those containing beryllium and nickel, initiate an allergic response in sensitive people. Skin diseases are the most common occupational diseases. Some skin diseases clear up quickly after exposure ceases, but others continue for a long time.

Many chemicals, especially solvents, permeate the outer protective layer of skin and are absorbed into blood vessels in underlying layers. Chemicals known to cross the skin barrier are given a "skin notation" by OSHA, ACGIH, and NIOSH. Some of these are hexane, benzene, and trichloroethane. Skin cells contain fat, and solvents dissolve fat. This can lead to skin damage and entry into the blood transport system. Once in the blood, chemicals are carried throughout the body. Figure 14-3 shows the structure of skin and its underlying blood vessels.

A small area of the body is not covered by skin, but by the corneas of the eyes. This covering is even more vulnerable than skin, as the eyes are always wet. This allows any soluble chemical to quickly dissolve, a pre-

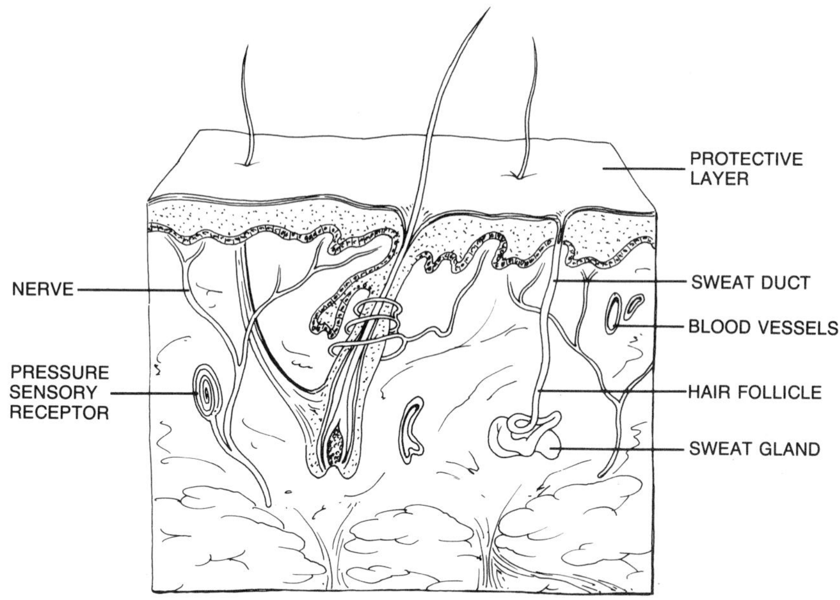

FIGURE 14-3. The structure of human skin.

requisite for crossing the body surface. Protection of the eyes is essential, because they are a route of entry and because they are so delicate. Irreversible eye damage can begin within a few minutes of exposure to some chemicals. Contact lenses, which may trap and hold chemicals against the cornea, are not suggested, or approved by OSHA, for use in contaminated environments.

Inhalation into the Respiratory System

All the parts of the human respiratory system are vulnerable to inhaled toxic chemicals. If the chemical is in the form of dust or other large particles, it is likely to be stopped in the nose, throat, or upper airway and cause damage there. A study of career firefighters in Seattle, Washington, showed a higher incidence of upper respiratory cancer in this population, which inhales smoke, than in the nonfirefighter population.

Another group of chemicals known to cause upper respiratory problems are those which are highly water soluble and dissolve out of inhaled air onto the damp surfaces there. Gases with high solubility, such as ammonia and hydrogen chloride, act on the upper respiratory tract within seconds and can cause fatal swelling. Moderately soluble gases like chlorine and sulfur dioxide cause both upper and lower respiratory distress within minutes. Low-solubility irritants like ozone and phosgene are more insidious; they may cause the lungs to fill with fluid 6 to 24 hours after exposure, without any irritant symptoms to serve as warnings.

Chemicals which reach the smaller branches of the bronchi or the far reaches of the lungs can cause a variety of health effects, ranging from damage of delicate membranes to lung cancer. Lung linings do not regenerate after traumatic injury, and damaged portions of a lung are no longer functional. The surface area of the lungs of an average 150-pound person has been estimated to be the size of a singles tennis court, and the amount of air inhaled daily is around 430 cubic feet, so the potential for exposure to inhaled toxins is great. Figure 14-4 shows the structure of the lower human respiratory system.

The terminal pockets (called "alveoli") inside the lungs are well adapted to perform their primary chore—the exchange of oxygen and carbon dioxide with blood inside tiny vessels surrounding these pockets. Many inhaled chemicals also cross these thin membranes and enter the blood, to be transported throughout the body.

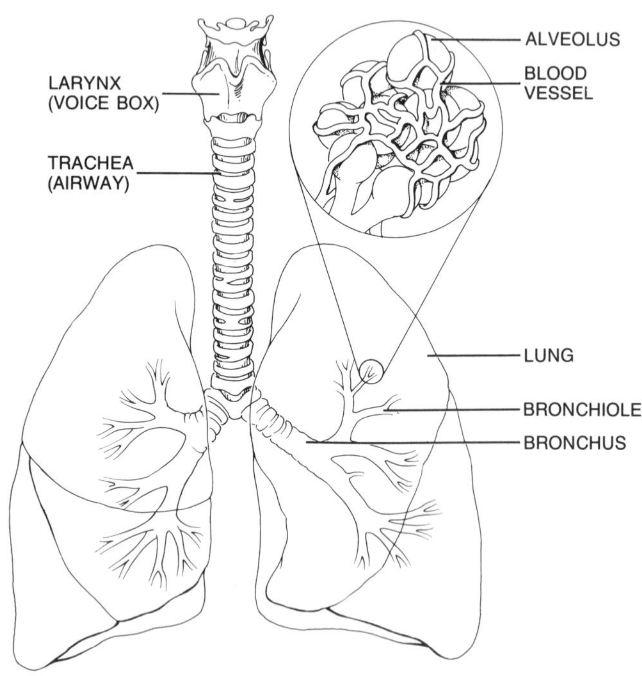

FIGURE 14-4. The human lower respiratory system.

Ingestion Into the Digestive System

Ingested chemicals are swallowed. It is very unlikely that responders will knowingly ingest harmful chemicals during or after an emergency response, so attention to proper protection and personal habits can greatly limit exposure to chemicals by mouth. Avoiding being splashed in the face will help, as will taking care to remove any contamination from the hands before touching the face or mouth. Hands should be washed before drinking, eating, smoking, or applying makeup. Some inhaled chemicals, particularly those in solid particulate form, may be brought up into the back of the mouth by mechanisms in the respiratory system which keep such particles from reaching the lungs. These are reflexively swallowed.

Chemicals which reach the stomach are not detoxified by the body's natural defense system there, since these are designed primarily to kill bacteria. Swallowed chemicals can generally pass through to the small intestine, another body organ with an extremely large surface area (almost the size of a football field) and an excellent system for absorbing ingested materials. This organ is where almost all of the body's digested food is absorbed into the bloodstream for transport to cells throughout the body. Chemicals absorbed here are also available for transport. Figure 14-5 shows the structure of the human digestive system.

You have seen many routes through which chemicals may be absorbed into the blood transport system in the body: across the skin, the linings of the lungs, and the digestive tract surfaces. What happens next depends upon where they go and how the body's natural detoxification mechanisms handle them.

Blood travels everywhere. There are several barriers inside the body through which some chemicals cannot pass, for example the blood-brain barrier, which prevents the entry of some chemicals into brain tissues. Most organs, however, are not protected in any way from materials carried in the blood. All the cells in the body are near enough to the blood supply to be reached by whatever the blood carries, since all cells need oxygen and nutrients to remain functional. It is possible, however, to look up a chemical and find that it has listed "target organs" where damage is most likely to occur. How do these tissues happen to get singled out for harm?

Each cell is bounded by a membrane which surrounds its contents. Cell membranes are selective about what they let into cells, and different kinds of cells let in different kinds of body chemicals. Ovary cells, for example, admit certain kinds of hormones which regulate the daily business of the ovary. Muscle cells admit calcium, and nerve cell membranes allow sodium and potassium to cross when messages are transmitted. The cell

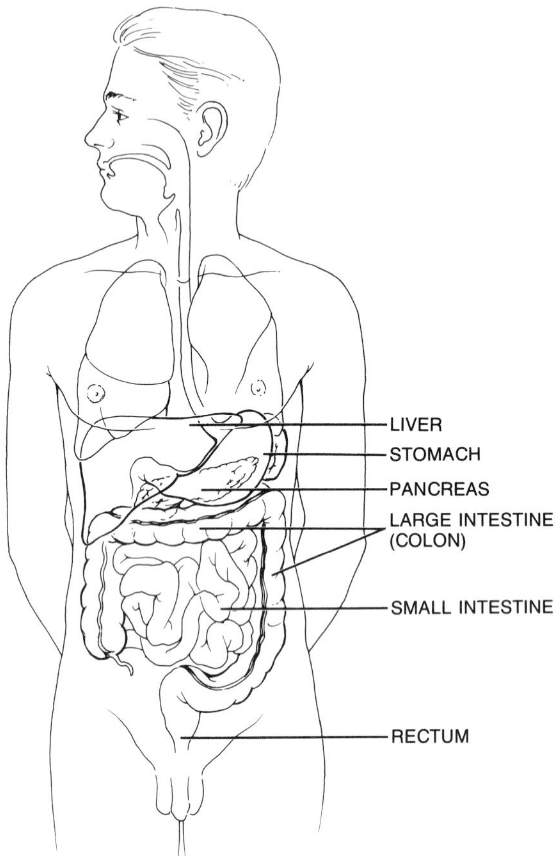

FIGURE 14-5. The human digestive system.

membrane around each of these cells has entry ports which are opened by a sort of lock-and-key system; the entering material is the key which fits the lock on the cell membrane.

Since the shape of the key depends on the chemistry of the entering compound, a toxic chemical which is similarly shaped can also open the lock. In this way, toxic chemicals get inside certain cells and cause damage there. Table 14-7 shows examples of toxic chemicals which mimic the shapes of materials certain cells need and get inside target tissue cells.

The liver is a large organ which has a number of important duties in the body's functional operations. One of these duties is to break apart molecules of nutrients, put them back together in useful forms, and then send

TABLE 14-7. Toxic Chemicals and Their Target Tissues

Agent	Target Tissues
Asbestos	Lungs
Carbon monoxide	Cardiovascular system, lungs, blood cells, central nervous system
Chlordane	Central nervous system, eyes, lungs, liver, kidneys, skin
Methyl ethyl ketone	Central nervous system, lungs
Hydrogen chloride	Respiratory system, skin, eyes
Hydrogen peroxide	Respiratory system, skin, eyes
Hydrogen sulfide	Respiratory system, eyes
Lead	Gastrointestinal tract, central nervous system, kidneys
LP gas	Respiratory system, central nervous system
Nitric acid	Eyes, skin, teeth
Toluene	Central nervous system, liver, kidneys, skin
Mineral spirits	Skin, eyes, respiratory system, central nervous system

Source: Adapted from NIOSH. 1990. *Pocket Guide to Chemical Hazards*. Washington, DC: U.S. Government Printing Office.

them back into the bloodstream and on their way to cells where they are used. The liver produces and uses a variety of certain large protein molecules, called "enzymes," for these operations.

Enzymes are made and preprogrammed to break molecules apart at certain specific spots. If we think of nutrients reaching the liver as a string of different-colored pop beads, we can think of enzymes as being programmed to separate the string between certain color combinations.

For example, Enzyme A may always separate the string between red and yellow adjacent beads. If a toxic chemical containing this same combination reaches the liver in your bloodstream, Enzyme A, which cannot tell toxic chemicals from nutrients, breaks it between the red and yellow beads. There is no way you or your liver can control this enzymatic action; it happens automatically wherever red and yellow beads are connected.

The result of Enzyme A's action may be beneficial or it may be harmful to you. If the broken sections are less toxic to your cells as they float around the body than the original chemical would have been, the chemical has been detoxified to your advantage. Unfortunately, this is not always the case. If the broken bits may combine in different ways with other bits to form chemicals more toxic than the original, you will suffer more serious health effects than if Enzyme A had not done its job.

Exposure to several chemicals at the same time may have unexpected results because of the effect of the detoxifying enzymes. Occasionally these results are advantageous; more often, they are not.

Groups of similar chemicals are metabolized by the same enzyme, so exposure to two similar chemicals at once may result in a shortage of detoxifying enzymes. For example, ethyl and isopropyl alcohols are metabolized by the same enzyme which breaks down chlorinated organic chemicals like chloroform and carbon tetrachloride. Exposure to one of the alcohols together with chloroform or carbon tetrachloride leaves the emergency responder short of the necessary detoxifying enzyme and puts him at greater risk of harmful effects. This interaction is called "synergism." The ethanol exposure may be by vapor inhalation, by skin absorption, or by recent ingestion of alcoholic beverages.

Chemicals which are synergistic with each other are sometimes noted in references materials. Synergism results in harmful effects that are more serious than if the chemicals acted separately. Some off-work factors cause synergistic effects when combined with chemical exposure. Cigarette smoking increases the harmful effects of exposure to a number of chemicals. Alcohol and prescription drug use can also increase damage.

Sometimes a chemical which is normally not toxic becomes so when someone is exposed to it at the same time exposure to another chemical occurs. This happens much the same way in which synergism occurs and is called "potentiation." For example, acetone, which does not cause liver damage by itself, increases the liver damage caused by carbon tetrachloride if an exposure includes both of these chemicals.

Rarely, an unexpected beneficial response occurs following multiple exposures. If one of the chemicals acts to increase the formation and action of enzymes, a lessened toxic response results. This action is called "antagonism." An example of this activity is combined exposure to phenobarbital and benzo(a)pyrene: Phenobarbital increases the body's production of the enzyme which detoxifies benzo(a)pyrene.

Two further body mechanisms may channel toxic chemicals after they leave the liver. If the chemical is water soluble or has been made water soluble by enzyme activities, it will be eliminated from the blood as it goes through the kidney tubules and will be excreted from the body in urine. Other methods of excretion include exhalation of gases dissolved in the blood (we know the odor of alcohol in a drinker's exhaled breath) and excretion of fat-soluble materials in mothers' milk. Nursing infants have been harmed by exposure to pesticides like DDT and hexachlorobenzene in mothers' milk.

Many fat-soluble chemicals seek out body fat cells and are stored there; this continues to be detrimental to the body, since fat cells are recycled continuously, even by people who are not losing weight. Each time the fat is burned for energy, any toxic chemicals stored there are released to circulate through the body and cause problems. Many pesticides are good examples of toxins stored in fat; their fat solubility also allows them to accumulate in animals living in ponds and streams which receive runoff from treated land.

ASSESSING HUMAN EXPOSURE TO CHEMICALS

There are two primary methods which can be used to determine if emergency responders have been exposed to toxic chemicals. These are recognizing the symptoms of exposure, and medical evaluation through monitoring and surveillance.

Recognizing Symptoms

The first and most obvious step in recognizing the symptoms of chemical exposure is knowing what they are. In every hazardous materials incident the chemical must be identified for a number of reasons, not the least of which is the gathering of human exposure data. The agency having jurisdiction should keep records regarding an individual's role and possible exposure in each incident. Once the chemical has been identified, part of the reference material that should be checked is the section describing the symptoms of exposure. The safety officer should inform all responders and medical personnel what these symptoms are so that everyone can watch for them. During the post-incident meeting, responders should be told what long-term symptoms to watch for. Often the accurate attribution of symptoms to their causes is the first indication of exposure to a chemical agent.

Since some physicians are poorly informed about chemical exposure, responders should be prepared to provide them with written information about the symptoms of such exposure and the target organs of incident chemicals.

Medical Surveillance and Monitoring

Medical surveillance is the regularly scheduled examination of any health parameters considered to be important to the job of an emergency responder. Medical monitoring is a more specific, case-related examination of a person who may have been exposed to a toxic agent. In a case where this agent is a chemical, monitoring should be specific to the action of the chemical and what is known about the way the body processes it.

OSHA has written separate standards for a number of toxic chemicals, and each standard includes specifically required medical monitoring.

Some of the chemicals for which standards have been written are asbestos, lead, benzene, and vinyl chloride. These are found in 29 CFR 1910.1000–1101.

Medical monitoring for chemicals which are not regulated by a specific standard can be referenced through other sources of information and consultation with a trained industrial hygienist or physician.

Detoxification mechanisms are well known for some chemicals and poorly understood for others. In the case of the former, it can be determined whether to look for the chemical or its breakdown products (called "metabolites") in blood, urine, or feces, or whether to examine the patient for signs of the effects which may be caused by the chemical. Sometimes, for example in the case of nephrotoxic agents, other signs of organ damage can be assessed. When the mechanism of the chemical is poorly understood, signs of damage may be harder to find and may not be detected until extensive harm has been done.

In all of these situations, responders need to work closely with their physicians and keep them informed of all toxicological information, in written form, which may be related to the exposure.

SUMMARIZING THE STEPS FOR A SAFE (NONTOXIC) RESPONSE

Armed with the knowledge of how toxic chemicals may harm the human body, an emergency responder can follow these steps to avoid or lessen exposure:

- Know what the chemical is, and get good information about its routes of entry and toxicity. Protect those routes of entry.
- Write down the possible exposure level and duration after the incident.
- Look up and write down the acute and chronic symptoms to watch for. Report, and check out, any suspicion that these symptoms exist.
- Take advantage of medical surveillance for annual health assessment and medical monitoring for exposure after each incident where exposure may have occurred.
- Don't be a hero. Your employer does not expect you to risk your health by rushing into a toxic environment without protection.

REFERENCES

Calabrese, E.J. 1987. Animal extrapolation. *Environmental Science and Technology,* 21(7):618–622.

Cashman, J.R. February 1989. *Hazardous Materials Newsletter,* 5.

DOL. 1974. Title 29 CRF 1910 Part 1000–1101.

EPA. 1985. *Principles of Risk Assessment. Report of the Workshop on Risk Assessment.* Washington, DC: Environ Corporation.

FEMA/DOT/EPA. *Handbook of Chemical Hazard Analysis Procedures.* Washington, DC: U.S. Government Printing Office.

Hallenbeck, W.H., and K.M. Cunnginham. 1986. *Quantitative Risk Assessment for Environmental and Occupational Health.* New York: Lewis Publishing, Inc.

Hilyer, B.H. 1990. A carcinogen risk assessment for hexachlorobenzene. Unpublished.

Klaassen, D.C., M.O. Amdur, and J. Doull, editors. 1986. *Casarett and Doull's Toxicology,* 3rd ed. New York: Macmillan Publishing Co.

Levy, B.S., and D.H. Wegman, editors. 1988. *Occupational Health,* 2nd ed. Boston: Little, Brown and Co.

USPHS. 1989. *Draft Toxicological Profile for Hexachlorobenzene.* Washington, DC.: U.S. Government Printing Office.

15

Air Surveillance

Kenneth W. Oldfield, M.S.P.H., and
John E. Backensto

Environmental professionals responded to a call by a hotel chain about complaints of odors by hotel personnel and guests at one of their facilities. The hotel and the surrounding development had been built on the site of a closed petroleum refinery approximately 3 years earlier. Complaints of odors from surrounding areas received local media attention, and the hotel's management took seriously similar complaints among their employees and guests. The complaints followed a period of seasonal rain.

The responders surveyed the lodging suites, using a combustible gas indicator (CGI) and a flame ionization detector (FID) which detects a variety of organic compounds. While most of the suites did not show detectable contaminants, peak readings of over 1,000 units on the FID (calibrated to methane) and 2–3 percent of the lower explosive limit (LEL) on the CGI were taken in a few locations, especially near plumbing and other conduits to the ground. Occupants were moved out of the suites in which contaminants were detected, and periodic monitoring was conducted in all suites.

Air samples, using charcoal tubes, were taken from the units with the highest survey readings and from an area considered to represent background conditions. These samples were sent to a laboratory which had agreed to provide 24- to 48-hour turnaround for analytical results. The samples were analyzed for common petroleum constituents; the total contaminants in any of the samples did not exceed 1 ppm. However, readings of the FID taken concurrently with the air sampling ranged from 10 to 100 ppm. Colorimetric indicator tubes for the above contaminants exposed during air sampling failed to produce a color change, indicating that no contaminant was detected. Both the FID and the air sampling equipment were calibrated before and after use.

The discrepancy in readings and analytical data was later attributed to methane gas emissions that were forced out of the soil by the rising groundwater level after the rain. Methane gas is a by-product of the decomposition of hydrocarbons, and would be detected by the FID but would not adsorb to the charcoal in the air sampling tube. Gas bag samples collected later from an apartment near the site and analyzed in a laboratory yielded methane levels of 4,000 ppm.

This incident illustrates the importance of understanding the use of air surveillance during a response. For example, relying on a single means of monitoring airborne hazards may result in misinterpretation of data or insufficient information to make sound decisions during a response. Therefore, a hazmat responder should be as familiar as possible with the general principles and equipment of air surveillance in order to make quick, sound decisions during an incident. This chapter will describe the principles and equipment used in air surveillance.

OBJECTIVES

- Understand the importance of standard operating procedures for responding to hazardous materials incidents
- Understand the strategies of air surveillance that may be a part of the incident response
- Be familiar with the components, uses, and limitations of common direct-reading instruments (DRIs) and colorimetric indicator tubes for air monitoring
- Understand the concepts of relative response, calibration, and inherent safety as they relate to direct-reading instruments
- Be familiar with the equipment and principles of air sampling
- Be familiar with considerations for selecting air surveillance equipment

STANDARD OPERATING PROCEDURES

Hazmat teams should develop standard operating procedures (SOPs) that outline the steps to be taken in identifying and measuring airborne hazards. Obviously, many things about a hazardous materials emergency cannot be predicted when an SOP is developed, but it should provide general guidance for a potentially chaotic situation. Air surveillance guidelines should be integrated into the team's overall SOP and should reflect the capabilities and limitations of the team's equipment and personnel. (See Chapter 7 for a discussion of SOPs.)

Ideally, the SOP would establish specific procedures for conducting air surveillance during any type of incident (e.g., transportation accidents, leaks/spills in an industrial setting) that may be encountered by the team. This may not be practical for public sector teams called to a wide range of hazardous material incidents, and a general SOP is more appropriate. To the extent possible, specific air surveillance tasks should be listed and prioritized for different site and environmental conditions. The SOP should be flexible enough to allow for the use of judgment at the site while providing consistency and guidance in surveillance procedures.

The SOP should also outline procedures for maintaining the equipment and supplies used in surveillance and calibration of instruments and pumps prior to use. The equipment and instruments that are battery-powered may require periodic maintenance such as charge-discharge cycles. If the equipment is not used regularly, it should be tested and inspected periodically to ensure that it is ready for use. Special maintenance and storage procedures specified by the manufacturer should be referenced or incorporated into the SOP. Sensors, sampling media, colorimetric tubes, and other items with a shelf life or service life should be checked regularly and replaced as needed.

The training of personnel in air surveillance techniques and the use of equipment should be an integral part of the air surveillance SOP. Hazmat personnel who are likely to conduct air surveillance activities should receive training in general principles and procedures of air monitoring and air sampling. Whenever a piece of equipment is acquired, all personnel who may be asked to use it should be fully trained in its operational characteristics and limitations. Most vendors or manufacturers of monitoring equipment can provide excellent training in the use and care of their products. This training should be considered when new equipment is purchased. If the equipment is used only occasionally, periodic refresher training should be given. A hazmat incident is not the time or place to become acquainted with the equipment.

SURVEILLANCE STRATEGIES

In an industrial setting, the identity of the hazardous material may be relatively easy to determine because the list of possibilities is limited. Therefore, before air surveillance is initiated, preliminary information gathered from visual observations, the materials stored in the location of the release, container labels, or other clues should be used to plan the air surveillance activities. If possible, witnesses to the release, plant managers or engineers, or other individuals with information about the identity and specific hazards of the released material (e.g., chemical supplier representatives) should be interviewed. If the chemical or physical hazard can be identified before monitoring begins, valuable time can be saved and the proper equipment can be chosen. If the hazard cannot be identified exactly, preliminary information such as the physical state and approximate size of the release is useful in selecting the appropriate level of protection for personnel entering the site to gather more information.

The first decision to be made in planning the air surveillance strategy is whether air monitoring should be attempted. Air surveillance involves time and effort that, during a response, must be invested wisely. In some cases, the monitoring equipment available to the hazmat team is not appropriate for the hazard (e.g., the instrument will not detect the hazard). For instance, if perchloroethylene, a noncombustible liquid, is spilled in an open area and the only monitoring equipment available to the team is a combustible gas meter, the time and effort invested in air surveillance could be better used to control the release. Also, if the released chemical is identified and judged unlikely to generate high concentrations of vapors, it may be decided that air monitoring is not necessary. This decision must be made only by professionals with a thorough understanding of the chemical and physical properties of all of the chemicals involved.

Once the decision to monitor for airborne hazards has been reached, the strategy for monitoring must be established. Preliminary information about the nature of the hazard involved in the incident should be reviewed to determine what information is missing in order to evaluate effectively the potential risk to the health and safety of the responders and the public. Objectives should be established that provide the specific information needed.

As discussed previously, general monitoring procedures should be addressed in SOPs for the response team before the incident occurs. During the incident, the need is to determine how to adapt the SOPs to the specific conditions of the site. The air surveillance strategy may involve the types of monitoring discussed in the following sections.

Perimeter and Background Survey

As discussed in Chapter 6, the first responder to a release should establish an isolation perimeter large enough to include the hazard area and an additional buffer zone. Before response personnel enter the hazard area, an inspection of this area should be conducted from the isolation perimeter to gather information about the identity and extent of the hazard. This survey should use air monitoring equipment to detect air contaminants in two specific areas—upwind and downwind of the hazard area.

Upwind monitoring will provide data on background levels of contaminants from sources away from the site of the incident, such as emissions from other processes in the plant or sources located on nearby property. Background levels of contamination may be compared to readings taken on site to determine what portion of the air contamination originated with the released material.

Downwind monitoring can be used to determine if the isolation perimeter contains all of the hazard area. In other words, if air contamination greater than background levels is detected downwind of the hazard area, the isolation perimeter should be expanded to prevent exposure to unprotected people.

Air monitoring around the entire perimeter may not be necessary because most of the chemical contamination will be carried by the wind in a plume. The upwind and downwind monitoring should be sufficient to detect the plume. However, the release areas closest to the perimeter should be monitored regardless of the wind direction. Also, if radiation hazards are suspected, the entire perimeter should be monitored, since radioactive emissions are affected only minimally by air movement. Perimeter monitoring should be included in the periodic monitoring described later in this chapter.

Initial Entry

In order to monitor conditions in the actual hazard area, an initial entry must be made. The objectives must be clearly outlined and communicated before the entry so that the team can conduct the survey efficiently and leave the hazard area quickly. The entry objectives for air surveillance may include

- Establishing that airborne hazards exist or potentially exist at the site
- Locating and delineating areas of high air concentration of the released material(s)
- Verifying preliminary or existing information with respect to the nature of the release
- Establishing boundaries for the site control zones based on visual observations of the current location and potential movement of the released materials
- Collecting information related to the specific protective measures and equipment required for response personnel
- Collecting information useful in choosing response actions

Preliminary information should be used to determine what specific hazards to monitor during the entry. If the identity of the hazardous material is known, DRIs or colorimetric indicators capable of monitoring the material should be selected. If the material cannot be specifically identified, the hazard it represents can be classified in one or more of the following groups:

- Combustible gases or vapors
- Oxygen deficiency or enrichment
- Toxic gases, vapors, or particulates
- Radioactivity

Air monitoring equipment capable of monitoring one or more of these hazards may be selected to provide general, initial detection of the possible hazards.

The initial entry should be a brief survey of the hazard area to identify and outline areas of high air concentration or other immediately dangerous to life and health (IDLH) conditions such as confined spaces or potentially explosive environments. The team should note any visual indicators of the existence of a hazard such as pooled liquids, discoloration or degradation of the ground surface, condensed vapor clouds, peeling or staining of painted surfaces, and the location of any victims.

If the approximate location of the release is known, the entry team should approach from the upwind direction and monitor the air on a continuous basis. Since chemical air contaminants are typically dispersed by wind, approaching the incident from upwind should allow the team to approach the release without entering the area of highest air contamination. Continuous monitoring is especially important if radiation hazards are suspected to be present, since the release of radioactivity is not significantly affected by wind direction.

Site conditions may be used to prioritize the hazards to be monitored if the chemical has not been identified. For instance, in areas with restricted air movement, the presence of an explosive and/or oxygen-deficient environment may pose the most immediate threat to the entry team. Because the entry team should be using at least Level B protection if the hazards are unidentified, the investigation of toxic atmospheres might be conducted after these hazards have been assessed and determined not to pose a life-threatening situation. On the other hand, if the area around the release is open and seems to be well ventilated, the accumulation of sufficient explosive vapors may be less likely and monitoring may focus on the comparatively lower-concentration toxic atmospheres.

The actual use of monitoring equipment should follow the general guidelines set forth in the team's SOPs. It is important to remember that some air contaminants are heavier than air and may accumulate near the ground. Air monitoring should be conducted near the ground surface and at the level of the breathing zone. Monitoring should emphasize areas where air movement is reduced, such as diked areas, ditches and low areas, buildings, or confined spaces, since vapors tend to accumulate in these areas. If radiation is being monitored, any debris or other materials that may be covering (i.e., shielding) the radioactive source should be monitored without being moved, as this may expose the team to the unshielded radioactivity.

The team should periodically test to ensure that the instrument is functioning properly by using procedures recommended by the manufacturer. These procedures may include visual indicators such as lights or audible signals that sound at regular intervals or special alarms that sound in the event of instrument failure. Alarm sounds or lights that warn of hazardous conditions should always be functional and adjusted to the appropriate limits for the hazard of concern (e.g., below 19.5 percent oxygen).

Periodic Monitoring

Air surveillance is an ongoing activity that does not end with completion of the initial entry survey. Even if the air concentrations in the hazard area are well defined, the changing nature of emergency response activities and site conditions makes periodic monitoring necessary. The objective of this type of monitoring is to detect changes that may positively (e.g., consistently lower concentrations allowing a downgrade to a lower protection level for responders) or negatively (e.g., detection of air contaminants leaving the hazard area, leading to an enlarging of the hazard zone) affect the emergency response activities. Some factors which may cause a change in air concentration during an emergency response include

- Response actions such as accidental releases while moving drums or containers, handling spilled materials, or conducting containment or confinement procedures
- Changes in weather conditions such as rain, decreased cloud cover leading to hotter conditions, changes in wind direction and speed, or sudden changes in atmospheric pressure
- Accumulation of spilled material due to the continued release before containment procedures can be completed

Periodic monitoring may take the form of regular surveys of all or part of the site similar to the initial entry survey, or it may be performed as area monitoring at a fixed position near the area of highest hazard. Fixed-area monitoring is not practical if only one instrument is available and a large hazard area may be involved. Survey-style monitoring dedicates at least one person to conducting the survey at the expense of other activities in the response. The frequency and method of monitoring should be site specific and based on the degree of hazard present and the potential for changing conditions to affect the air concentrations of contaminants.

Termination Monitoring

Emergency response termination procedures described in Chapter 7 may include using air monitoring data as criteria for ending response activities. Air contamination may remain in the hazard area even after the source of the contaminant has been contained or confined (see Chapter 11). Air monitoring should be performed throughout the hazard area to confirm that all spilled material or sources of contamination have been contained. Also, prior to incident termination, air contamination should be reduced to background levels. At that time, normal traffic (nonresponse personnel and vehicles) is allowed into the response area. This is especially true of incidents in confined spaces or areas of poor ventilation where hazardous gases or vapors may accumulate.

Air Monitoring Versus Air Sampling

Air surveillance may be accomplished by air monitoring with direct-reading instruments (DRIs) or colorimetric indicators for real-time measurements and/or air sampling. These two methods are not synonymous, and each has advantages and limitations. During a fast-paced emergency response, the need for rapidly available information is critical. Air monitoring with DRIs has the advantage of providing exposure data almost instantly. However, speed has its price. Most DRIs are not truly selective, but will detect a variety of chemicals and provide the total amount of all detectable substances present. Colorimetric indicators can provide concentration data on specific chemicals, but environmental conditions in the field can affect the accuracy of the results.

Air sampling is a more accurate means of measuring air contaminants and can often identify specific constituents of a mixture or an unknown substance. Unfortunately, air sampling takes valuable time, often requiring days or weeks to get the results. Also, it can be quite expensive, especially if a rush analysis is required. On-site analytical equipment can reduce the turnaround time, but can be expensive and requires experienced operators to provide meaningful data.

The circumstances of the incident, the variety of available air surveillance equipment, and the competence of the response personnel are all factors affecting the method or combination of methods selected during an emergency response. The following sections discuss equipment that may be used to conduct air surveillance.

DIRECT-READING INSTRUMENTS

While DRIs are usually not complicated to operate, the user must know more than just how to turn the unit on and take a reading. All DRIs have limitations and unique operating characteristics that must be understood for a correct interpretation of the data. For example, combustible gas indicators (CGIs) require sufficient oxygen in the atmosphere to support combustion in the detector. In low-oxygen conditions (<19.5 percent oxygen in air), the CGI may show a lower combustible gas level than is actually present. A user who is not aware of this limitation may have a false sense of security regarding the flammability hazard based on inaccurate data. Therefore, familiarity with the instrument is necessary to operate it correctly and interpret the results.

Components of DRIs

DRIs typically consist of a sensor, and amplifier, a display, and possibly a pump. The sensor is that part of the instrument that actually detects the presence of a contaminant in the air by means of physical principles related to the chemical or physical properties of the contaminant. Although each chemical has a unique set of properties, the sensor typically responds to general properties or principles that are common to a wide variety of chemicals. For example, CGIs detect combustion, which may be caused by methane or xylene. Sensors are unable to distinguish between different chemicals that cause this response. For this reason, DRIs are not able to identify unknown chemicals.

The sensor generates an electrical signal which is proportional to the concentration of the chemical in the air. In most cases, the signal produced by the detector is a small electric current, insufficient to produce a meter reading or display. For this reason, DRIs include an amplifier that receives the signal from the detector and sends the appropriate electric current to the display or readout. In instruments that can be field calibrated, the amplifier is where the adjustment is made. This will be discussed in the later section on calibration.

Even the most sophisticated and amazing instrument is useless if the data it generates are not communicated to the user. It is the function of

TABLE 15-1. Common Instrument Displays

Instrument Type	Units	Range
Combustible gas indicator	Percent (%) of LEL[a]	0–100%
Oxygen meter	% Oxygen in air	0–25%
Toxic gas meters (PID, FID, etc.)	Parts per million (ppm) in air[b]	0–1000 ppm
		0–2000 ppm
Radiation detectors	milliRoentgens/hour (mR/hr)	0–50 mR/hr

[a] LEL = lower explosive limit.
[b] IR Spectrophotometers may report data as absorbance units, which can be converted to ppm with information provided by the manufacturer.

the instrument display to communicate the data in a useful form. Typically, DRIs have either an analog (needle) or a digital display. The data should be displayed in units appropriate to the hazard being monitored. Table 15-1 describes some common instrument types and the display units.

Some instruments are capable of measuring over a wide range of concentrations. To accomplish this with an analog display, the display is divided into the appropriate increments, and there is a switch which changes the value of each increment. Figure 15-1 shows a meter display with two types of switches designed to provide both a wide range of measurements and sensitivity to change at low concentrations. In both meters, the first switch position allows more precise measurement but is limited to a small range of concentrations (0–5 ppm). The third position allows measurement that is less precise but covers a wider range of concentrations (0–500 ppm).

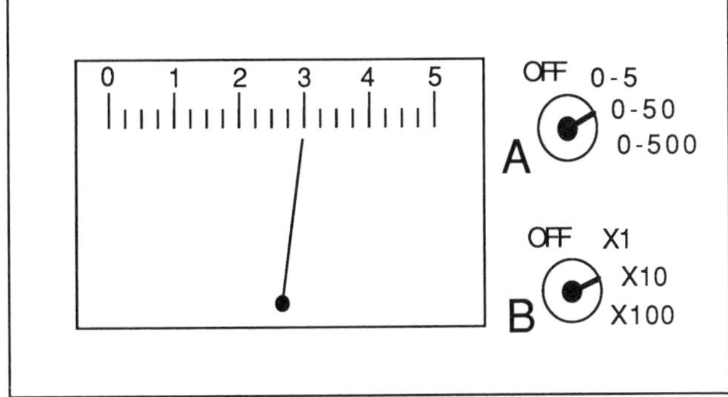

FIGURE 15-1. Instruments may have multiple concentration ranges.

If the range of concentrations is not known, the user should start with the widest range (the third position) and, as possible, switch back to smaller, more precise ranges (the second and then the first position). If a high concentration causes the meter set to the narrow scale (the first or second position) to go to full scale, the needle must recoil to provide the appropriate reading on the wider scale (the third position). During this recoil time, a peak reading may be missed. This recoil time is eliminated if the meter is set on the broadest range.

Many instruments contain a pump that will draw the air sample into the instrument detector. The pump will allow the instrument to be used for remote sampling if a length of tubing is attached. Instruments without a built-in pump may provide a bulb aspirator or other mechanical means of drawing air into the sensor area. The alternative is to allow the air to diffuse naturally into a sensor area located on the exterior of the instrument. Instruments designed to monitor the atmosphere continuously and give an alarm when dangerous concentrations exist often use this system to conserve battery power.

Commonly Used Instruments

Combustible Gas Indicator (CGI)

The CGI uses a hot-wire or catalytic sensor to detect the presence of flammable gases or vapors. A catalytic filament carrying an electric current burns flammable or combustible gases or vapors in a sample chamber. An identical filament which is isolated from the contaminated air carries the same current. The heat from the combustion in the sample chamber causes a change in the current passing through the filament. The difference in current between the two filaments is detected in a bridge circuit which produces the signal.

The CGI displays the data as a percentage of the LEL of the calibration gas. A full-scale reading (100 percent LEL) means that the minimum concentration of gas necessary to support a flame is present. The reading should not be interpreted as a direct reading of the concentration of the gas, but rather as the percentage of the LEL of the calibration gas. The actual concentration can be approximated by multiplying the meter reading (as a decimal value) by the LEL of the gas (usually a percentage) if the instrument is calibrated to the gas being measured. For gases other than the calibration gas, the meter reading must be converted first by the use of a relative response factor, as described later in this chapter.

If a flammable gas is present in air concentrations less than the LEL, the CGI will provide a reading of the concentration as a percentage of the LEL. In conditions within the flammable range of a gas (concentrations between the lower and upper explosive limits), the needle of an analog meter will remain above the 100 percent LEL mark. If the upper explosive limit (UEL) is exceeded, the needle may go to full scale and then quickly drop to zero. This movement could be missed if the operator is not looking at the instrument display at the time. Therefore, many instruments have an alarm system that must be reset by the operator each time the set point is exceeded, so that temporary peaks or concentrations over the UEL are not overlooked.

The CGI is designed to detect the potential for a flammable or explosive atmosphere before one actually exists. That is why a full-scale reading (100 percent LEL) represents only the minimum concentration necessary to support a flame. Response actions recommended by EPA are triggered at readings much lower than the LEL to provide a safety factor. These guidelines are:

- <10 percent LEL: continue work with caution
- 10–25 percent LEL: continue work with continuous monitoring
- >25 percent LEL: explosion hazard; withdraw from the area

The hot-wire detector is not selective and will only detect combustible or flammable gases or vapors, not particulates. Some chemicals burn hotter than others, so the sensor responds differently to the various chemicals. Therefore, a relative response factor must be used if a chemical other than the calibration gas is measured. The detector requires oxygen for the combustion process, so it may not function properly in an oxygen-deficient atmosphere.

Oxygen Meters and Electrochemical Sensors

Oxygen meters typically use electrochemical sensors which involve a chemical reaction between a chemical substrate in the sensor and a contaminant in air. The sensor chemical is contained behind a semipermeable membrane that allows the oxygen to diffuse into the solution. The reaction produces a small electric current which is collected by an electrode and is proportional to the concentration of the contaminant in air. By using different chemicals in the sensor, instruments can be made selective for certain other chemicals such as carbon monoxide, hydrogen sulfide, and chlorine.

The normal concentration of oxygen in air is approximately 20.7 percent. If the concentration detected during an incident is less than 19.5 percent, the atmosphere is considered oxygen deficient and should not be entered without atmosphere-supplying respiratory equipment such as an SCBA or a SAR. Atmospheres containing more than 25 percent oxygen are considered oxygen enriched and represent an increased combustion hazard.

The chemical reactions of electrochemical sensors may be affected by changes in temperature, humidity, and atmospheric pressure. Other chemicals may interfere with the action of the sensor, creating a signal when the contaminant is not present. Electrochemical cells may deteriorate or be exhausted and must be replaced periodically. High concentrations of carbon dioxide and carbon monoxide will shorten the service life of oxygen sensors.

Photoionization Detector (PID)

This sensor uses ultraviolet (UV) light to break the contaminant into ions which move to electrodes and create an electric current. The PID will detect many organic and some inorganic compounds at concentrations ranging from 1 ppm up to 1,000–2,000 ppm. The instrument is used to monitor toxic atmospheres in which low concentrations (parts per million range) can be harmful.

Though PIDs are not selective (i.e., will detect a wide variety of chemicals), some selectivity may be achieved on the basis of differing ionization potentials. The strength of the UV light necessary to ionize a compound (known as its ionization potential, or "IP") differs for various chemicals and is reported in units called electron volts (eV). For example, the ionization potential for 1,1,1-trichloroethane is 11 eV. If a 9.7-eV lamp is used with the PID, most of the 1,1,1-trichloroethane present in the environment will not be detected, while a chemical with an IP of 9.7 eV or less would be. Thus, the PID becomes selective for chemicals with an IP less than the strength of the UV lamp used. (Some ionization does occur at strengths less than the IP of a compound, but it is usually a very small fraction of that which occurs at or above the IP.)

High humidity and high electromagnetic energy may interfere with the instrument's response. If the UV lamp is dirty, the UV light will be blocked and unable to ionize the chemical, causing the instrument to give an inaccurate reading. Charged particles in the air other than the ions of the contaminant (e.g., dusts, particulates from diesel engines) will collect on the electrode and cause a false reading.

Flame Ionization Detector (FID)

The FID uses a hydrogen flame to burn the contaminant, resulting in the release of charged ions. As with the PID, the ions are collected on electrodes and cause an electric current signal. The FID will ionize organic compounds with an IP of 15.4 eV or less. It is not significantly affected by humidity and can measure over a wide range of contaminant concentrations.

The FID detects only organic compounds. Because the flame requires oxygen, the unit may not function in oxygen-deficient atmospheres. Impurities in the hydrogen used as fuel will be ionized and detected just like contaminants in the air, causing a false reading. As with the PID, other charged particles may collect on the electrode, interfering with the instrument readout.

Infrared (IR) Spectrophotometer

The IR sensor takes advantage of the fact that chemicals will absorb infrared light at certain discrete wavelengths. Therefore the IR sensor can be fairly selective (i.e., monitor one chemical at a time) on the basis of the wavelength of IR light used. The spectrophotometer consists of twin parallel chambers through which IR light is projected. Some of the light in the sample chamber is absorbed by the contaminant molecules, and the remainder is transmitted. The instrument displays the difference in the amount of IR light transmitted in the chamber either as the percent transmitted (%T) or as absorbance units (AU). The two units are related (inversely and directly, respectively) to the amount of contaminant in air and can be compared to data provided by the manufacturer to calculate the concentration expressed as parts per million. Some units provide microprocessors to make this calculation automatically for selected chemicals.

Many compounds absorb IR light at the same or nearly the same wavelength; therefore, interference may occur when multiple chemicals are present. In other words, a signal indicating absorption at one wavelength cannot be attributed definitively to a single chemical without other supporting evidence. Also, these instruments often require flat, level surfaces free of vibration or movement for proper operation.

Using Toxic Exposure Data

Results obtained from the PID, FID, IR spectrophotometer, or other instruments monitoring exposure to toxic chemicals should be compared to standards for the specific chemical present if the chemical has been identified. It is important to compare the results to the right standard for the kind of data generated. For instance, OSHA's time-weighted average permissible exposure limit (TWA-PEL) is intended to protect workers who are exposed to the chemical for 8 hours a day, 5 days a week over a normal working life. Since the emergency responder will be exposed to the chemical for only a brief period during a hazmat incident, exposure to a higher concentration may be acceptable.

Selection criteria for PPE may be based on the results of DRIs. If the chemical is identified, the selection is straightforward. If the chemical is unknown or if a mixture is being monitored, the reading should be considered a total reading, and the appropriate level of protection must be based on different criteria. EPA has established selection criteria for levels of protection based on total readings; these are discussed in Chapter 9.

Radiation Detectors

Radioactivity can be either in particle (alpha, beta) or wave energy (gamma, X ray) form. When radioactivity encounters molecules, it causes the molecules to become ionized and/or to release energy in the form of heat or light. Radiation detectors containing an ionizable detection medium are commonly used. The ionization of this medium by radioactivity is quantified and a signal is produced. Table 15-2 lists some common radiation meters.

Because of its ability to detect both beta and gamma radiation at relatively low levels (<1 mR/hr), the Geiger counter is often used to survey when unidentified sources may be present. However, the Geiger counter is not an effective instrument for measuring the amount of radioactivity of beta sources. Ion chamber instruments are also used as radiation meters, though they may not detect low-level radiation and therefore may not be effective as survey instruments for unidentified sources. If alpha radiation is known or suspected to be present, an alpha detector should be used during the survey.

Background levels of radiation should be monitored away from the site of the incident. Some radiation occurs naturally in geologic formations in certain parts of the country. Readings of radiation taken on site with detectors should be compared to the background readings. If the level on site is more than twice the background level, a radioactive source is likely present and response personnel should withdraw from the hazard area.

TABLE 15-2. Common Radiation Detectors

Detector	Radiation Type
Ion chamber	Beta, gamma, x-ray
Proportional counter	Alpha, beta
Geiger counter	Beta, gamma
Scintillation counter	Alpha, beta, gamma

Response activities should proceed only under the direction of qualified radiation experts.

Typically, instruments are designed to detect specific types or levels of radiation. Radiation can be shielded from the instrument during the initial survey but may be uncovered during later response actions, causing unexpected exposure to response personnel. Also, the distance between the detector and the source may result in a lower reading.

Gas Chromatography

Identifying components of an unknown mixture is a difficult job at best. One technology that may be used to help in this process is gas chromatography. A gas chromatograph (GC) is not a detection device but is useful in separating out components of a mixture or, in some cases, identifying an unknown compound. Some instrument manufacturers have coupled this technology with detectors such as the PID or FID in field instruments. These instruments can operate in the survey mode (detector only) or the GC mode (GC and detector). Because of the time requirements of operating in the GC mode, these instruments cannot simultaneously survey an area and separate the components of a mixture with the GC. Nonetheless, the GC can be a valuable tool for identifying and quantifying the components of a mixture.

The GC is essentially a long, thin tube (usually coiled) which contains a medium. Contaminants are introduced into the column in a carrier gas. The contaminant adsorbs to the medium with different strength bonds and subsequently is released to flow out the other end of the column into a detector (e.g., FID, PID). The detector is usually attached to a chart recorder that records the concentration detected over time. Based on the strength of the bond, column temperature, nature of the contaminants, and so on, all of a particular chemical or group of chemicals will leave the column at the same time, and this time will be different from that of other chemicals.

The length of time from injection of the chemical into the column until it exits the column and is detected is called its "retention time." The retention time for a chemical under specific column conditions is constant and will vary from chemical to chemical. The temperature of a column greatly affects the retention time of a chemical. Put simply, the individual chemicals in a mixture injected at one end of a GC column will exit the column at different retention times, but all of each chemical will come out at approximately the same retention time. The chart recorder will show a peak for each chemical at the appropriate retention time. The area under the peak is proportional to the amount of the contaminant present.

The practical use of this technology is that if column conditions (temperature, medium type, carrier gas flow rate, etc.) are kept constant and the retention times of individual contaminants in a mixture under those conditions are known (two very big "ifs"), the quantity of each chemical can be determined as it is detected by the detector, and, based on its retention time, it can be identified. Figure 15-2 shows a chromatogram.

The technology is useful but does have limitations and drawbacks. Obviously, the person using a GC must be well trained. Many things must be controlled. Calibration standards of each chemical to be detected must be run through the GC under specific conditions to confirm the retention times and response of the detector to the chemical. If the response team has a large variety of potential contaminants and can afford the equipment and training costs to operate a field GC, it can quickly provide valuable information about specific contaminants.

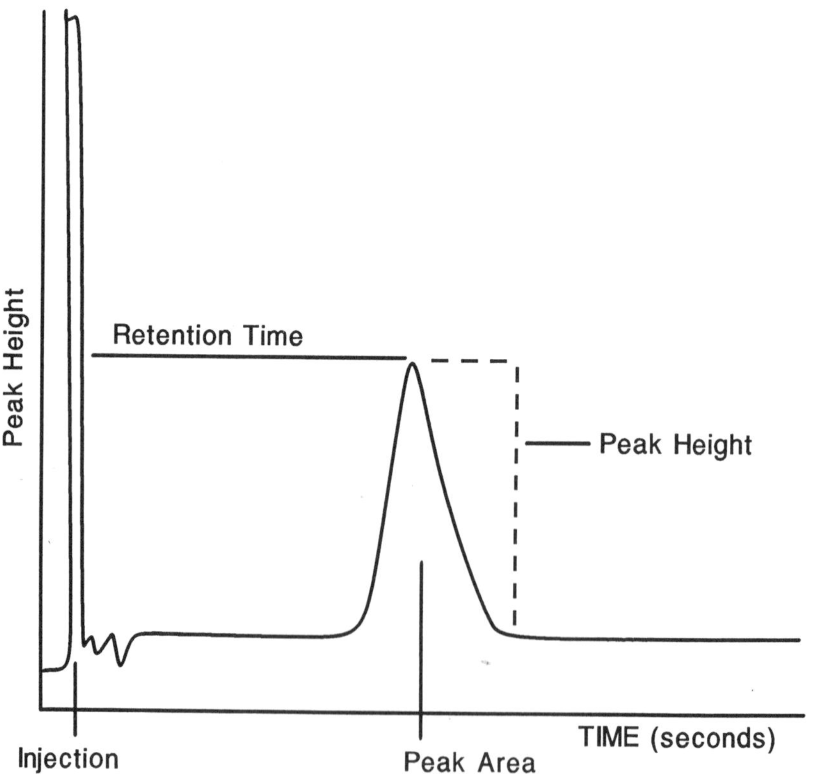

FIGURE 15-2. The GC shows the response of the detector over time.

Relative Response

In nearly all DRIs, the meter response is calibrated to one specific gas. If the instrument is used to detect a gas or vapor different from the calibration gas, the reading should be related to the calibration gas. The numerical value of the reading may under- or overstate the concentration of the contaminant gas, depending on how the instrument responds to that gas. The relationship of the contaminant gas to the calibration gas is referred to as the "relative response." Relative response represents the ratio of the meter reading to the actual concentration of the contaminant and is usually expressed as a percentage. For instance, the Foxboro OVA is calibrated to methane. The manufacturer's information states that the relative response to allyl chloride is 50 percent (.50). Therefore, a meter reading of 100 units in an atmosphere of allyl chloride would represent an actual concentration of

$$\text{Actual Concentration} = \text{Meter Reading} \div \text{Relative Response}$$
$$= 100 \div .50 = 200 \text{ ppm}$$

The OVA would understate the actual concentration of allyl chloride by half. As a rule, a relative response of less than 100 percent means that the meter reading will be less than the actual concentration present, and a relative response greater than 100 percent means that the meter reading will be greater than the actual concentration present.

The relative response may be plotted as a curve, comparing the meter reading to the actual concentration (Figure 15-3). These curves are created by the manufacturer by taking meter readings at several different actual concentrations. The manufacturer may be able to provide relative response factors or curves for specific contaminants that are identified at a hazmat incident.

Calibration

DRIs have electronics, sensors, and often moving parts that may wear down, become dirty, or stop functioning properly. The net result is that the instrument may no longer give accurate readings. If the sensor is detecting only part of what is present or the amplifier is only partially amplifying the signal, low readings will be given by the instrument even if the actual concentration is high. Obviously, this could have disastrous results. For this reason, DRIs must be calibrated or have their calibrations checked regularly. Calibration is the process of adjusting the instrument readout to correspond to a known concentration of a gas.

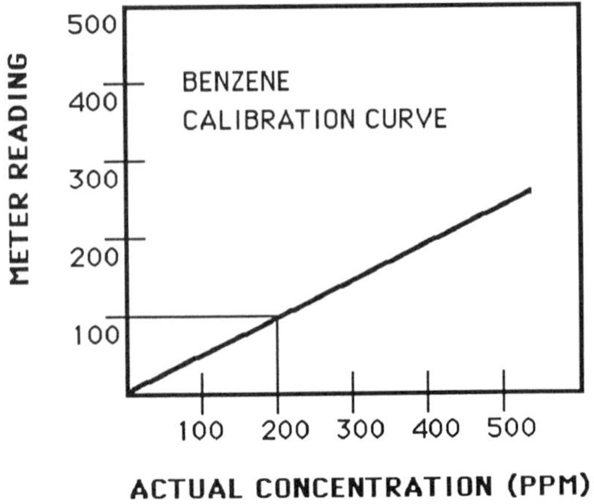

FIGURE 15-3. Relative response curves show the meter's response to a gas compared to the actual concentration of a gas.

Ideally, each instrument should be calibrated before and after each use. As mentioned previously, the manufacturer of an instrument will recommend a specific gas that should be used to calibrate it. For instance, the Foxboro Century OVA is calibrated to methane, and the MSA Model 260 combination (combustible gas and oxygen) meter is calibrated to pentane. The H·Nu PI-101 is factory calibrated to benzene, but the manufacturer recommends that field calibration be done to an isobutylene standard.

If an instrument is to be used for an extended period to monitor an atmosphere other than the calibration gas, it may be helpful to calibrate the instrument to respond directly to that gas. This can be done by adjusting the instrument readout to correspond to a known concentration of the suspected contaminant. Specialty gas suppliers recommended by the instrument's manufacturer may be able to supply calibration gases for a variety of contaminants.

Two general types of calibration are "zeroing" the instrument and adjusting the span potentiometer. The first adjustment simply tells the instrument what is considered baseline or zero contamination and is analogous to adjusting the idle on a car. The zero adjustment can be used to eliminate readings caused by electronic "noise" in the instrument or background levels of the contaminants. The second adjustment sets the amount of amplification that the instrument should give a signal from the sensor so that the readout corresponds to the actual concentration. The meter reading should be adjusted to match the actual real-time concentration.

DRIs may provide one or both of the above adjustments for the user. It is important to follow the manufacturer's instructions in the operation and calibration of the instrument. Some instruments are not designed to be calibrated by the user but should be sent back to the manufacturer for this purpose. Even in these cases, however, the user should be able to check the instrument's response against a known concentration of a calibration gas.

Inherent Safety

DRIs must be safe to use even in hazardous environments. Because an instrument contains electronics and possibly other ignition sources, it must be constructed in such a way that it will not cause ignition of an explosive atmosphere. The National Electrical Code of the NFPA describes minimum criteria that an instrument must meet to be considered "intrinsically safe." Instruments are typically tested by the Underwriters' Laboratory (UL) or Factory Mutual (FM) and must be marked to indicate the hazardous atmosphere for which they are certified. The code classifies hazardous atmospheres by class, group, and division. Class and group are used to describe the type of flammable material present as follows:

- *Class I* is flammable vapors and gases, and is further divided into groups A, B, C, and D based on similar flammability characteristics. Examples include gasoline and hydrogen.
- *Class II* is combustible dusts and is divided into groups E, F, and G. Examples include coal, grain, and metals such as magnesium.
- *Class III* is ignitable fibers such as cotton.

Divisions are used to describe the likelihood that the flammable contaminant will be present in a concentration sufficient to pose an explosion or combustion hazard. *Division I* atmospheres are considered most likely to contain the hazardous substance in flammable concentrations. *Division II* atmospheres have flammable or combustible substances present, but they are typically handled or contained in closed systems that are not likely to generate hazardous concentrations under normal conditions.

A typical marking on an instrument is that it is "intrinsically safe for Class I, Division I, Groups A, B, C, and D as approved by FM." This means that the instrument can be used in an atmosphere which potentially contains flammable concentrations of combustible or flammable gases or vapors. Approval of an instrument for use in one hazard class does not mean that it can be used in all hazard classes. This approval assumes that the instrument will be used according to the manufacturer's directions and that it is not modified by the user.

COLORIMETRIC INDICATORS

Colorimetric indicators provide a means of quantifying air contamination with a reasonable degree of selectivity. Three types of indicators may be used: liquid reagents, chemically treated papers, and glass tubes containing chemically treated solids. The principle of operation is that a contaminant in air reacts with a chemical reagent to cause a change in color, which is proportional to the amount of contaminant in the air. Liquid reagent devices are available that produce color changes that are fairly easy to observe. But handling liquids in the field may be awkward and inconvenient. Chemically treated papers are easy to use but do not typically have a means of controlling the volume of air contacting the paper. Therefore, their accuracy may be affected by factors such as the amount of air movement over the paper.

Because of their ease of use and quick results, glass tubes containing chemically treated solids—known as "colorimetric" or "detector tubes"—are popular. They typically consist of a glass tube containing a granular carrier solid which has been impregnated with a specific chemical reagent. As contaminated air passes through the tube, the contaminant reacts with the reagent on the carrier solid to produce a color stain. The air is drawn through the tube by either a hand pump or a battery-powered pump. Thus, the volume of air sampled can be controlled. It should be noted that detector tubes are designed and calibrated to be used exclusively with the manufacturer's pump. To use a detector tube with another manufacturer's pump, even if the pump volume is the same, can lead to significantly inaccurate results.

Operation of Colorimetric Indicator Tubes

The tubes may operate in one of three ways (Figure 15-4). First, the pump may be operated until the length of the stain reaches a certain preset point. The number of pump strokes (i.e., the volume of air) required to reach this full stain is compared to a chart to determine the concentration of the contaminant in air. In this case, a high concentration of contaminant in air would require fewer strokes to reach the full stain. Second, a set number of pump strokes (volumes) of air will be drawn through the tube, and the length of the stain will be compared to a calibration scale, often printed on the tube, to determine the concentration. For a set number of pump strokes, a high concentration will cause a longer stain. Third,

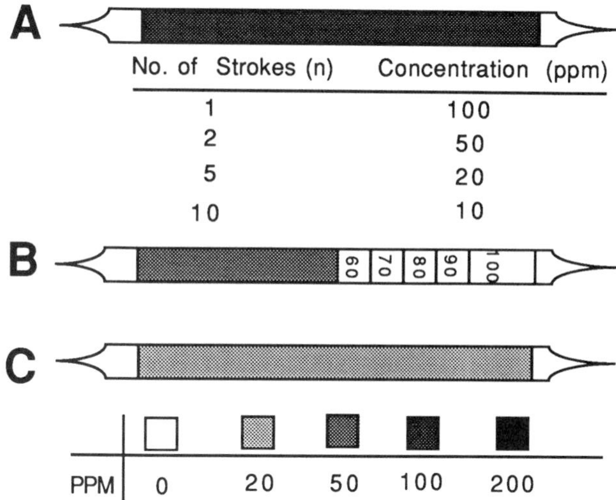

FIGURE 15-4. Colorimetric indicator tubes show results in one of three ways.

a predetermined number of pump strokes of air are drawn through the tube, and the degree or tint of the color change is compared to a chart to determine the concentration. For these tubes, a high concentration causes a deeper or darker color change after a set number of pump strokes. It is critical that the operator of the tube be familiar with the manufacturer's directions and know which mode of operation is used.

Sample pumps must be checked for leaks to ensure that the appropriate volume of air is drawn through the tube. Also, the pump must be allowed to complete every pump stroke. Incomplete strokes or leaks in the system will cause less than the appropriate volume of sample air to pass through the tube, potentially resulting in a lower reading than is actually present.

The tubes may be specific for a certain gas or vapor or may detect groups of chemicals, such as alcohols or aromatic hydrocarbons. Tubes designed to detect one chemical will usually react with certain other chemicals (known as "interferences") to produce a similar color change. The manufacturer will provide a list of known interferences. In general, detector tubes provide an opportunity to make a qualitative and quantitative determination of the presence of a particular chemical.

Limitations of Colorimetric Indicator Tubes

Some limitations must be considered when detector tubes are used. Detector tube systems have rather poor accuracy, with errors ranging from

25 to 50 percent for many tubes. NIOSH tested and certified detector tubes at one time, but it has since discontinued this practice. Manufacturers generally provide information about the accuracy of the tubes with the instructions.

Because a chemical reaction is involved, detector tube accuracy may be affected by such factors as temperature, humidity, and atmospheric pressure. Where temperature significantly affects the performance of the tube, the manufacturer will include compensation factors in the instructions. Colder temperatures will usually slow down the reaction, so if detector tubes are to be used in cold weather, they should be stored in a warm place and carried next to the body. High temperatures may affect the rate of the chemical reaction and may reduce the detector tube's shelf life, as described below.

One source of error in the use of colorimetric tubes is the visual interpretation of the length or degree of color change. The leading edge of the stain may be uneven or may be lighter than the rest of the stain, calling for a judgment on the part of the operator as to what constitutes the end of the stain. The same difficulty applies to judging the degree or tint of color change even when comparison charts are provided. When in doubt, it is advisable to use the most conservative (i.e., highest) reading so that more protection is provided to the responder and the public.

Detector tubes have a specific shelf life. Chemical reagents will deteriorate over time, even if the tube is not opened and exposed to air. Also, high temperatures may cause degradation of the reagent. The manufacturer will stamp an expiration date on each pack of tubes. Storing the tubes in a refrigerator may maintain or extend their shelf life, but expired tubes should not be used.

AIR SAMPLING

Air sampling typically involves collection of the contaminant from a known volume of air on an appropriate sample medium (e.g., activated charcoal, liquid medium in an impinger, filters). The contaminant remains associated with the medium until it is removed at the laboratory for analysis. The result of this analysis may be reported as the total mass (or volume) of the contaminant and can be divided by the total volume of air sampled to calculate the average air concentration for the sample period. Laboratory analytical equipment can provide more accurate qualitative and quantitative analysis of air contaminants than most field air monitoring equipment.

The unpredictable nature and short duration of hazardous materials incidents make the use of air sampling difficult. In order to accomplish this during an emergency response, planning before an incident is required. Some of the equipment and media described below should be kept on hand. Personnel responsible for air monitoring should be able to deploy the equipment rapidly. A qualified laboratory should be available to provide rush analysis. Laboratories are usually very helpful in providing the information necessary to plan sampling events.

Sample Period

The length of time over which the sample is to be collected ("sample period") will largely be determined by (1) the requirements of the analytical method and (2) the type of exposure of concern. Most analytical equipment and methods require that a minimum amount of contaminant be present on the sample medium ("minimum detection limit") to be detected. The amount of contaminant deposited on the sampling medium is a function of the concentration of the contaminant in the air and the amount of air sampled. Therefore, sampling a low-concentration atmosphere will require a longer sample period to collect enough contaminant for analysis.

The other consideration for the sample period is the type of exposure of concern. If the suspected contaminant produces short-term (acute) effects after brief exposure to a high concentration, a short sample period should be used at times of expected highest exposure in order to minimize the effect of averaging over time. On the other hand, the contaminant may cause long-term or cumulative effects after longer exposure to lower concentrations. In this case, a long sample period would provide data on the average exposure for a longer period of time, perhaps the entire life of the incident. Planning and training before an incident occurs are necessary to make good decisions about the type of sampling to be used during an incident.

Sampling Systems

Two sampling systems are used for the collection of air samples: active samplers which draw contaminated air through a sample medium by means of a pump, and passive samplers which rely on natural forces such as diffusion or permeation to collect samples. These systems are described in the following sections.

Active Samplers

Active sampling systems draw air mechanically through a sampling medium which collects the contaminant. In this way, the contaminant contained in a large volume of air is concentrated on the medium, which is then analyzed in the laboratory. An active air sampling system typically consists of the following components:

- A sampling pump to move air
- Inert tubing to carry the air
- Sampling medium which collects the contaminant

Active sampling systems typically rely on battery-powered pumps to draw air through the sampling medium or into a sampler container. Most pumps now have some means of adjusting the airflow rate, which may be specified by the analytical method. Flow rates may vary from a few cubic centimeters per minute (cc/min) to over 10 liters per minute (10,000 cc/min). Pumps with the broadest range of flow rates will offer the most flexibility for air surveillance of a variety of contaminants.

The sampling pump must be calibrated before and after each use to ensure a constant flow rate. Unlike DRIs, active air samplers concentrate the contaminant on a sampling medium for later analysis. In order to get a concentration value, the amount of contaminant analyzed by the laboratory must be divided by the total volume of air sampled. Therefore, it is essential that the flow rate be kept constant during the sample period so that the total volume of air can be calculated. Calibration of the pump flow rate is the means of ensuring that the flow rate is right and has not changed over the sample period. Because the sampling medium and the tubing may affect the air flow rate, calibration should be performed with the whole sampling system intact.

Passive Samplers

Passive samplers or dosimeters are becoming increasingly popular as alternatives to active sampling methods. These samplers have no moving parts, and therefore are relatively easy to use and do not require calibration or maintenance. Some may be read directly, similar to colorimetric tubes, and others must be sent to a laboratory for analysis.

Passive samplers may be classified as either diffusion or permeation samplers. Diffusion samplers have a sorbent medium which is separated from the contaminated air by a grid section. This grid creates a layer of stagnant air between the contaminated air (high concentration) and the sorbent material (low concentration), forming a concentration gradient. If given the opportunity, chemicals will move from areas of high concentration to areas of low concentration by natural molecular forces. Thus, the air contaminant will naturally move to the sorbent medium from the air.

Permeation samplers take advantage of similar natural forces, but the sample medium is covered by a membrane. The membrane may be permeated (passed through) by certain chemicals and not by others, thus screening out unwanted contaminants. As the chemical permeates the membrane, it is collected by the sorbent medium on the other side.

Sampling Media

Air sampling methods for gaseous and vapor contaminants make use of a variety of sample collection media, including solids, liquids, long-duration colorimetric tubes, and sampling bags. When solid sorbents are used, contaminants in the air adsorb to the solid medium itself or to a chemical coating on the medium. Two widely used solid sorbents are activated charcoal and silica gel. Other solid media include porous polymers such as Tenax and Chromosorb, and specialty sorbents for unique uses. The sampling and analytical methods will specify the medium to use.

Sampling with a liquid medium typically involves drawing contaminated air through a liquid which absorbs the contaminant. The liquid is contained in an impinger or a bubbler that allows contact between the contaminant, which is either reactive or soluble, and the liquid reagent. The liquid is then sent to a laboratory for analysis.

Airborne particulates, including liquid and solid aerosols, are typically sampled using filter media. The two most common types of filters are fiber mesh and membrane (thin polymer membranes or sheets with tiny holes). As contaminated air is drawn through the filter medium, the particulates are impacted and trapped on the filter, which is sent to the laboratory.

In some cases, it may be desirable to collect samples of the air in a gas bag for analysis outside the hot zone. This may be the case when an unidentified contaminant is present in high concentration. The use of multiple colorimetric indicator tubes takes time, and the procedure could easily be conducted by personnel outside of the hot zone, where a lower level of protection is appropriate. Analysis on site by gas chromatography may also be desired. Shipment of the bag off site to a laboratory may not be practical because (1) the bag may leak, (2) the chemical may permeate the wall of the bag, or (3) the chemical may degrade over the time required for shipment. Therefore, any analysis must be conducted on site.

The sample can be collected by attaching inert tubing to the exhaust port of a sampling pump and to the inlet of the gas bag and placing the pump inlet in the area to be sampled. This procedure draws contaminated air through the pump, so only pumps certified as inherently safe should be used if explosive gases or vapors may be present. Also, the potential for incompatibility between the contaminant and the materials used in the pump must be considered.

Laboratory Analysis

Once air has been drawn through a collection medium, the sample container is sealed and sent to an analytical laboratory. Gases or vapors are then desorbed (removed) from the collection medium and run through analytical equipment (often a GC with an appropriate detector) for analysis. The analysis may be both qualitative (what is present) and quantitative (how much is present), depending on whether the contaminant has been identified.

The results are usually reported in mass units (milligrams, mg, or micrograms, μg) or in volume units (microliters, μl). If the total amount detected is divided by the total volume of air sampled (expressed as liters or cubic meters), an average concentration for the sample period is obtained. The units often used for air concentrations are milligrams of contaminant per cubic meter of air (mg/m^3). This is a mass (mg) to volume (cubic meters) relationship and is not the same as parts of contaminant per million parts of air (ppm), which is a volume to volume relationship. Results given as milligrams per cubic meter may be converted to parts per million based on the molecular weight of the contaminant and the temperature and atmospheric pressure during the sampling. Laboratory personnel can make this conversion if it is desired.

Samples of particulates may be analyzed in two ways. First, the exposed filter medium is weighed on a very sensitive scale, and that weight is compared to the preexposure weight of the filter. The added mass is the mass of the particulate in the air that was sampled. The sample mass is divided by the total volume of air sampled to obtain an average concentration. This is called "gravimetric analysis" and is usually used for dust particles.

The other method of particulate analysis involves fiber counting. A filter which has been exposed to contaminated air is turned clear, and the fibers on a portion of the filter are counted under a microscope. Using mathematical formulas, the number of fibers per cubic centimeter of air can be calculated. This type of analysis is most commonly used for determining airborne asbestos concentrations.

Using Air Sampling Data

The purpose of air surveillance is to gather information that can be used to make decisions during an emergency response. Air monitoring with DRIs provides immediate information, but this information is usually not chemical specific and the accuracy of the instruments varies. Air sampling provides concentration data that are chemical specific and accurate at lower levels.

If the samples can be analyzed during the response, the data may be compared directly to the appropriate criteria to make decisions about protective measures. For instance, evacuation decisions can be made by comparing the concentration data on a specific chemical to the short-term public exposure guidance level (SPEGL) described in Chapter 14. Selection of proper PPE for emergency response can be made based on the specific hazardous characteristics and known concentration of the released material. Potential health effects can be predicted and monitored if the chemical's identity and exposure data are available.

Even if laboratory results are obtained too late for use during the incident, the data may be useful in cleanup efforts to select the appropriate PPE based on actual exposure data. Also, the data may be used to help prepare for future responses to similar incidents. Emergency responders could estimate air concentrations during a future incident involving the same material based on air sampling results obtained during the initial incident. In order to make the comparison, sufficient information about site conditions must be reported to allow knowledgeable individuals to assess the impact of differing conditions (e.g., higher temperature, larger quantity release, higher humidity) on the concentration.

SELECTING AIR SURVEILLANCE EQUIPMENT

General Considerations

Technological advances in air monitoring and air sampling equipment have resulted in a wealth of choices for the consumer. As in any other market, enterprising manufacturers present their products in the best possible light and make impressive claims about how they perform. The hazmat team must decide which equipment to purchase only after careful consideration. This section will describe some important factors in making the decision.

Budget

This area is mentioned first not because it is most important, but because it is the consideration most likely to limit the options of the team. Air surveillance equipment ranges in cost from detector tube kits that can be purchased for under $500 to field GC units costing over $15,000. Add to that the costs for accessories, maintenance, and training of operators, and it becomes apparent that equipment can represent a significant investment of money which should be made carefully.

Hazards

The list of hazardous materials present at the facility should be reviewed to determine what specific materials are most likely to be involved in a release and the type of hazard they represent. The hazmat team should be capable of detecting each type of hazard at the facility (e.g., oxygen-deficient, flammable, toxic, or radioactive atmospheres). This may involve a single piece of equipment or several different types, depending on the variety of materials used.

Selectivity

"Selectivity" refers to the ability of equipment to monitor one chemical or group of chemicals to the exclusion of others. Selectivity is useful when only a few chemicals are used at a site or when one specific chemical poses a higher hazard than others. Selectivity allows the user to focus on a particular hazard with a greater degree of confidence. However, the inability to detect other chemicals would not be practical in a situation where multiple hazards are likely to be encountered, such as when public sector hazmat teams are called to transportation incidents.

In considering selectivity, it is important to remember the problem of interference, described in the discussion of colorimetric indicators. The sensors of single-gas monitors will usually respond to certain other chemicals if they are present. If a single-gas detector is purchased, potential interferences should be determined from the manufacturer's information.

Operating Range

The range of concentrations that can be measured by an instrument is referred to as its "operating range." An important part of the operating range is the lower detection limit or the lowest concentration of chemical in air that will cause a response by the instrument. The appropriate range will depend on the nature of the hazard to be monitored. For example, a combustible gas meter, which measures in the percentage range (parts per hundred), would not be appropriate for monitoring exposure to gases which are toxic at concentrations in the range of parts per million. A concentration of the gas that is high enough to be detected by the combustible gas meter would be much higher than the toxic exposure limit. On the other hand, an instrument designed to monitor toxic exposures may not have a sufficient upper range to monitor the flammability of a gas. Some instruments are capable of operating in both the percentage and parts per million ranges.

Ease of Operation

Instruments should be easy to operate and the results easy to read. Instruments that are difficult to operate will slow down responders or require too much of their attention, perhaps causing them to miss important visible information during an entry. A bulky or hard-to-operate instrument may limit the activities the responder can perform during an entry, forcing additional entries. The buttons, switches, and knobs should be easy to operate, even while wearing gloves.

Planning Purchases

As mentioned previously, the hazmat team will probably not have an unlimited budget with which to purchase equipment. This usually means that the air surveillance equipment will have to be purchased and accumulated over time. Therefore, the hazmat team should use the above considerations to prioritize purchases.

Before purchasing equipment, the team should investigate the resources available locally. Equipment used by the plant health and safety department to monitor for compliance with OSHA chemical exposure limits may be available for use in emergency situations. As facilities purchase new equipment, the old equipment may be available for purchase by hazmat teams at other facilities. Industrial facilities with similar potential hazards that are located in the same area may decide to pool their resources and purchase equipment to be shared in the event of an emergency at any facility.

Another option is to rent air monitoring equipment, thus eliminating purchase and maintenance costs. The major drawback of renting is delivery time unless the equipment is available locally. Typically, rental companies can ship equipment by overnight carriers for use the following day, but that may be too late for use during the response. Also, rental charges may be high, depending on the equipment and any minimum rental periods.

Planning is an important part of the decision to purchase equipment for air surveillance. The materials present at a facility which would represent the greatest and most immediate airborne hazard should be determined based on factors such as quantity stored, handling procedures, flammability, and toxicity. Equipment capable of detecting or monitoring these conditions should receive first priority when considering equipment purchases. After high-priority hazards are addressed, equipment for other potential hazards can be added as budgets permit.

In some cases, the most immediate threat will be the potential for fire or explosion, and a CGI would be the first priority. Because the CGI requires sufficient oxygen to function, combination detectors that simultaneously monitor oxygen content and combustible gases should be considered. These instruments are available from many manufacturers, and costs range from $1,000 to $2,500. Since the instrument can be used to detect most combustible or flammable gases or vapors, it is a good initial investment, especially if multiple flammable chemicals are present on site. The instrument should be able to be calibrated in the field, as discussed in the section on calibration in this chapter.

Radioactive sources such as those used in gauges or testing equipment in industrial facilities are usually tightly contained and do not represent an exposure hazard to nearby workers. However, in an emergency such as a fire or explosion, containment may be compromised and emergency responders may unknowingly be exposed to harmful radioactivity. Radioactivity can be detected only with monitoring equipment. Therefore, proper radiation detection equipment such as a Geiger-Mueller or an ion chamber detector, should be available to responders called to an incident where radioactive material may be present. Radiation survey meters are available for about $1,000. Be sure that the survey meter selected is capable of monitoring the type (alpha, beta, or gamma) of radiation used at the site.

If a particular chemical poses a high or unique toxic threat at low concentrations, an instrument capable of monitoring that chemical should be considered. Hand-held instruments which monitor chlorine, hydrogen sulfide, carbon monoxide, and other specific chemicals are available for costs starting at several hundred dollars. These instruments can detect concentrations as low as 1 ppm and provide some degree of selectivity. As discussed above, the potential that other chemicals may be detected by the instrument must be considered.

Another option for detecting specific chemicals is the use of colorimetric indicator tubes. Though the accuracy of the data varies by plus or minus 25 percent or more, these tubes provide "quick and dirty" detection of hundreds of specific chemicals. Many manufacturers have assembled hazmat kits that contain the hand-held pump and a variety of tubes for commonly encountered contaminants and sell for up to $1,000. A facility with a limited number of contaminants present may decide to purchase the pump kit ($200–$500) and a few boxes of tubes ($30–$40 each) for only those contaminants present on site. It is important to remember that the tubes have a limited shelf life. They should be checked regularly and purchased in relatively small quantities to avoid throwing away a large number of outdated tubes.

If many different toxic chemicals are potential hazards at a facility or if somewhat higher accuracy is needed, more expensive instruments such as a PID, FID, or IR spectrophotometer are available. These instruments can be calibrated to provide a direct reading of the air concentration of the chemical and can show fluctuations in the concentration over time. They are capable of monitoring a wide variety of different chemicals and, when properly calibrated, provide a more accurate reading than colorimetric indicators. The costs of the instruments start at about $4,500 for a PID and can go to well over $15,000 for field GCs with one of these detection methods.

SUMMARY

Exposure to hazards present in the air is difficult to control, so information about the extent of airborne hazards can be very important to decisions about protecting the public and response personnel. The principles and equipment available to gather this information can vary widely in complexity and usefulness. Understanding these principles and equipment can mean that valuable information is available to the emergency responder. On the other hand, failure to understand or plan for air surveillance during a response can lead to a lack of information or, worse yet, misinterpretation of information and bad decisions. Training, planning, and resources are the keys to successful implementation of air surveillance during an emergency response.

REFERENCES

ACGIH. 1983. *Air Sampling Instruments for Evaluation of Atmospheric Contaminants*, 6th ed. Cincinnati: American Conference of Governmental Industrial Hygienists.

EPA, Office of Emergency and Remedial Response. *Air Surveillance for Hazardous Materials*. Cincinnati: Environmental Protection Agency.

EPA, Office of Emergency and Remedial Response. 1984. *Standard Operating Safety Guides*. Washington, DC: Environmental Protection Agency.

Gollnick, D.A. 1986. *Basic Radiation Protection Technology*. Altadena, CA: Pacific Radiation Corporation.

Leidel, N.A., K.A. Busch, and J.R. Lynch. 1977. *Occupational Exposure Sampling Strategy Manual*. NIOSH Publication No. 77-173. Cincinnati: National Institute for Occupational Safety and Health.

National Safety Council. 1988. *Fundamentals of Industrial Hygiene*, 3rd ed. Chicago: National Safety Council.

NIOSH/OSHA/USCG/EPA. 1985. *Occupational Safety and Health Guidance Manual for Hazardous Waste Site Activities*. Washington, DC: U.S. Government Printing Office.

16

Computer Use in Emergency Response

W. Donald Fattig, Ph.D.

The focus of SARA Title III legislation is on planning for response to emergencies arising in the course of operations with hazardous materials. It is not surprising that a number of computer applications have become available to assist responsible parties in meeting the compliance-driven requirements of the law.

In using computers in emergency response, two broad categories are evident: (1) on-site use at industrial facilities subject to the law and (2) use by off-site emergency responders summoned to the scene of a spill. The first focuses upon advance planning for emergency response and the periodic and incident-driven reporting requirements of the law, and the second is concerned with helping firefighters make decisions at the scene regarding appropriate firefighting procedures, isolation of the incident, and possible evacuation of the public in the vicinity of the incident. Clearly, there is considerable overlap between these functions, but there are obvious differences of focus.

In reviewing possible computer usage in this area, the stated legal objectives and imperatives may be divided into several categories.

OBJECTIVES

- Be aware of computer software which supports emergency response activities.
- Encourage incorporation of computer applications in health and safety planning and training.
- Understand criteria for evaluating emergency response software.

- Understand technical considerations in choosing hardware and software for emergency response activities.

RISK ASSESSMENT

A focal point of computer programs dealing with risk assessment is providing quick access to a chemical data base that is searchable by several criteria, such as CAS (Chemical Abstracts Service) number, chemical name and synonyms, and specific health hazards. These include systems that are available on diskette or CD-ROM, as well as on-line systems that can be accessed via modem. In addition to the usual considerations of cost and compatibility with computer systems in use at the site, important features that should be evaluated in choosing a particular application include

- Accuracy and completeness of the information provided
- Provisions for timely updating of the data base by the vendor
- For first responders, linkage to other modules such as air dispersion and plume modeling, inventory, and site storage mapping information

Accuracy and completeness of technical information are best ensured by the use of established data bases provided by an agency having the resources to give continuing attention to the product. Many commercially available software products incorporate data bases such as *RTECS* (Registry of Toxic Effects of Chemical Substances) from NIOSH, *CHRIS* (Chemical Hazard Response Information System) from DOT (Coast Guard), *OHM/TADS* (Oil and Hazardous Materials/Technical Assistance Database) from EPA, and the *CAMEO 3.0* (*Computer-Aided Management of Emergency Operations*) *Chemical Database* from the National Safety Council. The last has recently been expanded to include over 3,300 chemicals, and over 14,000 chemical synonyms and trade names have been added.

On-line systems provide for continuous updating by addition of entries, but so do some stand-alone packages. A majority of the programs available for use on site include modules featuring storage locations, Material Safety Data Sheet generation, and reports.

Some commercially available software packages for chemical hazard assessment include *CHEMBANK* (CD-ROM), *CHRIS PLUS* (diskettes), *MicroCHRIS* (diskettes), and *CAMEO* (diskettes). On-line systems include *CHEMLINE* and *TOXLINE* from the National Library of Medicine and *Chemical Information System* (*The CIS*). Firesoft *EPA Hazmat* has data on 1,400 compounds (including brand names and trade names), selected for spill history, production volume, and toxicity; it gives information on hazard levels, fire protection and disposal information, storage procedures, and chemical and physical properties.

OPERATIONS, DATA MANAGEMENT, AND REPORTING

In addition to a chemical data base, many software packages feature modules that provide for other compliance needs. These focus on SARA Title III Sections 311–312 (community right-to-know reporting requirements) and include the information discussed below.

Inventory Control

A custom data base that tracks the inventory of hazardous substances is an essential feature of emergency planning. While this kind of software may once have been the province of large industrial facilities, requiring mainframe or minicomputers, many microcomputer applications now incorporate this sort of data management. For CERCLA hazardous substances (RQ chemicals) also subject to reporting under SARA Title III Section 304, applications like CHEMASYST and SARATRAX automatically generate the required report and remind the user of reporting deadlines when inventories exceed the reportable quantities specified. SARATRAX determines whether a spill needs to be reported and to whom. Reports produced by these programs may also be incorporated into planning by emergency responders.

Site Location Maps

Site location maps showing storage facilities, including buildings and room or bay locations within buildings, underground and aboveground tanks, quantities of substances stored, access points, telephone locations and numbers, firefighting equipment placement, and personnel evacuation routes, are all necessary for both effective incident prevention and response. Ideally, these should be available as graphic maps capable of being displayed on the computer monitor. Quantities stored should be updated via linkage to the inventory module of the program; changes in response equipment type and/or storage location should be entered as needed. This may require separate graphics or CAD/CAM (computer-aided design/computer-aided manufacturing) software, in addition to the hazardous materials management program itself.

Prevention of storage of incompatible materials, changes in response procedures, and current lists of substances by chemical name and hazard class are important considerations in selecting software. The data should be searchable by chemical name and synonyms, hazard class, and location (address, building name or identification, room, bay, telephone numbers, etc.).

Reporting

SARA Title III Section 312 requires submission of an emergency and hazardous chemical inventory form, covering the same chemicals covered by Section 311, to the LEPC, the SERC, and the local fire department. The inventory form incorporates a two-tier approach.

Tier I

Facilities must submit the following aggregate information for each applicable OSHA category of health and physical hazard:

- An estimate (in ranges) of the maximum amount of chemicals for each category present at the facility at any time during the preceding year
- An estimate (in ranges) of the average daily amount of chemicals in each category
- The general location of hazardous chemicals in each category

Tier II

Upon request of an LEPC, SERC, or local fire department, the facility must provide the following information for each substance subject to the request:

- The chemical name or the common name as indicated on the Material Safety Data Sheet
- An estimate (in ranges) of the maximum amount of the chemical present at any time during the preceding calendar year
- A brief description of the manner of storage of the chemical
- The location of the chemical at the facility
- An indication of whether the owner elects to withhold location information from disclosure to the public

Many of the comprehensive computer programs available commercially include modules that will generate Tier I and Tier II reports; ordinarily this is not a difficult computing task, since it is derivable from inventory control and site location data mentioned above.

Form R Reports Under Section 313

Section 313 of Title III requires EPA to establish an inventory of toxic chemical emissions (annualized) from certain facilities. Owners and operators of facilities subject to this reporting requirement must submit Form R for specified chemicals. The specification of which facilities must file these reports is based on several criteria, including

- Ten or more full-time employees
- Industries listed in Standard Industrial Classification Codes (SIC) 20–39
- Whether or not the industry manufactured, processed, or otherwise used a listed toxic chemical in excess of specified threshold quantities

Several computer applications generate Section 313 Form R reports. This feature is included here because of the obvious implications for reporting of a spill or other release incident requiring emergency response.

Computer applications that incorporate Tier I and II reporting and Form R reporting include *313 Advisor, Hazmat Manager, HAZ-MAT II, HAMIN, SARA!,* and *SARATRAX*. The program *313 Advisor* may also suggest engineering alternatives for minimizing Section 313 releases, important in emergency response preplanning activities.

Employee Safety

For employees in general and especially for emergency responders, an accurate record of exposure to hazardous substances, together with a history of injury or illness, is an important part of emergency response planning. Some applications, such as the *SDMS Injury/Illness Surveillance System* and the *TOMES Plus System,* are dedicated to this task, while others (*HAZARD, HazMat Manager, Haz-Mat II*) are more comprehensive packages including options for producing accident reports, OSHA 101 accident reports, OSHA 200 logs, employee exposure data, and injury/illness reports.

Appendix B of the NIOSH *Guide for Evaluating the Performance of Chemical Protective Clothing* mentioned in Chapter 9 includes information describing several Macintosh and MS-DOS computer applications and data files (*CPCbase, Forsberg's Chemical Protective Clothing Performance Index, GlovES+*) pertaining to the choice of proper chemical protective clothing. Breakthrough times, permeation rates, and degradation data allow for site-specific choices; update services for registered owners of software may be available. *Hyper CPC* is a Macintosh Hypercard version of Krister Forsberg's Chemical Protective Clothing Permeation Index.

A recently released program, *3M Select Software,* assists in the selection of respirators. A menu-driven program that is easy for computer novices to use requests contaminant and concentration input data from the keyboard and returns generic respirator recommendations for most contaminants found in industrial settings. This software includes over 1,000 chemical synonyms and acronyms and is able to return recommendations for commonly found chemical combinations. Use of the program meets the OSHA audit trail requirement. The vendor plans to update the data base annually, and purchasers will receive renewal notices.

Event Tracking and Recording

Not all spills or incidents fall under the reporting requirements of Title III Section 313 or EPA reporting requirements for 360 chemicals comprising the EHS (extremely hazardous substances) list. It is desirable to use software that provides for recording of these incidents as well. The data obtained may be of assistance in planning emergency response, changes in storage and/or inventory, and may become a part of employee medical records.

For example, *CAMEO 3.0* allows creation of an Incident Report specifying the chemical name (including whether or not it is a RCRA chemical), quantity released, accident type, record of notification, and so on. The data base thus created is searchable by facility, chemical, and year and allows the creation of reports.

Process Tracking

Label Generation

Many of the software packages will generate appropriate shipping labels for specified chemicals.

Manifest Production and Storage (RCRA)

HAZARD, Hazardous Waste Tracker, Haz Waste Tracker, HAZWASTE, and *HazMat Manager* provide for the generation of shipping manifests and/or their storage in electronic form. Some also give shipping container type and size information. Transportation routes are included in some software items.

EMERGENCY RESPONSE TRAINING

Training for emergency response is an integral part of emergency response planning. Training needs vary considerably, but those most likely to be directly involved in responding to and managing, on-scene, an incident are likely to benefit most from the use of software designed specifically for responders. The most widely used computer application in this area, available in both MS-DOS and Macintosh versions, is *CAMEO 3.0* from the National Safety Council. The Macintosh version is a linked group of Hypercard stacks; a Hypercard-like version, under *WINDOWS*, is available for MS-DOS users; both are user-friendly modes easily mastered by those previously unfamiliar with computers.

CAMEO 3.0

CAMEO 3.0 is appropriate for use by both in-house emergency response teams and municipal or other fire departments and emergency response personnel. From a training standpoint, its great virtue is that it allows for comprehensive simulations ranging from generic to highly site-specific situations. Version 3.0 represents a major improvement over previous versions, especially in the thoroughness and accuracy of its Chemical

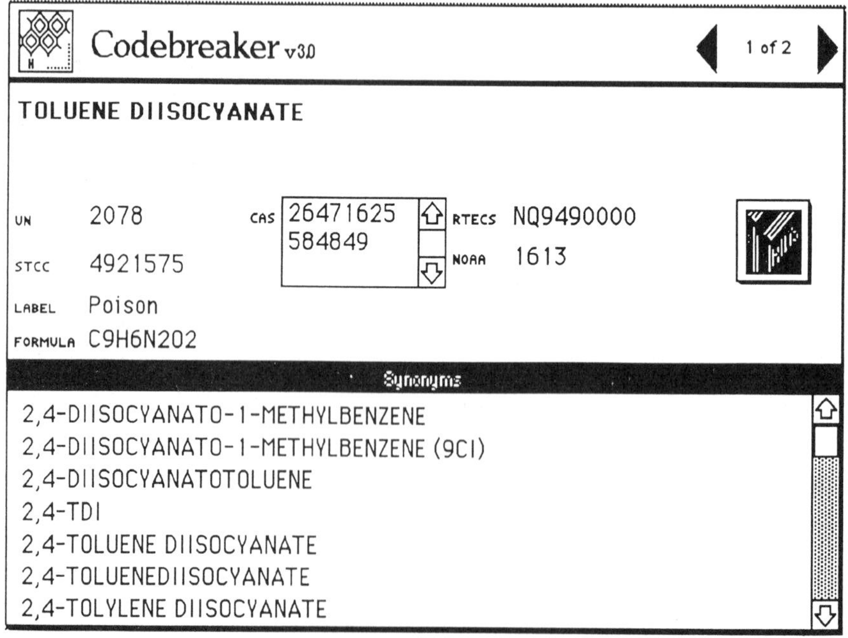

FIGURE 16-1. Results of a search of the *CAMEO 3.0 Chemical Database*. The search words "toluene" and "cyanate" were combined.

Database and the companion *ALOHA* (Areal Locations of Hazardous Atmospheres) plume modeling program.

The searchable Chemical Database is the heart of *CAMEO 3.0,* and familiarization with searching it is a key feature in assessing risk. Knowing the nature of a spill is the first step in response. Trainees may achieve a high level of proficiency in risk evaluation by searching using chemical names and synonyms, trade names, UN numbers, label identification, and RTECS categories. It is possible to conduct chemical searches using specific words and numbers that may be joined (AND/OR) in Boolean combinations (see Figures 16-1, 16-2, and 16-3). Information found on each chemical's card includes the NFPA diamond, physical properties, fire and health hazards, firefighting recommendations, protective clothing, and first aid (see Figure 16-4). Results of searches may be printed.

Facilities data and chemical inventory information are also part of *CAMEO 3.0;* these may include customized building plans. An information-tracking checklist allows one to keep the list current and linked to updated site storage mapping data.

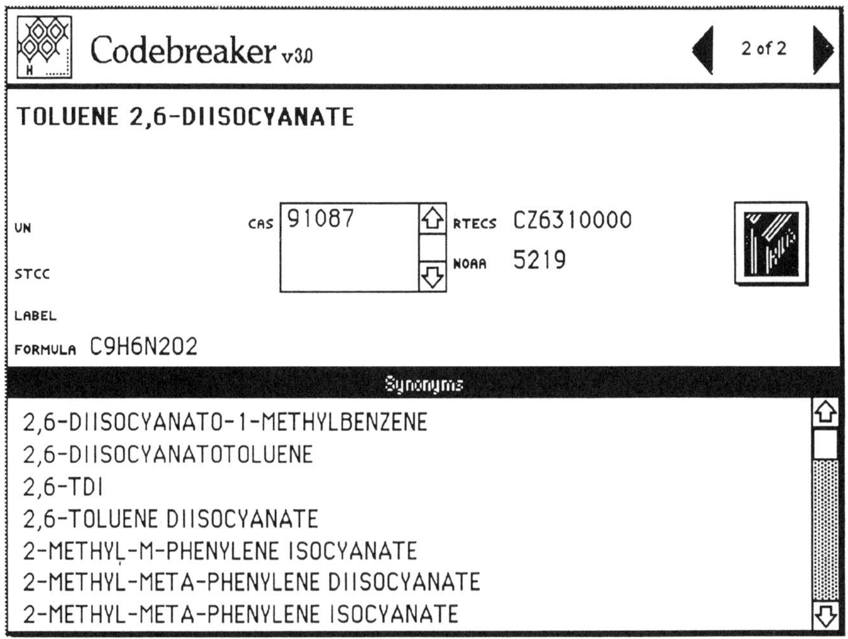

FIGURE 16-2. Second of two cards returned by the search in Figure 16-1.

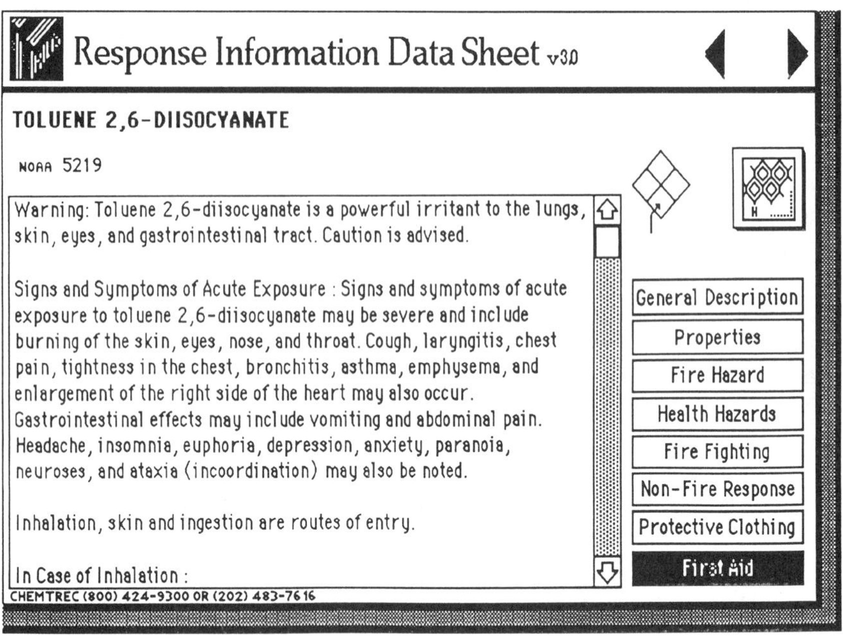

FIGURE 16-3. Response Information Data Sheet for the chemical in FIGURE 16-2.

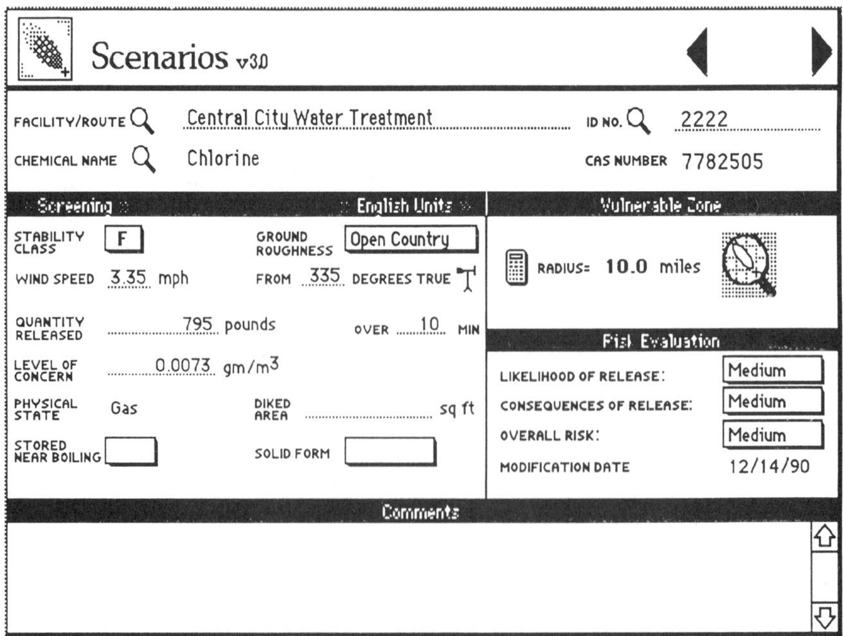

FIGURE 16-4. A scenario card for *CAMEO 3.0*.

The program allows for:

- Facilities data and chemical inventory information, including customized building plans and an information-tracking checklist for keeping the list current and linked to updated site storage data
- Creation of Incident Reports (see above)
- Transportation data, including routes used to transport priority chemicals (EHS chemicals), congested areas, etc.
- Data on schools and other special populations whose density may pose special problems in the event of a spill; this includes information on average population present, seasonal minima and maxima, and like information
- Scenarios (including plumes) for training and response, including vulnerable zone determinations for EHS chemicals, calculated conservatively using guidelines from the *Technical Guide for Hazard Analysis* (see References)

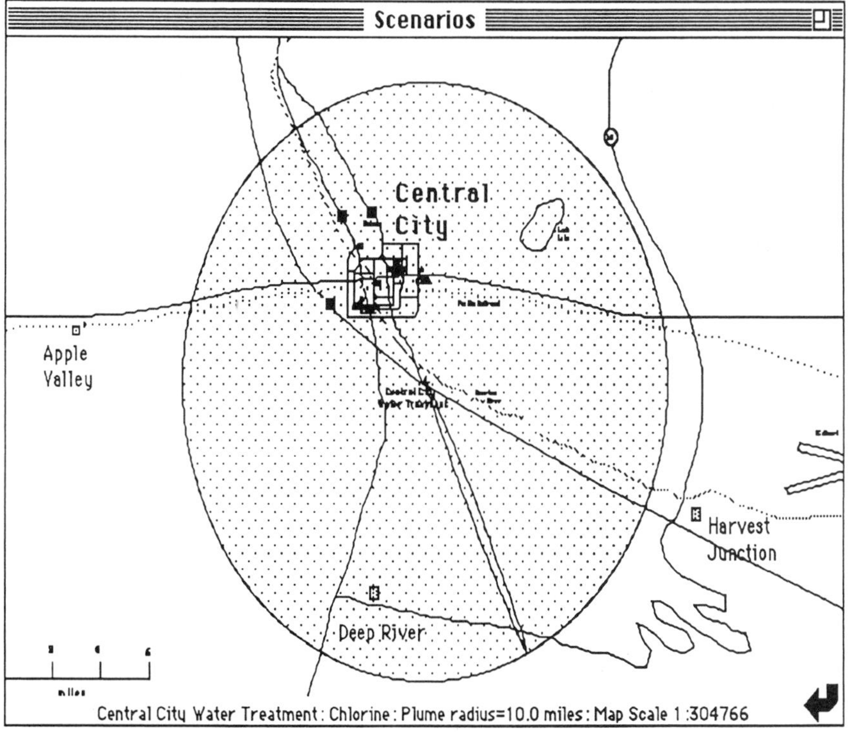

FIGURE 16-5. Vulnerability map created by *CAMEO 3.0* for the scenario in Figure 16-2.

All of the above features may be linked to maps, greatly enhancing the presentation of data. (See Figure 16-5.)

ALOHA

CAMEO's companion program, *ALOHA,* originated as a tool to aid in response situations. It is a plume modeling program for estimating the movement and dispersion of gases via a model that, given initial conditions, can generate a time-dependent plot of a spill plume. Exposures at downwind locations may be estimated over the time and dose calculated, and infiltration of air into buildings can be modeled.

In summary, while *CAMEO 3.0* and *ALOHA* have been seen widely as fire department–oriented programs, an increasing number of industrial facilities are recognizing their value as on-site emergency response tools. The concerns addressed by such programs on site are the same as those for off-site responders. Especially attractive is the ability to construct site-specific graphic maps for response planning and to generate, with appropriate software, detailed local maps from Census Bureau TIGER-Line files. The programs are user friendly, enabling them to be incorporated quickly and effectively into emergency response training; an important capability is the construction and evaluation of scenarios. These features contribute to the planning process by refining safety precautions before an incident occurs.

First Aid and Cardiopulmonary Resuscitation

An especially well-conceived and executed computer package for training in the first aid area is Andrew S. Kanter's *What to do before the Ambulance Arrives . . .* Two Hypercard stacks for the Macintosh (*First Aid* and *CPR Stack*) provide a wealth of information on symptoms and procedures.

ARCHIE

ARCHIE (*Automated Resource for Chemical Hazard Incident Evaluation*) is an MS-DOS program for evaluating hazard risks associated with vapor dispersion, fire, and explosion impacts from the release of hazardous materials. It consists of modeled outcomes for releases from specified tank types and pipelines, under various conditions, for hazardous gases and liquids. Intended as a training and planning tool, *ARCHIE* introduces useful concepts related to the physical chemistry of substances. Extensive documentation for the program may be found in the *Handbook of*

Chemical Hazard Analysis Procedures (see References); both the program and the book are available from the FEMA.

General Bulletin Board Systems

The HMIX (Hazardous Materials Information Exchange) bulletin board system, operated and maintained by FEMA and DOT, affords a mechanism for emergency response professionals to communicate up-to-date concerns and to share information. It includes a conference section supporting *ARCHIE,* a message section for exchange of ideas and information, details about *CAMEO* regional user groups, and text files of regulations, interpretations, and pending legislation. A user's manual is available. HMIX may be reached at 1-800-PLAN-FOR nationwide and 1-800-367-9592 in Illinois.

The National Safety Council supports *CAMEO* questions via Compuserve (ID No. 72730, 1035), HMIX, NOAA's *CAMEO* bulletin board system (account NSCCAM), and the International Association of Fire Chief's (IAFC) ICHIEFS bulletin board system under *CAMEO* (Connect ID: NSCCAM).

EVALUATING EMERGENCY RESPONSE SOFTWARE

The great variability found in computer software for emergency response is a reflection of differing primary concerns of the users. This is often correlated with the user's level of responsibility and the time frame within which his or her responsibilities are executed. Beyond the level of the software users themselves, others depend on data generated with the systems.

Those who respond directly to a spill are concerned primarily with problems of quick and accurate hazard identification; safe containment of the spilled substance and any resulting fires; and prevention of exposure and concomitant health hazards to themselves, other workers, and the public. Software such as *CAMEO* can aid their response decisions and actions on the scene.

Managers tend to be more concerned with the compliance-driven aspects of emergency response. Are there adequate provisions for hazard recognition? Are the Material Safety Data Sheet forms current and complete? Have the requirements for storage locations, inventory control, and reporting been met?

Officials at the local fire department, LEPC, SERC, and EPA have an interest in receiving the computer-generated reports mandated under the law. Each of these agencies depends on accurate reporting to carry out its comprehensive planning, enforcement, and public policy roles.

Finally, the public is another indirect consumer of the computerized information. Individuals, organizations, communities, and political jurisdictions have an interest in the outcomes of computer use in emergency response.

All of these concerns overlap considerably, and the reactions to the products of computer usage feed back in the form of computer program design and implementation.

The following are some general guidelines for evaluating computer software; the points apply to emergency response software, too.

- Match the computer system and the software. Plan for future software releases to require more computer memory and/or storage.
- Read reviews. Ask for a satisfied customer's name.
- Try before you buy. Are demonstration disks available? Does the software vendor provide a free trial period?
- What are the vendor's software upgrade policies? Are upgrades free? If not, what will they cost? How often is the software upgraded?
- Is site licensing available for installation of software on multiple computer systems or on a computer network?
- Is training or consulting available from the vendor or from other sources? Is it included in the purchase of the software? If not, what does it cost?
- Buy from a reputable vendor. Don't be a software pirate. Purchased software often assures availability of future releases or upgrades.
- Expect to have to upgrade software and hardware from time to time and include this aspect in your budget to avoid expensive surprises.

COMPUTER HARDWARE: TECHNICAL CONSIDERATIONS

The sophistication of computer hardware and software has increased hand in hand in the past decade, and the trend will continue in the next decade. In choosing software for emergency response, the user needs to be aware that some hardware features have become essential requirements. These include fast microprocessors for searching data bases; math coprocessors to speed calculations such as air plume dispersion patterns; large RAM memory to accommodate ever larger program releases; graphics capability for displaying site maps, air plumes, etc.; increased size of hard disk storage to accommodate data bases; CD-ROM storage ability;

network compatibility for multiuser operations; and a fast modem (2,400-baud minimum) for accessing on-line services.

Current developments indicate the increasing utility of capabilities for graphics scanning and FAX modems for transmission of reports. Advances in optical character recognition (OCR) are likely to make this feature a widely used mode of data input within a few years. The standard modem will become a 9,600-baud device. Speech recognition is already a feature of some programs, and its incorporation as an optional feature of hazmat software will likely increase as speech recognition hardware becomes more widely incorporated in systems.

Accompanying software developments, such as the ability to generate map graphics from Census Bureau TIGER-Line files on CD-ROM, data compression-expansion algorithms, and improved network management software, are already having an impact on the use of computers in emergency response. As these new features become incorporated into hazmat software, memory and storage hardware will have to be increased in order to incorporate the new features.

REFERENCES

Bishop, J. July 1991. ER Planning. *Hazmat World,* 48–52.

Donahoe, M.L. July 1991. National Chemical Response and Information Center and Hazardous Material Emergency Response. *Hazmat World,* 58–60.

Donley E., editor. 1991. *1990/91 Environmental Software Directory*. Garrisonville, VA: Donley Technology.

National Response Team/EPA/FEMA/DOT. 1987. *Technical Guidance for Hazards Analysis. Emergency Planning for Extremely Hazardous Substances.* Washington, DC: U.S. Government Printing Office.

FEMA/DOT/EPA. 1989. *Handbook of Chemical Hazard Analysis Procedures.* Washington, DC: U.S. Government Printing Office.

EPA/Information Management Staff/Office of Solid Waste and Emergency Response. December 1990. *Report on the Usage of Computer Models in Hazardous Waste/Superfund Programs: Phase II Final Report.* Washington, DC: U.S. Government Printing Office.

Hazmat Team Planning Guidance. April 1990. U.S. Environmental Protection Agency Directive 9285.3-05. Washington, DC: U.S. Environmental Protective Agency.

Newsom, D.E. July 1991. *ARCHIE Promotes Dialogue Between Industry, LEPCs. Hazmat World,* 53–57.

Schroll, R.C. 1990. *Hazardous Materials Emergency Response.* Firecon. East Earl, PA.

Title III Fact Sheet: Emergency Planning and Community Right to Know. Washington, DC: U.S. Environmental Protection Agency 1987.

APPENDIX: SOURCES OF EMERGENCY RESPONSE COMPUTER SOFTWARE MENTIONED IN THIS CHAPTER

3M Select Software
3M Occupational Health and Environmental Safety Division
3M Center Bldg. 220-3E-04
St. Paul, MN 55144-1000

313 Advisor
E.I. Du Pont de Nemours and Company
Engineering Test Center
P.O. Box 6094
Newark, DE 19714-6094

CAMEO 3.0
National Safety Council/Environmental Health Center
1050 17th Street, NW, Suite 770
Washington, DC 20036

CHEMBANK
Silver Platter Information, Inc.
One Newton Executive Park
Newton Lower Falls, MA 02162

Chemical Information System (The CIS)
Chemical Information Systems, Inc.
7215 York Road
Baltimore, MD 21212

CHRIS PLUS
HRD Software
22 Amherst Road
Amherst, MA 01002

Firesoft EPA Hazmat
Firesoft/Public Service Computer Software
1220 L Street, NW, Suite 200
Washington, D.C. 20005-4018

Haz Waste Tracker
Resource Consultants, Inc.
P.O. Box 1848
Brentwood, TN 37024-1848

HAZARD
North American Software, Inc.
P.O. Box 3309
Tustin, CA 92680

Hazardous Waste Tracker
HRD Software
22 Amherst Road
Amherst, MA 01002

HazMat Manager
Ray & Associates
1502 Harbins Road
Norcross, GA 30093

HAZMIN
Logical Technology, Inc.
1422 W. Main St. PO Box 3655
Peoria, IL 61614-0655

HAZWASTE 5.0
HazMat Control Systems, Inc.
3409 Lakewood Blvd., Suite 2C
Long Beach, CA 90808

HAZ-MAT II
Management and Communication Consultants
P.O. Box 14372
Oklahoma City, OK 73114

Hyper CPC
Michael J. Blotzer, CIH, MS
RR6 Box 1045A
Augusta, ME 04330

MicroCHRIS
Fein-Marquart Associates
7215 York Road
Baltimore, MD 21212

SARA!
OSHA-Soft Corporation
P.O. Box 668
Amherst, NH 03031

SARATRAX
IIT Research Institute
2719 Pulaski Highway
Edgewood, MD 21040

SDMS Chemical Management System
EcoAnalysis, Inc.
Arcade Plaza Suite A
221 E. Matilija Street
Ojai, CA 93023

TOMES Plus System
Micromedex, Inc.
600 Grant Street
Denver, CO 80203-3527

What to do before the ambulance arrives . . .
Andrew S. Kantor
41 University Road
Brookline, MA 02146

17

Mental Stress in Emergency Response

Ralph A. Johnson, Ph.D.

Emergency response work has many of the characteristics associated with high-stress occupations. Though firefighters in a large urban setting will have different problems than a maintenance worker in a remote rural hazardous materials facility, both may suffer from job and job-related pressure. One may go to work afraid of some unknown danger, while the other may dread the dreary routine, but both can claim that the work causes them to feel stress. In order to see and understand the stressful nature of these jobs, we can start by defining stress and by examining work stress in general. By identifying the special characteristics of emergency responder jobs, we should be able to locate some specific factors that create or compound stress. This method can also help to identify measures to remove some of the stressful factors in the workplace or at least reduce some of the effects of a stressful job.

OBJECTIVES

- Understand the nature of workplace stress
- Be aware of the factors that contribute to stressfulness in emergency response work
- Understand the methods to avoid, reduce, and manage workplace stressors

DEFINING STRESS

Stress is hard to define. Many people are well acquainted with its effects. Some feel confident in their ability to identify the causes of it in their own lives or the lives of others, but a precise definition is difficult to find. In our everyday environment, including family relationships, commercial transactions, and work conditions, factors commonly referred to as "stressors" create mental or emotional pressure for an individual. In a general sense, stress is the nonspecific result of a demand placed upon the body or mind. (Fiedler 1990) It is the biological or psychological reaction to a stressor. In one sense, all human interactions with nature and other humans produce stress. While stress per se is not harmful, too much stress can create problems for a person. Workplace stress theories are built on the central premise that "working conditions and other environmental factors present in the workplace produce a given stress load on the person" (Workplace Health Fund 1991). These stressors affect a person in two critical ways. When there are too many stressors, or when they last too long or are too intense, they may produce a direct stress effect in a person, or they may reduce the worker's tolerance to other stressors (Workplace Health Fund 1991) For our purposes, when the person's body or mind cannot adequately cope with the pressure, we can say that the individual has a problem with stress. Stress is also hard to define because each individual suffers or accommodates the pressure according to his or her personal capacity (Fiedler 1990). So, similar pressures or stressors produce different levels of reaction and different problems for each individual. Headaches, tension, irritability, nervousness, nausea, fatigue, and muscle pain are among the common symptoms of excessive stress.

STRESS AT WORK

Work is stressful. The conventional wisdom of a few decades ago predicted that the future workplace would consist of meaningful, creative careers. Each worker would have pleasant surroundings during working hours and plenty of leisure time. The reality of contemporary life, however, is that many workers feel squeezed by work. They do not feel that they have leisure time. They bring home mental and physical symptoms of work stress. They feel busy, behind, dissatisfied, uncertain, and unable to master either their life or their work. They complain that they are "stressed out." Karasek and Theorell (1990) present a significant number of studies indicating that stress is actually increasing.

Common symptoms of excessive workplace anxiety can be found from the highest levels of corporate management to low-status, minimum wage jobs. The workplace has generally become physically safer throughout this century, even as it has become more dangerous mentally. According to a 1988 study by the National Council on Compensation Insurance, "Psychological disorders have become one of the fastest growing occupational ills of this decade. Stress now accounts for about 14% of occupational disease claims, up from less than 5% in 1980" (McCarthy 1988).

Americans in past generations had to manage jobs and live under less than ideal conditions. Why do our current workplaces still foster the kinds of stressors that we experience? Of the many explanations, perhaps the most compelling is that work and the work experience are rapidly undergoing an unprecedented rate of change. This revolution has increased the insecurities and competitive pressures for bosses and subordinates alike (Bennett 1988).

In the past generation, many of the traditional production jobs in agriculture and manufacturing have disappeared or have undergone a radical transformation. Of the new jobs that were created, the professional and technical ones require specialized education. The more numerous sales and service jobs have less stringent entry requirements, but the rewards seldom match the wages and benefits of the manufacturing sector. One message of the new economic era is clear: Hard work alone is not likely to provide the path to prosperity. Education, training, a pleasant personality, and good customer service are all required.

Emergency responder occupations emerged as part of the urbanization and industrialization of America. Developments in communication, transportation, and lifesaving technologies made the job possible. Speed, crowding, dangerous materials, untested projects, and increasing interdependence made the job necessary. The annual production of hazardous waste in the United States has grown from 1 billion pounds at the end of World War II to more than 100 billion pounds today (Epstein, Brown, and Pope 1982, 7). Risking one's life and health can create overwhelming pressure. Studies show a clear link between dangerous work and stress (Poulton 1978). Workers who witness great tragedies may find that their identification with the victims gives them an overpowering sense of grief or despair. A single critical incident may be the source of job stress. Other emergency responders may never see a dramatic incident, but their work with hazardous materials may create a constant feeling of ambiguity and uncertainty (Bureau of National Affairs 1987). In either experience, emergency responders are likely to feel the stress of their job.

Technological Change

In many cases, work changes, especially technological changes, have made work more comfortable and safer. Sometimes the changes have freed humans from drudgery, allowing them to perform more interesting and creative tasks. In other cases, workers have lost a valued job or function to a machine or production method. As new technology continues to transform the workplace, the human resources are required to adapt, to learn, or to leave. Clearly, technological change and workplace stress are related (International Labour Office 1984).

The process of change creates anxiety because workers know that their lives will never be the same as before, but they do not know how they will actually be. Will life at work be easier or more complex? Will it be more interesting or boring? Will they be able to understand the work? Will they be able to influence the changes? Even if they know that change is imminent, they still may not know how it will affect them.

Emergency responder jobs have changed and will continue to be susceptible to new and rapid transformations. The nature of the changes falls into two major categories: (1) changes in the types of emergencies and the types of hazardous substances and (2) changes in the procedures and equipment used for responding to the emergency situation.

In the first category, new chemicals and combinations of chemicals, new fire hazards, and new pressures to protect densely populated areas may all be considered stressors unique to the job of emergency responder (Pelletier 1984). The second type of change requires training and adaptation. Keeping up with the latest hazards, techniques, and equipment may seem to be a never-ending chore. In cases where a responder is required to wear PPE, the discomfort may be increased. Valued as a potential lifesaver, PPE may be resented because of its weight, heat retention, or restrictiveness. (For a more detailed look at PPE, see Chapter 9; for more information about physical stress, see Chapters 18 and 19.) As the workplace has changed, emergency response work has probably changed more than other occupations, and the continuing change and rapid rate of change mean that it is difficult to master the skills, knowledge, and techniques, necessary to feel comfortable and secure.

Social Change

Historically, work on the farm was likely to be performed by people from the same culture, the same tribe or ethnic background, often the same family. Skilled trades were transferred from one family member to the next. The workplace brought similar kinds of people together. Diversity might mean the integration of young and old workers, but more often it meant a workplace that accommodated the values of the landowner or enterprise operator, as well as those of the hired hands. As large-scale manufacturing developed, workers with different backgrounds were likely to work together. While workplace loyalties might run along ethnic or family lines, generally the major workplace distinction was between the boss and the workers.

Job segregation may have provided a sense of satisfaction for the workers with top jobs and a feeling of frustration for those with undesirable work, but it also provided a sense of predictability. Sons could plan for a life very much like those of their fathers, and daughters could resign themselves to the limits imposed upon their mothers. After the Second World War, job segregation became the subject of national public protest, but major disruption of segregated work practices did not begin for several more decades. The civil rights movement and the women's movement of the 1960s and 1970s focused on job inequality. Subsequent legislative and rhetorical pressure promoted workplace change. As a result, when people go to work today, they are more likely to interact with people of other races and backgrounds than before.

Social change is necessary. Fairness, equal opportunities, and removal of racial, age, and gender restrictions promote American values at their best. But social change is hard for individuals, especially when the economic and social stakes are high. Groups who enjoyed favored treatment in the past are likely to regard the newer types as intruders (Serrin 1984). Newcomers are likely to resent the superior attitudes of old-timers (Lauffer 1985). White males who have always worked for other white males may experience some difficulties if they are assigned to work for a black female. At the same time, the tension of the situation will probably cause her to suffer as much as or more than her employee (Davidson and Cooper 1983).

Emergency responder jobs are more likely than most to have undergone significant social changes. Much of the emergency response work was once viewed as macho as well as heroic. In many cases, emergency response jobs were held exclusively by white males. With such rigid racial and gender exclusion, many of the changes were public and dramatic. In addition to the work itself, the social changes in the work environment have added to the pressure and tension. With the great amount of change that has already occurred and the changes that are still likely to come, it may be difficult for the emergency response worker to determine the nature of the changes or envision any way to influence the types or schedule of change.

Political Change

American history suggests a country where men and women with ordinary backgrounds could work hard and distinguish themselves. The Founding Fathers outlined a country in which government was deliberately limited in order to allow maximum freedom for individuals. Many citizens who started with nothing were able to convert hard work into a successful life and a comfortable standard of living. For others, though, hard work produced fatigue and disappointment for themselves and success for someone else. Still, a productive life was a constant American ideal. As technological and social changes took place, political changes began to emerge.

As early as the Great Depression, federal administrations saw that even though there was a surplus number of workers available and ready to produce, there was a shortage of persons with the means to purchase and consume the abundance of goods on hand. Government work programs were created not because production was the primary desire but because, through this method, workers could become economically valuable consumers. This national recovery legislation did, however, reinforce the idea that work was the path to success.

The subsequent decades continued to promote consumerism. Government programs were created to subsidize housing, education, and the purchase of private automobiles. Cheap money promoted individual spending. Post–World War II mass media transformed advertising into a major national industry. The new image makers also changed the image of the worker. The productive ideal was overshadowed by the consumer ideal. Success meant enjoying more, enjoying better, and enjoying now. Work was pushed into the shadows; it seemed less valuable. The conspicuous consumption and record levels of consumer spending of the past few decades seem to be a cultural statement that work and the need to produce are not especially valuable. The debates about major issues often expose strong public opinions toward work. Promoting consumerism aims to make consumers enjoy lower prices and better products in a free trade environment. What that means, though, is that those who produce the products will receive less consideration. Those who make high-quality, useful products are vulnerable to those who can make a substitute quicker or cheaper. In this environment, work and workers appear to be less valued and less valuable than consumers. Under these circumstances, when work is difficult or boring, it is likely to be more stressful (Everly and Girdano 1980). It may not seem to be a valuable activity; it may not seem to lead anywhere; it may not seem worth the trouble.

Emergency response work is highly susceptible to stress from these factors. Since much of the work is in the public sector and the remainder is contracted to or tightly regulated by public agencies, the work and the workers are extremely vulnerable to public opinion and political changes. Once again, the working conditions of emergency responders may not provide the predictability, understanding, and control necessary to minimize stressors on the job.

Acute and Chronic Stressors

One of the many ways to categorize workplace stress for emergency responders is acute and chronic.

"Acute stress" describes a human response to a single powerful event or overwhelming situation. Pain, horror, outrage, humiliation, jealousy, and conflict may all produce a sense of unbearable pressure. The acute stress originates in situations that are clearly beyond the normal experience or expectation. Most workers do not face the high risk of potential danger that is possible. Emergency response workers must live with that burden. Even heroic work can bring about acute stress. Responders may experience tremendous pressure when they see injury or death. Victims with any resemblance to a worker's family members may induce an overwhelming depression (Bryant, 1991).

"Chronic stress" is associated with the workplace pressure that accumulates day after day. General conditions, social structures, and interaction patterns, feelings of uncertainty, incompetence, or information deprivation may contribute to an individual's feeling unable to control his or her life and worried about what might happen next. Chronic stress originates in the daily routine, and the accumulation of those responses becomes more than the individual can bear without damage. Emergency response work that requires shift work, sleep interruption, and/or meal interruption can add significantly to accumulated stress.

SYMPTOMS OF STRESS

Stress is a part of living and working. It becomes a threat to the worker and the work organization when there is too much or too often, or when it lasts too long. The criteria of intensity, frequency, and duration help to distinguish between acceptable and unacceptable levels of stress. There is some difficulty in describing the nature of the symptoms of stress because each individual may have a different tolerance level to a specific stressor. The types of reactions will also vary by individual. What seems to happen is that the stress symptoms may be related to the strengths and weaknesses of the particular person. "Stressed out" may mean headaches, stomach pain, sleeplessness, tension, fatigue, weight gain, or even weight loss. Workers who find that they are undergoing changes in their body, mind, emotions, or habits may want to examine their work and life to determine whether or not they are suffering from excess stress. Anyone who can say "I just do not feel like my old self anymore" may want to find out why the change has occurred.

REACTION TO STRESSFUL SITUATIONS

If stress is the body's or mind's response to those factors which produce too much pressure, what are the likely responses? Early studies of behavior in stressful situations identified two general responses known as "fight" and "flight." The observation of animal behavior led to the conclusion that in times of great danger the quasi-instinctual drive propelled the animal to attack the danger or run from it. The animal's body provided chemical fortification to fight harder or run faster than usual. In stressful situations, human organisms react similarly to their animal counterparts (Karasek and Thorell 1990). When facing acute pressure, the human inclination should be to flee the situation or to engage immediately in aggressive action to overcome the danger. The human body provides adrenocorticotropic hormone, which stimulates the glands to increase energy resources (Workplace Health Fund 1991). This hormonal stimulus may provide temporary energy, but not without costs to the body. Too much stress may result in physical deterioration.

Many stress problems originate with the accumulation of pressure over a period of time rather than in a single traumatic incident. Humans with problems of accumulated stress are just as likely to withdraw and become passive or helpless as they are to fight or flee. If they exhibit aggressiveness, it may not be aimed at the problem but rather at a less formidable substitute.

Fortunately, humans seem to have other ways to handle stressors. They can take steps to increase their personal resistance to the stressors they are likely to encounter, or they can find ways to identify and remove or reduce the stressors that are harmful.

SHAPING UP TO MANAGE A STRESSFUL JOB

Emergency responders should constantly seek out stressors in the workplace and take steps to eliminate or reduce their effects. Since a low-stress workplace does not seem a likely possibility in the near future, workers need to protect themselves from the disabling effects of stress (Ellis 1978). There are several practical steps that help reduce the effects of working in a stressful environment (Ross 1985).

Relaxation

The first step in combating stress is learning how to relax. There are several books and training techniques that can assist persons in learning how to relax. Relaxation techniques often include slow, deep-breathing responses to stressors with control that signals the body to slow down and the emotions to cool off. Shutting out distractions, sitting comfortably, gently stretching muscles, and instructing the mind to relax seem to help counter the body's stress reaction.

Exercise

Exercise is recommended as a measure to counter the harmful effects of stress. Regular moderate exercise helps to keep the body fit, but it also helps to induce relaxation. It helps to bring balance to a person who suffers from mental fatigue along with physical restlessness. A third major advantage of exercise is that it gives the participant a sense of control and concrete accomplishment.

Structure

People who are susceptible to job stress may find that imposing a concrete structure on their time may help overcome some negative effects. A self-imposed structure gives the person a sense of control. Building a personal structure should take time and analysis. Generally, the individual will want to devote sufficient time to find some sense of accomplishment, as well as allowing time for activities that are enjoyable. Even though it sometimes seems as if there is never enough time, most people who scrutinize their lifestyle will find that they may be spending too much of their time in ways that are neither productive nor enjoyable, or perhaps too much time with people who are both counterproductive and unpleasant.

Meditation

The word "meditation" may conjure up an image of an Eastern religious practice. In reality, meditation is simply a method for shutting out distractions and finding a sense of quiet and control in a noisy, frantic world. Though one can spend time and money learning complex meditation techniques, it is possible to meditate without outside help. Simply finding a quiet, comfortable place for a period of 15 or 20 minutes, once or twice a day, is the major requisite. Some may find that concentration on a complete blank or an empty space may clear their troubled minds, while others will want to focus on a pleasant, calming picture.

Biofeedback

Those who feel that they are disabled by the effects of stress may want to seek medical attention. Expensive instruments capable of monitoring and reporting the vital signs of the body under certain conditions can help them understand their reactions, as well as the conditions that bring about negative reactions. Armed with this information, they can take steps to avoid or control the major stressors in their lives. The principles of this process, known as "biofeedback," may be used on a simpler level. Those who may feel some negative effects of stress but do not need medical attention can still benefit from monitoring their blood pressure, cholesterol level, weight, appetite, amount of sleep needed, and physical or emotional changes. By observing and evaluating basic indicators, individuals may be able to give themselves greater control.

Personal practices, including poor nutrition, tobacco, drugs, too much alcohol, or too little sleep, may be increasing the negative effects of stress. These habits can also be changed.

REDUCING STRESSORS ON THE JOB

Someone devoted to staying in shape can make his or her life a little easier, but the long-term goal of combating the negative effects of workplace stress needs to address the factors in the workplace that are causing stress. Stressors must be identified. Ideally, a comprehensive plan for removing or reducing the effects of those stressors should be developed. Finally, the plan should be implemented as soon as possible. In general, small, autonomous public responder teams will have more flexibility to make changes, while industrial hazmat teams may find the process a long one.

In the attempt to build a workplace that is less stressful, the workers will want to give some sense of *predictability, understanding,* and *control* to each member of the work team.

Change is the factor that is most likely to rob the emergency responder of any ability to predict the daily routine and envision the future on that job. Since work is not likely to stop changing and the rate of change is not likely to slow, stress can be reduced by including broad workplace representation in decisions about introducing change. Giving the workers a voice in the decision-making process makes them an active part of the process rather than a passive receptor of the change. Comprehensive training related to the changes should also help alleviate some of the associated stress. Fear of the unknown, and particularly of the unknown dangers, adds to the possibility of traumatic stress. Training preparation and team-building exercises help to create the confidence useful for facing danger.

Some of the major stressors on the job can be eliminated by helping each member of the work team understand the organization, the nature and procedures of the job, and the needs and expectations of co-workers. Perhaps the best tools for combating the effects of stress are groups, techniques, and instruments that help each individual to understand the self. Many emergency responders have reduced some of their burdens by creating peer counseling structures.

Workers who have little control over their lives and their work appear to be among those with highest stress levels. Scientific management theories of the past century have suggested that management should centralize control of the work organization. However, there is strong evidence that decentralized organizations which encourage worker participation in decisions not only produce higher morale but also provide the organization with previously untapped resources—the skills, attitudes, and commitment of the workers.

Wise planning, strong training, and discerning organizations will probably never eliminate stress from emergency response jobs. In order to reduce the stressors that cannot be eliminated, work teams should examine the individual worker's exposure to inherent stressors of the particular job. By examining the frequency, intensity, and duration of stress exposures, measures can be created to provide more breaks, rotate stressful duties, and reduce the time that the worker will be subjected to high stressors. Organizational accommodation and job design can change stress levels.

Kenneth R. Pelletier (1984) suggests that in order to bring about a significant reduction of stress in the workplace, a major overhaul of the workplace and the worker must include "a new work ethic based upon cooperation, empowerment of the individual and nondestructive lifestyles." Danger, boredom, unpredictable situations, restrictive equipment, and frequent procedural change often contribute to the stressfulness of emergency response work. Structural change can remove many of the stressors. Organizations and individuals can also develop practices to help manage stress. In any case, for emergency responders to lead healthy, happy, and productive lives, reducing and managing the stressors must become an important priority.

REFERENCES

A BNA Special Report. 1987. Stress in the workplace: costs liability and prevention. Washington, DC: The Bureau of National Affairs.

Bennett, A. April 22, 1988. Is Your Job Making You Sick? *Wall Street Journal Reports:* Medicine and Health, p. B1.

Bryant, W. June 16, 1991. Voluntary Counselors Helping Firefighters Battle Job Stress. *Birmingham News,* p. 25A.

Davidson, M., and C. Cooper. 1983. *Stress and the Woman Manager.* New York: St. Martin's Press.

Ellis, A. 1978. What People Can Do for Themselves to Cope with Stress. In *Stress at Work,* ed. C.L. Cooper and R. Payne, pp. 209–222. New York: John Wiley & Sons.

Epstein, S.S., L.O. Brown, and C. Pope. 1982. *Hazardous Waste in America.* San Francisco: Sierra Club Books.

Everly, G.C., Jr., and D.A. Girdano. 1980. *The Stress Mess Solution: The Causes and Cures of Sress on the Job.* Bowie, MD: Prentice-Hall Co.

Fiedler, N. March 1990. Understanding Stress in Hazardous Waste Workers. *Occupational Medicine* 5(1):101–108.

International Labour Office, 1984. *Automation, Work Organization and Occupational Stress.* Geneva: ILO Press.

Karasek, R., and T. Theorell. 1990. *Healthy Work: Stress, Productivity, and the Reconstruction of Working Life.* New York: Basic Books.

Lauffer, A. 1985. *Careers, Colleagues, and Conflicts: Understanding Gender, Race, and Ethnicity in the Workplace.* Beverly Hills, CA: Sage Publications.

McCarthy, M.J. April 7, 1988. Stressed Employees Look for Relief in Workers' Compensation Claims. *Wall Street Journal,* p. B1.

Pelletier, K.R. 1984. *Healthy People in Unhealthy Places: Stress and Fitness at Work.* New York: Delacorte Press.

Poulton, E.C. 1978. Blue Collar Stressors. In *Stress at Work,* ed. C.L. Cooper and R. Payne, pp. 51–80. New York: John Wiley & Sons.

Ross, K.C. 1985. *A Guide to Managing Stress.* Daly City, CA: Krames Communications.

Serrin, W. December 9, 1984. White Men Discover It's a Shrinking Market. *New York Times,* p. 2E.

Workplace Health Fund. 1991. *Ergonomics Training Manual.* Washington, DC: U.S. Government Printing Office.

18

Temperature Stress in Emergency Response

Barbara M. Hilyer, M.S., M.S.P.H., and
Kenneth W. Oldfield, M.S.P.H.

The Houston, Texas, hazmat team is one of the best-trained and most highly experienced hazardous materials response teams in the United States. In 1990 they responded to 682 incidents involving 190 different chemicals. The team reported only five lost-time injuries: two resulted from a vehicle accident en route to a scene, and three were the result of heat stress. Preparation and planning enabled the team to work safely and efficiently in 681 chemical incidents, but they could not completely control the temperature stress factor inherent in one release.

Stress from temperature extremes is often a concern during emergency response. Work mut be done no matter what conditions of temperature, wind, and humidity are present at the scene. Work often must be done while wearing protective clothing which interferes with the effeciveness of the body's thermoregulatory system. Hazmat responders stay at the scene until the incident is terminated, often without regard for extremely uncomfortable, perhaps even dangerous, temperature conditions. This chapter will discuss potentially dangerous hot and cold incident conditions, and give suggestions for reducing the hazard presented by heat stress and cold stress during emergency response.

OBJECTIVES

- Be able to identify conditions likely to lead to heat and cold stress
- Recognize the symptoms of dangerous physiological responses to heat and cold
- Know how to prevent and treat heat-induced and cold-related illnesses

HEAT STRESS

Four thousand people die in the United States each year from illnesses brought on by heat stress. Although stress from cold temperatures is sometimes encountered during hazmat response, heat stress is more prevalent, more dangerous, and more readily adapted to by a fit, acclimatized individual. Heat-related illnesses are sometimes serious and can be fatal; emergency responders should prepare for work in hot environments by learning the prevention, symptoms, and treatment of heat-induced illnesses.

Sources of Heat

Two sources of heat can cause an increase in body temperature: environmental and metabolic heat. Both can be controlled to some extent during a hazardous materials response.

Environmental Heat

Environmental heat comes from sources outside the body. It may be present in the form of high ambient air temperature, infrared radiation from the sun or heat-generating equipment, or heat from a fire or chemical reaction. Not all of the suggestions made by industrial hygienists and governmental agencies for controlling environmental heat are feasible in emergency response, since responders cannot choose the time and location of the incident. Procedures suggested by health professionals which can be used to reduce environmental heat during hazmat incidents include providing shade for rest breaks and selecting PPE which is not unnecessarily burdensome.

Metabolic Heat

Metabolic heat is produced by the body during cellular activities. It is increased when the body's workload increases. In work situations where tasks are repetitive and can be anticipated, the workload can be measured and the amount of metabolic heat produced predicted so that schedules can be adjusted in hot weather. Workload reduction is based on a formula that estimates total heat stress using combined environmental heat and predicted metabolic heat. Emergency responders do not work in highly predictable situations, and this sort of estimate would be difficult, if not impossible, for their duties. Responders can, however, implement good work practices, since metabolic heat production can be decreased by working more slowly and utilizing more human or mechanical help on heavy tasks.

Workers who are in good physical condition can perform tasks more efficently with less effort. Since the energy expended in work efforts generates heat and adds to the metabolic heat load, physically fit workers are able to work harder and for longer periods, with less metabolic heat production, than workers who are not physically fit.

Thermoregulation in Hot Environments

The human body has a thermoregulatory system which maintains body temperature at approximately 98.6°F. The "thermostat" is located deep within the brain, in an area called the hypothalamus. The hypothalamus receives input from specialized temperature-sensing cells throughout the body and, if the temeprature drops or rises, this part of the brain directs other organs to initiate processes to warm or cool the body. Two cooling responses occur when they are stimulated by the brain following an increase in body core temperature: vasodilation (enlargement of blood vessels) in the skin and sweating.

Vasodilation

Vasodilation is the enlargement of tiny blood vessels called capillaries. Enlargement of capillaries in the skin allows the volume of blood at the body surface to increase. The blood, having been warmed as it circulated throughout the body, can now dissipate heat by radiation into the air. Heat radiation is effective whenever the temperature of the air is below 95°F.

Sweating

If surface vasodilation is not effective in lowering body temperature, the sweating response is stimulated. Sweat glands begin to produce sweat and deliver it to the skin surface via ducts. There it evaporates, changing from liquid to vapor. Heat is used to change water into water vapor in the evaporative process; the heat is taken from the body. If the surrounding air is very humid, at or approaching saturation with water vapor, sweat will not evaporate and no cooling takes place. Since the message from the sensory cells to the brain still reads "hot," sweat continues to be produced even though it does not evaporate.

Heat-Induced Illness

In situations where the body's thermoregulatory mechanisms are unable to reduce the core temperature, illness will result. The illness may be heat rash, heat syncope (fainting), heat cramps, heat exhaustion, or heat stroke.

Heat Rash

Heat rash occurs when sweat is prevented from evaporating, usually by clothing. The outer skin layer of dead cells becomes saturated, swelling and clogging the sweat ducts. Sweat glands continue to produce sweat (remember that temperature reduction is the only thing that will turn sweat production off), and the sweat seeps into the deep skin layers and causes irritation. Tiny red, raised blisters appear on the surface of the skin, and a pricking sensation is felt. Allowing the skin to dry under cool, dry conditions will stop heat rash. Since clothing is required for skin protection during emergency response, it may not be possible to prevent heat rash.

Heat Syncope

Heat syncope usually occurs when a person is standing still in a hot environment. Skin vasodilation results in an increase in the volume of blood in these vessels; with blood also collecting in the legs and feet, the resulting shortage of blood to the brain may lead to dizziness and fainting. This is not a serious heat illness unless it is accompanied by symptoms of heat exhaustion; it can be prevented by simply walking around. Resting in a reclining position allows blood to circulate to the brain again, and full recovery is prompt and complete.

Heat Cramps

Heat cramps occur in working muscles that are deficient in sodium, potassium, calcium, or other electrolytes. Profuse sweating may deplete the body of electrolytes, especially in an unacclimatized person, causing electrolyte-deficient muscles to contract involuntarily. Treatment involves immediate infusion of intravenous electrolyte solutions. Orally administered electrolyte replacement drinks or salted water will probably resolve the problem quickly. To prevent muscle cramps, a balanced diet should be eaten, and responders should acclimatize themselves to heat.

Heat Exhaustion

Heat exhaustion is a more serious state characterized by fatigue, nausea, headache, dizziness, pallor, and profuse sweating. It usually occurs in the setting of sustained exertion in hot conditions when an individual is dehydrated from deficient water intake. Oral temperature may be elevated, but not in every case. Heat exhaustion can progress to heat stroke if the victim is not cooled. The individual suffering from heat exhaustion should rest in a cool place and drink water until the urine volume indicates that body fluids have been replaced. She should not return to active work that day.

Heat Stroke

Heat stroke is a serious condition requiring immediate medical treatment. The sweat resources have been depleted, so the skin is dry and hot but may still appear red due to vasodilation. Heat stroke quickly leads to collapse, delirium, and coma. Eighty percent of people who suffer full-blown heat stroke die, and one-half of the surviving 20 percent suffer brain damage.

The onset of heat stroke may be rapid, with very little warning to the victim that a crisis stage has been reached. Since the first symptoms include confusion and impaired judgment, the buddy system is vital in hot environments.

Heat stroke victims must be cooled immediately while waiting for EMS personnel. Clothing should be removed and the victim continuously sprinkled with water or wrapped in a thin, wet sheet. Fanning to increase air movement and evaporation should be initiated. The patient should not be immersed in extremely cold water, since that action may trigger vasoconstriction in the skin and shut down the one cooling mechanism still operating.

Prevention of Heat-Induced Illness

Reducing the Heat Stress

Working in hot environments cannot be totally prevented in hazmat response, but it can be reduced. Isolation of the worker from hot environments is not possible, since a hazmat response is conducted at unpredictable and uncontrollable times. It is possible, however, to reduce the amount of time a responder spends in hot conditions by enforcing frequent rest breaks, and to reduce metabolic heat by providing help with tasks which necessitate heavy work.

Reducing the Individual's Response to Heat Stress

Once environmental and metabolic heat have been reduced as much as possible by the means suggested above, the hazmat responder depends on physiological responses to cool the body. Beneficial responses can be enhanced, and dangerous responses reduced, by consideration and control of the factors that influence an individual's response to heat stress.

Hydration

Hydration (providing water to body cells) is an important factor in one's response to heat stress. Drinking water must always be available to responders, and they should be reminded to drink water often, both on and off the job. It is possible to lose 1 liter of fluid each hour by sweating. Since the body's total fluid volume is only around 40 liters, it is easy to see why fluid must be replaced. If hypohydration (inadequate body fluid) exceeds 1.5 to 2 percent of body weight, tolerance of heat stress begins to deteriorate, heart rate and body temperature increase, and work capacity decreases. When hypohydration exceeds 5 percent, it may lead to collapse. Since the feeling of thirst is not an adequate guide for water replacement, responders working in heat should be encouraged to drink water every 15 to 20 minutes. The daily amount of hypohydration can be estimated by measuring body weight after work and comparing it to that morning's baseline weight; it should not exceed 1.5 percent.

One of the most important physiological factors causing hypohydration is alcohol consumption. Alcohol is a diuretic that depresses the body's production of a kidney hormone that prevents water loss. When the hormone is in short supply, more urine is produced and more water is lost from the body. Responders who consumed alcohol the night before are already on the way to hypohydration when they arrive at a hazmat incident.

Acclimatization

Acclimatization to hot conditions has been found to be effective in preventing heat-related illnesses. The process takes 5 to 7 days, during which work hours in the hot environment are gradually increased. During 7 days of acclimatization, heart rate and body temperature become lower while performing the same work. The major physiological mechanism involved in acclimatization is sweating; acclimatized persons sweat sooner and more profusely, and produce more dilute sweat, than do unacclimatized individuals. Unacclimatized individuals lose four times the volume of electrolytes in a liter of sweat that acclimatized persons lose.

Acclimatization schedules may be difficult to adhere to, given the sporadic nature of emergency response. Industrial hazmat teams whose members will be called on to respond in hot areas should limit membership to individuals who can acclimatize. It is inappropriate to assign persons who work entirely in air-conditioned environments to active positions on hazmat teams whose responses can be expected to include heat stress.

Aging

The aging process results in a more sluggish response of the sweat glands, which leads to less effective control of body temperature. One study of 5 years' accumulation of data on heat stroke in South African gold mines found a marked increase in heat stroke with increasing age of the workers.

Men over 40 made up less than 10 percent of the workers, but they accounted for 50 percent of the fatal and 25 percent of the nonfatal cases of heat stroke.

Drug Consumption

Many drugs prescribed or used for therapeutic purposes can interfere with thermoregulation. Almost any drug that affects central nervous system activity, cardiovascular reserve, or body hydration can potentially affect heat tolerance. Some over-the-counter medications, including those that lessen the symptoms of colds or allergies, fall into this group. A responder taking medications should seek the guidance of a physician in evaluating tolerance of heat.

Alcohol has been mentioned earlier regarding its effect on hydration. Alcohol has been commonly associated with the occurrence of heat stroke. Some drugs that are used on social occasions have been implicated in cases of heat disorder, sometimes leading to death.

Disease

Individuals suffering from nonheat-related disorders, such as degenerative diseases of the cardiovascular system and diabetes, are in extra danger when they are exposed to heat, particularly when a stress like hard work is imposed on the cardiovascular system.

Physical Fitness

In the absence of disease or the use of medications or drugs (including alcohol), the most effective way for an emergency responder to improve heat fitness is to improve physical fitness. All the components of overall fitness are important; the two most critical ones for heat fitness are body composition and aerobic capacity.

It is well established that overweight and obesity predispose individuals to heat disorders. Fat provides an insulating layer which prevents heat dissipation, adds weight requiring more energy to carry, and changes the body surface to body volume ratio, further reducing surface heat dissipation.

Trained individuals with relatively high aerobic capacity produce dilute sweat and sweat more profusely than untrained persons, thereby cooling their bodies better. They also generate less metabolic heat for a given amount of work because they can do it more easily and efficiently.

Individual Variation

Studies have been done to determine if gender affects heat acclimatization. Results of these studies indicate it does not; men and women can acclimatize equally well if they meet the other heat acclimatization criteria. However, individuals vary in their heat stress tolerance. In all experimental studies of the responses of humans to hot environmental conditions, wide variations in responses have been observed. These variations are seen not only between different individuals but also, to some extent, in the same individual exposed to high heat stress on different occasions. Such variations are not totally understood.

Screening for Heat Tolerance

It would be helpful to know which individuals are likely to succumb to heat stress before making assignments to an emergency response team. Making this determination without placing hazmat team applicants in potentially harmful hot conditions can, and should, be accomplished.

NIOSH recommends that all individual who will be required to work under hot conditions have a medical examination prior to assignment to the job. The examination should include

- A comprehensive work and medical history, with special emphasis on information about any previous heat illnesses or heat intolerance
- A comprehensive physical examination that gives special attention to the skin, liver, and kidneys, and to the nervous, cardiovascular, and respiratory systems
- An assessment of the use of therapeutic, over-the-counter, or social drugs, including alcohol
- An assessment of obesity
- An assessment of the worker's ability to wear any protective clothing and equipment that might be required on the job and may add to heat stress

The screening technique that has been found to correlate most strongly with heat fitness is the max VO_2 evaluation. This is an estimate of the maximum amount of oxygen an individual can supply to working muscles and of cardiorespiratory endurance. Methods for assessing max VO_2 are described in Chapter 19. It is well documented that people who train for cardiorespiratory fitness improve their max VO_2 scores. It follows that people can train in the same ways to improve their ability to withstand work in hot environments.

Monitoring for Heat Strain

Monitoring for heat strain should be part of the medical monitoring carried out during an incident response. Several physiological parameters can be measured to judge an individual's response to heat. The two measures which are easiest to monitor under incident conditions are heart rate

TABLE 18-1. Suggested Frequency of Physiological Monitoring for Fit and Acclimatized Workers[a]

Adjusted Temperature[b]	Normal Work Ensemble[c]	Impermeable Ensemble
90°F (32.2°C) or above	After each 45 minutes of work	After each 15 minutes of work
87.5°–90°F (30.8°–32.2°C)	After each 60 minutes of work	After each 30 minutes of work
82.5°–87.5°F (28.1°–30.8°C)	After each 90 minutes of work	After each 60 minutes of work
77.5°–82.5°F (25.3°–28.1°C)	After each 120 minutes of work	After each 90 minutes of work
72.5°–77.5°F (22.5°–25.3°C)	After each 150 minutes of work	After each 120 minutes of work

[a] For work levels of 250 kilocalories/hour.
[b] Calculate the adjusted air temperature (ta adj) by using this equation: ta adj °F = ta °F + (13 × % sunshine). Measure air temperature (ta) with a standard mercury-in-glass thermometer, with the bulb shielded from radiant heat. Estimate percent sunshine by judging what percent time the sun is not covered by clouds that are thick enough to produce a shadow. (100 percent sunshine = no cloud cover and a sharp, distinct shadow; 0 percent sunshine = no shadows.)
[c] A normal work ensemble consists of cotton coveralls or other cotton clothing with long sleeves and pants.

Source: NIOSH/OSHA/USCG/EPA. 1985. *Occupational Safety and Health Guidance Manual for Hazardous Waste Site Activities*. Washington, DC: U.S. Government Printing Office, p. 8–22.

and oral temperature. Baseline pulse and temperature should be taken at the beginning of every workday for comparison.

Table 18-1 provides suggested times for monitoring breaks. The environmental temperatures listed there were measured with a plain thermometer that does not take humidity into consideration. For responders who do not wear impermeable ensembles, times should be adjusted downward in humid weather, with monitoring and rest occurring more frequently.

Temperature is the best indicator of the response to heat stress. Rectal temperature is the most reliable indicator of body core temperature, but monitoring rectal temperature is not feasible in hazmat incident situations. Forehead patch devices should not be used, as they may be inaccurate and may be loosened by sweating. Oral temperature should be measured at the end of a work period, before drinking, by placing a clinical thermometer under the tongue. If the oral temperature exceeds 99.6°F, the next work period should be one-third shorter than the last and the rest period the same length. If the oral temperature still exceeds 99.6°F at the end of the next work period, the following work period should again be shortened by another third. Monitoring and adjustment of work periods should continue until the worker's initial temperature is less than 99.6°F. No personnel should be permitted to wear semipermeable or impermeable clothing if oral temperature exceeds 100.6°F.

Heart rate provides an additional indication of heat strain. Heart rate can be detected at several pulse locations; the most easily accessible are the radial pulse on the inside surface of the wrist and the carotid pulse on either side of the adam's apple. The pulse should be taken during the first minute of seated rest following work and should not exceed 110 beats per minute (bpm). If it does exceed 110 bpm, the following work cycle should be shortened by one-third. Monitoring and shortening of work cycles should continue until the responder's heart rate during the first minute of rest is lower than 110 bpm. A method which can be used to determine if an individual is recovering from an overheated condition involves taking the pulse during the first and third minutes of seated rest. If the initial heart rate is over 110 bpm but falls by 10 or more bpm by the third minute, one would conclude that the work is strenuous but the individual's thermoregulatory system is responding adequately to the stress.

COLD STRESS

Just as emergency responders work in hot environmental conditions, they also respond to hazardous materials incidents when the temperature is low. Water is often present at an incident, either in the form of rain or as part of the response, and wet conditions exacerbate cold injury. Responders may be injured by working in cold conditions and must take appropriate steps to reduce this hazard.

Conditions Leading to Injury

Three factors must be considered in assessing the danger of cold injury during a hazmat response: air temperature, wind velocity, and the presence of water in the work area.

Temperature

Although it is obvious that physiological damage from cold occurs more readily at low temperatures, it is impossible to define a temperature at which injury may begin. Assessment of the cold injury hazard must include consideration of the other important factors: wind velocity and the presence of water. Since the body's thermoregulatory mechanisms are most efficient when they lead to the creation of a layer of warm air around the exterior of the body, wind lessens the efficiency of the warming mechanisms when the warm air is blown away. When a person becomes wet, body heat is lost even faster; in fact, this is the reason sweating is an effective means of heat dissipation in hot environments. Hypothermia has been known to occur in air temperatures as high as 65°F, or in water at 72°F, especially when the individual is fatigued.

Wind Velocity

Wind chill is used to describe the chilling effect of moving air in combination with low temperature. For example, an air temperature of 10°F with a wind velocity of 15 miles per hour (mph) is equivalent in chilling effect to still air at −18°F. As a general rule, the greatest incremental increase in wind chill occurs when a wind of 5 mph increases to 10 mph. Table 18-2 shows the cooling power of wind.

The Presence of Water

Water conducts heat away from the body 240 times faster than still air does. An emergency responder whose clothing and skin are wet is at greater risk of cold injury than one who remains dry. Even inside boots and garments which are impervious to water, the body cools faster if the outer surface of the clothing is losing heat to water. A responder standing in water while wearing waterproof boots can loose a great deal of heat from the feet, as the interior air space of the boot is cooled by heat loss through the boot material.

Another source of water on the skin is perspiration if the responder has become overheated while working inside protective clothing. Removal of the clothing exposes the wet skin to cold air, resulting in rapid cooling through the evaporation of sweat. A responder who was overheated can rapidly become chilled by this means.

Thermoregulation in Cold Environments

The same portion of the hypothalamus of the brain that sets cooling mechanisms in motion when the core temperature rises also initiates heating responses when the core temperature drops. The two warming mechanisms are surface vasoconstriction and shivering.

Vasoconstriction

Since warmed blood loses heat at the surface of the body, reducing the flow of blood at the surface decreases heat loss. When body temperature drops, skin capillaries constrict; as they become smaller in diameter, they hold a smaller volume of blood. Heat is retained inside the body. One of

TABLE 18-2. Cooling Power of Wind on Exposed Flesh Expressed as an Equivalent Temperature Under Calm Conditions

Estimated Wind Speed (in mph)	Actual Temperature Reading (°F)											
	50	40	30	20	10	0	−10	−20	−30	−40	−50	−60
	Equivalent Chill Temperature (°F)											
Calm	50	40	30	20	10	0	−10	−20	−30	−40	−50	−60
5	48	37	27	16	6	−5	−15	−26	−36	−47	−57	−68
10	40	28	16	4	−9	−24	−33	−46	−58	−70	−83	−95
15	36	22	9	−5	−18	−32	−45	−58	−72	−85	−99	−112
20	32	18	4	−10	−25	−39	−53	−67	−82	−96	−110	−121
25	30	16	0	−15	−29	−44	−59	−74	−88	−104	−118	−133
30	28	13	−2	−18	−33	−48	−63	−79	−94	−109	−125	−140
35	27	11	−4	−20	−35	−51	−67	−82	−98	−113	−129	−145
40	26	10	−6	−21	−37	−53	−69	−85	−100	−116	−132	−148
(Wind speeds greater than 40 mph have little additional effect.)	LITTLE DANGER In chr with dry skin Maximum danger of false sense of security.				INCREASING DANGER Danger from freezing of exposed flesh within 1 minute.			GREAT DANGER Flesh may freeze within 30 seconds.				
	Trench foot and immersion foot may occur at any point on this chart.											

Source: U.S. Army Research Institute of Environmental Medicine, Natick, MA.

the dangers of becoming exhausted in a cold environment is that the vasoconstrictive protective mechanism becomes overwhelmed, resulting in sudden vasodilation and rapid heat loss.

Shivering

Metabolic heat is produced by working cells; working muscle cells are set into motion by the thermoregulatory system when body temperature drops. Shivering is the involuntary contraction of muscle fibers. These contractions produce energy in the form of heat and help the body maintain its optimal temperature. Shivering is an obvious indicator that body temperature has dropped.

Harmful Effects of Cold Stress

Responses to cold conditions range from discomfort to death. They include several stages of frostbite, may include trench foot under wet conditions, and can result in general hypothermia.

Frostbite

Frostbite is freezing of tissue due to exposure to extreme cold or contact with extremely cold objects. Wind chill can play an important role in accelerating frostbite. Frostbite is characterized by sudden blanching or whitening of the skin. If the tissues are cold, pale, and solid, deep frostbite has occurred; this is a serious injury that may lead to necrosis and loss of fingers, toes, or other affected parts.

Treatment of frostbite begins with slow, careful warming of the affected part. It should be immersed in water maintained at 102°–105°F (comfortably warm to the inner surface of an unchilled forearm). Warming should be discontinued as soon as flushing indicates the return of blood. The injured part should be elevated after being warmed, and contact between it and any surface except a sterile bandage should be prevented. The victim should not be allowed to walk on a frozen foot, but should exercise a thawed part of the body by moving it around.

Trench Foot

Trench foot, also called "immersion foot," is a condition resulting from long, continuous exposure to damp and cold while remaining relatively immobile. It can progress from a stage where the foot lacks an adequate blood supply to one in which gangrene occurs. Since emergency responders usually wear waterproof boots, and since they seldom stand still for long periods of time, they are unlikely to suffer from trench foot. The same condition can, however, affect the tip of the nose and ears under cold, wet conditions of immobility.

General Hypothermia

Individuals subjected to prolonged cold exposure and physical exertion are at risk for general hypothermia, the cooling of the entire body. Sweating and the fatigue of the vasoconstrictive response add to the hazard. Symptoms are usually exhibited in four stages: (1) shivering; (2) apathy, listlessness, sleepiness, and sometimes rapid cooling of the body to less than 95°F; (3) unconsciousness, glassy stare, slow pulse, and low respiratory rate; and (4) freezing of the extremities. Coma and death can result from general hypothermia if the cooling process is not reversed. The danger of hypothermia is increased by the consumption of sedatives and alcohol.

Prevention of Cold Injury

There is not a great deal a hazmat responder can do to acclimatize himself to working in cold temperatures. The body does not adapt physiologically to cold as well as it does to heat. Staying in good physical condition with an efficient cardiorespiratory system, and not ingesting alcohol and other drugs, will enhance an individual's thermoregulatory capability. Generalized prevention strategies focus on safe work cycles, with adequate rest in a warm place and keeping a warm air layer around the body.

Work and Rest Cycles

Work and rest cycles to prevent cold stress can be based on the individual's response to the cold. When shivering begins, the body's temperature has dropped below the optimum level. If shivering continues or grows stronger, an inability to warm the body is indicated, and the worker is warned that it is time to move to a warmer location. Some sort of warm area should be provided at an incident to enable chilled responders to warm up.

Protective Clothing

Many fabrics are available which greatly enhance body heat retention. The most efficient natural fibers for heat retention are silk and wool; however, new synthetic fibers are available which have excellent insulating capabilities and offer the additional advantages of light weight and rapid drying. In addition to holding a warm layer of air next to the body, most of the new synthetic insulating fabrics wick moisture away from the body to the outer surface of the garment. Garments made from these fabrics range in thickness from thin thermal underwear and glove liners to thick pile clothing. Many of the new garments, which were developed for mountain climbing and other cold-weather outdoor sports, are extremely lightweight compared to natural fibers. They are most effective when worn under a tightly woven or otherwise wind-impervious outer garment.

It should be noted that at temperatures below 59°F the hands and fingers become insensitive long before cold injuries take place, thereby decreasing manual dexterity and increasing the risk of accidents. Hands and feet may be left with an inadequate blood supply when vasoconstriction in the extremities takes place in the body's effort to reduce heat loss, requiring especially well-insulated coverings. Sock liners of silk or polypropylene under double thicknesses of wool or polypropylene socks and boots that include insulating liners will help protect feet from cold. Glove liners and gloves with several separate layers provide the same protection for hands. Some of these can be worn under chemical-resistant gloves. Gloves which are tight or elastic should not be worn, as they enhance cold by inhibiting blood flow. The head should always be covered with a warm hat, and as much of the face as possible should be covered, since these areas are well provided with blood vessels and are locations where heat loss can be extensive.

SUMMARY

When a hazmat team responds to an accident that involves an actual or potential release of a hazardous material, the biggest concern will probably be the behavior of the materials involved. Obviously hazardous materials present real dangers to responders; however, other factors, such as heat and cold, can be equally dangerous under certain conditions. Heat stress and cold stress are often enhanced by the equipment and duties of emergency responders. Since some controls are possible with the use of appropriate measures, the hazards presented by heat and cold stress must not be overlooked.

The suggestions for prevention and treatment of heat and cold stress made in this chapter are derived from the references listed below and from conversations with two physicians, one working in occupational medicine and one with a physical fitness research background. Individual response plans should be written in consultation with medical and safety personnel experienced in the prevention and treatment of temperature stress and knowledgeable about the conditions and personnel specific to the team for which the plan is written.

REFERENCES

ACGIH. 1985. *Threshold Limit Values for Chemical Substances and Physical Agents in the Work Environment and Biological Exposure Indices with Intended Changes for 1984–1985*. Cincinnati: American Conference of Governmental Industrial Hygienists.

Favata, E.A., G. Buckler, and M. Gochfeld. 1990. Heat Stress in Hazardous Waste Workers: Evaluation and Prevention. *Occupational Medicine: State of the Art Reviews* 5(1):79–91.

Levine, S., and W. Martin, editors. 1985. *Protecting Personnel at Hazardous Waste Sites*. Stoneham, MA: Butterworth Publishers.

Levy, B.S., and D.H. Wegman. 1988. *Occupational Health, Recognizing and Preventing Work-Related Disease*, 2nd ed. Boston and Toronto: Little, Brown and Co.

NIOSH. 1986. *Occupational Exposure to Hot Environments. Revised Criteria*. Washington, DC: U.S. Department of Health and Human Services.

NIOSH/OSHA/USCG/EPA. 1985. *Occupational Safety and Health Guidance Manual for Hazardous Waste Site Activities*. Washington, DC: U.S. Government Printing Office.

Strydom, N.B. 1971. Age as a Causal Factor in Heat Stroke. *Journal of the South African Institute of Mining and Metallurgy* 72:112–114. In; Favata, Buckler, and Gochfeld (1990).

19

Physical Fitness for Emergency Responders

James C. Hilyer, Ed.D., M.P.H., and
Lynn M. Artz, M.D., M.P.H.

Responding to a hazardous materials incident can be physically demanding. Responders may be required to move quickly in response to changing or unanticipated conditions. They may have to lift or move heavy objects or repetitively move light objects, sometimes for considerable distances. Responders may have to work hard for long periods of time without rest or work extremely hard for shorter periods of time without undue fatigue. These physical challenges often must be met while wearing heavy or hot protective equipment. The development of more sophisticated equipment and more efficient techniques continues to improve the safety outlook for responders, but the crucial factors in performance and personal safety are still the mental and physical capabilities of the human being performing the job.

Responding adequately and safely to the physical challenges of hazmat emergencies requires high levels of physical functioning and physical fitness. Special training and frequent assessment of physical fitness are necessary to ensure that appropriate response levels are maintained. Hazmat responders who are trained firefighters have probably experienced fitness testing and have been expected or required to maintain physical standards; however, private sector teams may not have been asked by their employers to meet these same kinds of standards. Physical fitness is specifically addressed by NFPA's Professional Qualifications Standards. The NFPA recognizes the importance of physical fitness in the successful job performance and personal safety of the firefighter. An annual job-related illness and injury rate of approximately 50 percent supports the rationale of the NFPA requirements.

OBJECTIVES

- Understand the different definitions of physical fitness and how each relates to physical functioning in job competence
- Be aware that physical fitness can improve job performance, decrease physical strain, and decrease the risk of injury
- Know how to attain and maintain a level of physical fitness appropriate to hazmat response
- Learn how physical fitness can be measured, and how these measurements can be used to screen potential members of the hazmat team and to monitor maintenance of quantified levels of fitness
- Be aware of the importance of maintaining appropriate levels of physical fitness, not only for successful job performance but also for personal safety and the potential for improved health and quality of life

HEALTH BENEFITS OF PHYSICAL ACTIVITY AND PHYSICAL FITNESS

Increasingly, research indicates that regular physical activity improves physical fitness and has health benefits for most people. Regular physical activity helps to prevent coronary heart disease and high blood pressure, important concerns for emergency responders. It can also help to prevent diabetes mellitus, osteoporosis, obesity, and mental health problems. Physical activity may be helpful in improving arthritis pain, low back pain, frequent headaches, and some digestive disorders. Regular physical activity may also reduce the incidence of stroke and help to maintain the functional independence of the elderly.

Because coronary heart disease is the leading cause of death in the United States, as well as the leading cause of on-the-job death and disability among emergency responders, the role of physical activity in preventing coronary heart disease is of special importance. The U.S. Department of Health and Human Services reports that physical inactivity is an independent risk factor for coronary heart disease, with a relative risk of 1.9. This means that physically inactive people are almost twice as likely to die from coronary heart disease as people who engage in regular physical activity. The relative risk is only slightly less than the relative risk for cigarette smoking, high blood pressure, or high blood cholesterol. Also, there are more people at risk for coronary heart disease due to physical inactivity than for any other single risk factor. Those who have other risk factors for coronary heart disease, such as obesity and hypertension, may particularly benefit from physical activity.

Few Americans and few emergency responders engage in regular physical activity, despite its potential health benefits. The guidelines established by the American College of Sports Medicine for exercise levels which give maximum protection against coronary heart disease recommend "exercise which involves large muscle groups in dynamic movement for periods of 20 minutes or longer, three or more days per week, and which is performed at an intensity of 60 percent or greater of an individual's cardiorespiratory capacity." Less than 10 percent of the American adult population exercises at that recommended level.

Approximately 26 percent of American adults are overweight. A major contributing factor to this alarming statistic is physical inactivity. Overweight is an independent risk factor for coronary heart disease. It also increases an individual's risk for high blood pressure, diabetes, and low back problems, and is associated with elevated serum cholesterol levels and noninsulin-dependent diabetes. Overweight increases the risk for gallbladder disease and some types of cancer, and is implicated in the development of osteoarthritis of the weight-bearing joints, particularly the knee.

Emergency responders are at risk for frequent muscle injuries, tendon strains, and ligament sprains, as well as joint injuries, especially to shoulders, knees and backs. Back injury is the most prevalent injury among the entire U.S. work force, and certainly among emergency responders. Injury statistics are not available for hazmat responders as a separate group, but shoulder injuries account for approximately 15 percent of all injuries to police officers and firefighters, who engage in activities similar to those of hazmat responders. Back injuries account for approximately 40 percent of all injuries to this same population. Considerable evidence suggests that increasing physical activity and improving levels of physical fitness can play a significant role in the prevention of musculoskeletal injuries. Data gathered during 1990 for firefighters and hazmat personnel in Birmingham, Alabama, show a significant correlation between fitness levels and injury rates (Table 19-1).

TABLE 19-1. Comparison of Fitness Rating and Injury Rates for the Birmingham, Alabama, Fire and Rescue Service, 1990

Fitness Rating	No. of Personnel	No. of Injuries	% of Injuries in Each Class
Excellent	106	15	14.1
Good	232	37	15.9
Average	136	37	27.2
Fair	54	37	68.5
Poor and very poor	6	11	183.3

DEFINING PHYSICAL FITNESS

Physical activity is defined as any bodily movement produced by skeletal muscles that results in energy expenditure. Physical fitness enhances the ability of an individual to perform physical activity to accomplish a predetermined objective or to reach a satisfactory level of performance.

Functional Physical Fitness

Functional physical fitness is the ability to perform physical activities necessary for daily living. When functional fitness is lost, a person usually requires human or mechanical assistance. Loss of the ability to walk because of trauma or disease to the spinal cord is an example of the loss of functional physical fitness. Maintaining good functional physical fitness is important to the quality of life.

Activity-Related Fitness

Activity-related physical fitness means different things for different people, depending on the activities they perform. Being fit for one activity does not necessarily imply that an individual is fit for a different activity. Having the physical capabilities of running 3 miles in 30 minutes, for example, does not necessarily mean that a person can successfully play par golf, since the physical activity requirements of the two achievements are vastly different.

There is a general transference of physical capabilities from one physical activity to any similar activity, but not to an activity that has dissimilar components. Each dissimilar activity and each level of performance within an activity has its own physical ability demands.

Job-Related Fitness

Just as physical fitness levels needed to perform certain activities at certain performance expectations are very specific, there are also levels of physical fitness that must accompany the successful completion of job tasks requiring physical activity. The level of activity and the degree of fitness required depend upon the job description and a task analysis of the job to be completed. Specific job analysis is relatively easy for jobs like assembly line jobs, where repetitive actions make up all the requirements of the job. There are specific physical tasks to be completed and there are measurable, predictable levels of physical activity required to do these tasks. This is not the case with emergency responder job descriptions. When emergency responses are initiated, a wide range of possible situations may greet the responder. The job description at the incident scene may require simple mop-up tasks, or it may require dealing with severe hazard conditions, weather extremes, and bulky equipment. The physical abilities needed to complete the tasks safely and successfully may require relatively low or extremely high fitness levels. The fitness level of the responder must be compatible with the most severe conditions. The responder must arrive at the emergency scene with the physical fitness capability to do all the tasks that may be necessary, and do them under extremely difficult circumstances.

The business of formulating job descriptions and measuring levels of physical fitness needed by emergency responders to complete their job tasks safely and successfully is difficult. At this time, a job task analysis for the determination of specific physical fitness criteria has not been completed for hazmat responders. However, the NFPA's professional qualifications for firefighters include specific minimum physical fitness requirements for entry-level firefighters, and these are probably useful guidelines for emergency responders of all kinds. These requirements are minimum standards only, and local departments may exceed them; however, they may not set lower standards if they are to be accredited by the NFPA. The following paragraphs quote these requirements.

NFPA 1001. 2-3. Minimum Physical Fitness Requirements

2-3.1 The candidate, after successfully completing the medical examination and with written authorization of the examining physician, shall complete one of the following:
- (a) Run 1½ miles within 13 minutes.
- (b) Walk 3 miles within 38 minutes.
- (c) Bicycle 4 miles within 12 minutes.
- (d) Swim 500 yards within 8 minutes and 20 seconds.
- (e) Run in place 75 steps per minute for 15 minutes.
- (f) Run on motorized horizontal treadmill at 10 miles per hour for 6 minutes.
- (g) Climb stairs consisting of 10 steps at 9 round trips per minute for 9 minutes.

2-3.2 The candidate will perform 35 bent-knee sit-ups within 2 minutes.

2-3.3 The candidate shall complete one of the following:
- (a) Flexed Arm Hang—Minimum Time: 8 Seconds (Palms Away)
- (b) Pull Ups—Minimum: 7 (Palms Away)
- (c) Push Ups (Standard)—Minimum: 25

2-3.4 The candidate, given a beam secured to a level floor and measuring 20 ft long by 3 to 4 in. wide, and given a length of fire hose weighing at least 20 lb, shall walk the length of the beam, carrying the length of hose, without falling off, or stepping off the beam.

2-3.5 The candidate, given a weight of 125 lb, shall lift the weight from the floor and carry the weight 100 ft without stopping.

2-3.6 The candidate, starting from an erect position with feet apart, the distance closely approximating shoulder width, shall move a 15 lb-weight in the following manner: bend over, grasp the weight with both hands while it is at a point on the floor between the feet, and lift weight to waist level, then place the weight on the floor approximately 12 in. outside the left foot, and without letting go, raise the weight to waist level and touch it to the floor about 12 in. outside the right foot. The weight shall then be moved alternately in this fashion from left foot, to waist level, to right; right to waist level to left until it has been moved 7 times in each direction with total horizontal distance of travel being at least 24 in. more than the space between the feet for each of the 14 moves. This shall be done in less than 35 seconds.

Health-Related Fitness

The most important concept of physical fitness relates to the acquisition and maintenance of good health. The health-related components of physical fitness include cardiorespiratory fitness, muscular strength and endurance, flexibility, and body composition. The scientific literature strongly supports the health benefits of cardiorespiratory fitness and body composition. An increasing amount of research also supports the importance of muscular strength and endurance and flexibility to good health.

Health-related physical fitness includes the basic functional abilities that support a good quality of life and provides the base for job- and activity-specific abilities. Because of the importance of the health-related fitness components to overall well-being, freedom from disease and disability, and quality of life, each component will be examined separately.

COMPONENTS OF HEALTH-RELATED PHYSICAL FITNESS

Cardiorespiratory Endurance

Cardiorespiratory endurance describes the ability to work or exercise over a period of time without fatigue, as a result of the ability of the heart, lungs, and blood vessels to supply adequate oxygen to working muscle cells. This physical ability is commonly called "stamina" or "aerobic capacity."

Having good aerobic capacity can yield the following benefits:

- Having more energy
- Being able to work hard and exert oneself for a long time without tiring
- Help with stress management
- Help with weight management
- Help in preventing and controlling diabetes
- Help in preventing osteoporosis (bone degeneration)
- Reducing the chances of developing heart disease, thereby preventing heart attack and untimely death

Emergency responders with good aerobic capacity can also enhance their job-related fitness and be

- Better able to spring into action rapidly and without injury
- Better able to maintain exertion over time
- More alert and able to respond quickly and safely to threats and hazards in their jobs

Cardiorespiratory endurance testing measures the fitness of the heart and lungs. Cardiorespiratory endurance is expressed by the term "maximum oxygen uptake (max VO_2)." This term describes the amount of oxygen utilized by the body during a measured amount of activity for a specific period of time. The max VO_2 value is expressed in milliliters of oxygen consumed each minute for every kilogram of a person's body weight.

Max VO_2 can be measured by a variety of methods. The most accurate method is the graded exercise test on a treadmill or a stationary bicycle with a calibrated ergometer. Other methods, called "submaximal testing," are much quicker and do not require a physician's presence. Submaximal testing does not yield an actual max VO_2 but rather a predicted one.

While treadmill and stationary bicycle ergometers are often used for max VO_2 testing of individuals, the 1.5-mile run and the 3-mile walk are submaximal tests commonly used with large groups. Group testing for aerobic capacity is inexpensive and easily administered by an individual with some technical expertise and background in fitness. Research by Kenneth Cooper, M.D., at the Aerobic Research Institute in Dallas, Texas, has provided normative data by which completion times on the 1.5-mile run and the 3-mile walk can be equated to a predicted max VO_2 value. All norms are age and sex adapted. These norms can be adjusted to

TABLE 19-2. Points Assigned for Aerobic Fitness, Based on the Time (in Minutes) Required to Run 1.5 Miles or Time Required to Walk Three Miles

Running—1.5 Miles

Men

Points	20–29	30–39	40–49	50–59	60+
100	<9:45	<10:00	<10:30	<11:00	<11:15
95	9:45–10:14	10:00–10:29	10:30–10:59	11:00–11:44	11:15–12:34
90	10:15–10:45	10:30–11:00	11:00–11:30	11:45–12:30	12:35–14:00
85	10:46–11:22	11:01–11:45	11:31–12:15	12:31–13:30	14:01–15:06
80	11:23–12:00	11:46–12:30	12:16–13:00	13:31–14:30	15:07–16:15
75	12:01–13:00	12:31–13:35	13:01–14:15	14:31–15:45	16:16–17:36
70	13:01–14:00	13:36–14:45	14:16–15:35	15:46–17:00	17:37–19:00
65	14:01–15:00	14:46–15:37	15:36–16:33	17:01–18:00	19:01–19:30
60	15:01–16:00	15:38–16:30	16:34–17:30	18:01–19:00	19:31–20:00
50	16:01–17:00	16:31–17:30	17:31–18:30	19:01–20:00	20:01–21:00
40	17:01–18:00	17:31–18:30	18:31–19:30	20:01–21:00	21:01–22:00
30	18:01–19:00	18:31–19:30	19:31–20:30	21:01–22:00	22:01–23:00
20	19:01–20:00	19:31–20:30	20:31–21:30	22:01–23:00	23:01–24:00
10	20:01–21:00	20:31–21:30	21:31–22:30	23:01–24:00	24:01–25:00
0	>21:00	>21:30	>22:30	>24:00	>25:00

Women

Points	20–29	30–39	40–49	50–59	60+
100	<12:30	<13:00	<13:45	<14:30	<16:30
95	12:30–12:59	13:30–13:44	13:45–14:49	14:30–15:29	16:30–16:59
90	13:00–13:30	13:45–14:30	14:50–15:55	15:30–16:30	17:00–17:30
85	13:31–14:40	14:31–15:30	15:56–16:42	16:31–17:45	17:31–18:30
80	14:41–15:54	15:31–16:30	16:43–17:30	17:46–19:00	18:31–19:30
75	15:55–17:11	16:31–17:45	17:31–18:30	19:01–19:30	19:31–20:00
70	17:12–18:30	17:46–19:00	18:31–19:30	19:31–20:00	20:01–20:30
65	18:31–18:45	19:01–19:15	19:31–19:45	20:01–20:15	20:31–20:45
60	18:46–19:00	19:16–19:30	19:46–20:00	20:16–20:30	20:46–21:00
50	19:01–20:00	19:31–20:30	20:01–21:00	20:31–21:30	21:01–22:00
40	20:01–21:00	20:31–21:30	21:01–22:00	21:31–22:30	22:01–23:00
30	21:01–22:00	21:31–22:30	22:01–23:00	22:31–23:30	23:01–24:00
20	22:01–23:00	22:31–23:00	23:01–24:00	23:31–24:30	24:01–25:00
10	23:01–24:00	23:31–24:30	24:01–25:00	24:31–25:30	25:01–26:00
0	>24:00	>24:30	>25:00	>25:30	>26:00

Walking—3 Miles

Men

Points	20–29	30–39	40–49	50–59	60+
100	<33:00	<34:00	<35:30	<38:00	<40:00
95	33:00–33:30	34:00–34:30	35:30–36:00	38:00–38:30	40:00–40:30
90	33:31–34:00	34:31–35:00	36:01–36:30	38:31–39:00	40:31–41:00
85	34:01–36:15	35:01–37:30	36:31–39:15	39:01–42:00	41:01–44:30
80	36:16–38:30	37:31–40:00	39:16–42:00	42:01–44:30	44:31–48:00
75	38:31–40:15	40:01–42:15	42:01–44:30	44:31–47:00	48:01–51:30
70	40:16–42:00	42:16–44:30	44:31–47:00	47:01–49:30	51:31–54:00
65	42:01–44:00	44:31–46:45	47:01–49:30	49:31–52:30	54:01–57:00
60	44:01–46:00	46:46–49:00	49:30–52:00	52:31–55:00	57:01–60:00
50	46:01–47:00	49:01–50:00	52:01–53:00	55:01–56:00	60:01–61:00
40	47:01–48:00	50:01–51:00	53:01–54:00	56:01–57:00	61:01–62:00
30	48:01–49:00	51:01–52:00	54:01–55:00	57:01–58:00	62:01–63:00
20	49:01–50:00	52:01–53:00	55:01–56:00	58:01–59:00	63:01–64:00
10	50:01–51:00	53:01–54:00	56:01–57:00	59:01–60:00	64:01–65:00
0	>51:00	>54:00	>57:00	>60:00	>65:00

Women

Points	20–29	30–39	40–49	50–59	60+
100	<35:00	<36:30	<38:00	<41:00	<44:00
95	35:00–35:30	36:30–37:00	38:00–38:30	41:00–41:30	44:00–44:30
90	35:31–36:00	37:01–37:30	38:31–39:00	41:31–42:00	44:31–45:00
85	36:01–38:15	37:31–39:15	39:01–41:30	42:01–44:30	45:01–48:00
80	38:16–40:30	39:16–42:00	41:31–44:00	44:31–47:00	48:01–51:00
75	40:31–42:15	42:01–44:15	44:01–46:30	47:01–49:30	51:01–54:00
70	42:16–44:00	44:16–46:30	46:31–49:00	49:31–52:00	54:01–57:00
65	44:01–46:00	46:31–48:45	49:01–51:30	52:01–54:30	57:01–60:00
60	46:01–48:00	48:46–51:00	51:31–54:00	54:31–57:00	60:01–63:00
50	48:01–49:00	51:01–52:00	54:01–55:00	57:01–58:00	63:01–64:00
40	49:01–50:00	52:01–53:00	55:01–56:00	58:01–59:00	64:01–65:00
30	50:01–51:00	53:01–54:00	56:01–57:00	59:01–60:00	65:01–66:00
20	51:01–52:00	54:01–55:00	57:01–58:00	60:01–61:00	66:01–67:00
10	52:01–53:00	55:01–56:00	58:01–59:00	61:01–62:00	67:01–68:00
0	>53:00	>56:00	>59:00	>62:00	>68:00

indicate levels of cardiorespiratory endurance required to perform specific job tasks safely.

The chart in Table 19-2 was developed for a metropolitan fire department's fitness testing program based on an extensive task analysis of the job descriptions and actual work done by firefighters in Birmingham, Alabama. This program, called FIT✔CHECK, is administered annually to over 700 fire and rescue service personnel, including members of several hazmat units. The numerical values in Table 19-2 are easily converted into estimated max VO_2 values. Task analysis and literature review determined that a score of at least 75 points in aerobic capacity was needed for safe and effective firefighter performance.

Activities best suited for cardiorespiratory endurance training include running, rapid walking, swimming, biking (Figure 19-1), jogging, stair

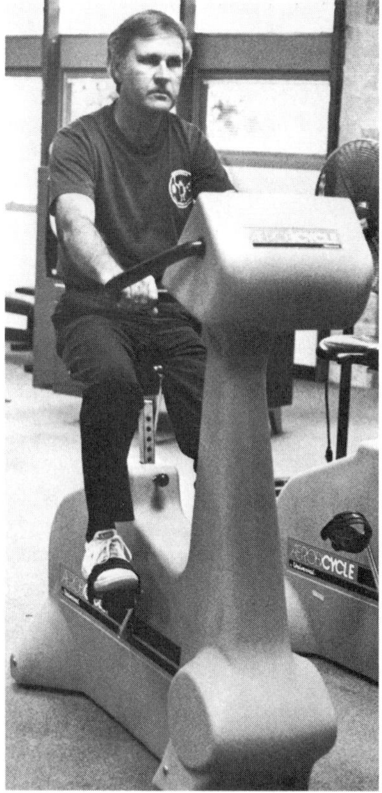

FIGURE 19-1. Riding a stationary bicycle is a convenient way to train for cardiorespiratory endurance. (Barbara Hilyer)

climbing (Figure 19-2), aerobic dance, and cross-country skiing. To improve cardiorespiratory endurance, an individual should select an aerobic exercise he or she enjoys and exercise for at least 20 to 30 minutes three times per week at an intensity of 65 to 90 percent of the maximum heart rate. Maximum heart rate (in beats per minute) can be estimated by subtracting one's age from 220. Multiply this number by 0.65 and 0.90 to determine the lower and upper limits, respectively, of one's recommended heart rate range during exercise (Figure 19-3).

Start out slowly and build up gradually. Exercise at least 20 to 30 minutes, three times each week. Exercising more frequently or for longer

FIGURE 19-2. A stair climbing machine that can be adjusted to control the workload is a good aerobic training device. (Barbara Hilyer)

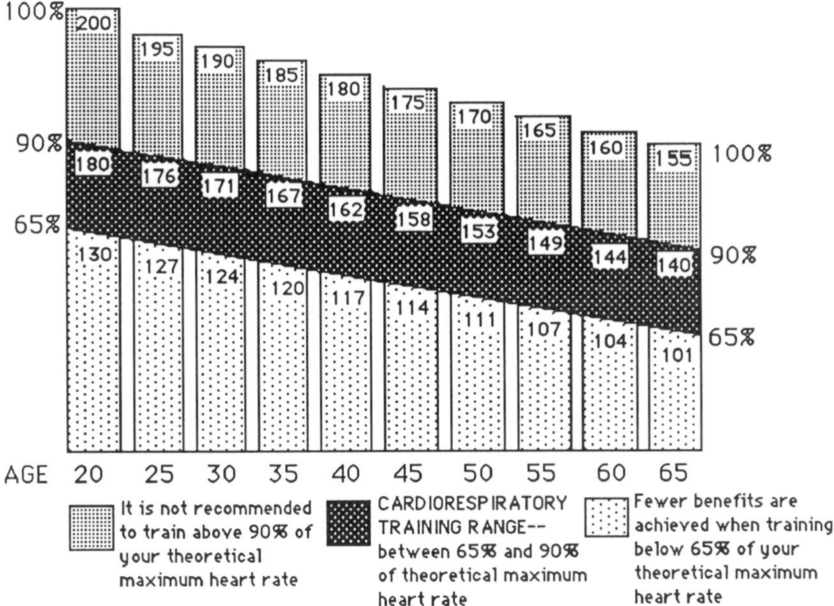

FIGURE 19-3. Training intensity should attain a heart rate between 65 and 90 percent of an individual's theoretical maximum heart rate.

periods of time is even better. Monitor the heart rate, and exercise within the target heart rate range. Warm up gradually, cool down slowly, and stretch before and after exercise.

Body Composition

Body composition refers to the percentage of the body that is fat tissue and the percentage of body tissue that is not fat (muscle, bone, and organ tissue). Percentage of body fat is second only to cardiorespiratory endurance in importance to health and fitness. Some individuals with excessive body fat are clinically obese. Obesity is diagnosed when an individual's fat tissues make up 20 percent more of the body than is desirable. Desirable body fat percentages are approximately 15 percent for adult males and 25 percent for adult females.

Having a desirable body composition can

- Help to prevent heart disease, high blood pressure, diabetes mellitus, and other diseases

- Reduce the risk of back injury and disability
- Reduce the risk of premature death
- Help to prevent chronic fatigue
- Improve appearance and enhance self-esteem

Emergency responders with a healthy ratio of body fat can also

- Get into and out of tight places more easily, reach spill release points, and move equipment
- Exert themselves strenuously and for long periods of time without tiring

Body fat measurements are considered much more important for health than is body weight. Body weight, especially when used as a measurement for classification of overweight, does not take into consideration muscular development. Many individuals who have highly developed skeletal muscle are classified as overweight if height and weight charts are consulted; however, these same individuals may have acceptable body fat measurements. Another method of defining overweight is the body mass index (BMI). BMI is calculated by dividing weight in kilograms by height squared in meters. For men over 20 years of age, a BMI of 27.8 or greater is considered overweight. For women over 20 years of age, a BMI of 27.3 or greater is considered overweight. Like height and weight charts, the BMI does not consider body composition in determining the overweight status. Many athletes and workers who do hard physical tasks are classified as overweight by height and weight charts or BMI, but body fat measurements may determine that they are not overly fat.

Body fat can be measured in several ways. The most sophisticated and accurate method is underwater weighing, which involves submerging the subject in water and measuring the amount of water that is displaced. Underwater weighing is time-consuming and expensive, and is not practical for large groups of people.

The use of electronic devices to measure body fat has become popular in the last few years. Electronic impedance systems, using sound waves and electrical currents, have been developed and have some degree of reliability for certain classifications of subjects.

The most commonly used method for determining body composition uses skinfold measurements taken by calibrated calipers (Figure 19-4). Since fat is stored just under the skin, as well as in other places, skinfold measurements can be used to estimate body fat. For men, skinfold measurements are commonly made at the chest, abdomen, and thigh. For

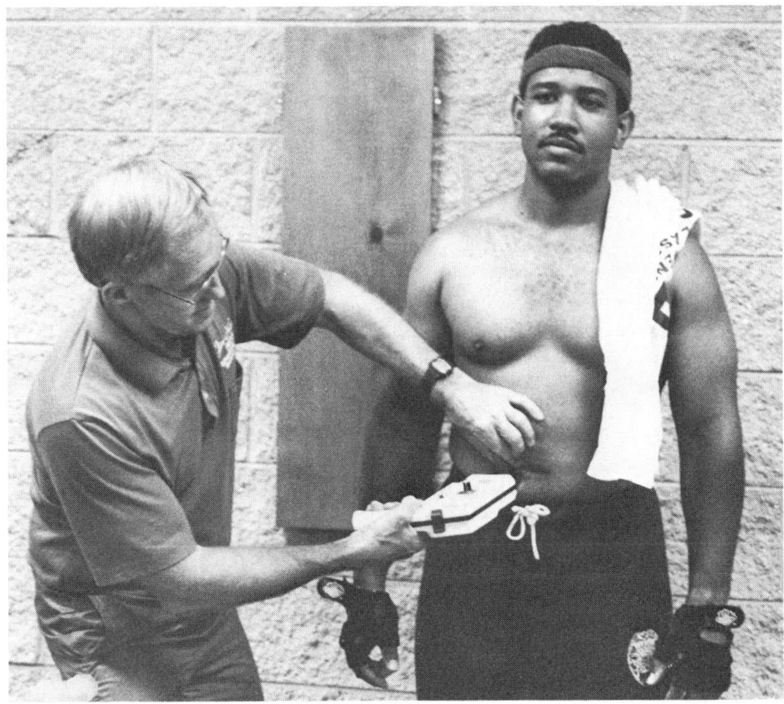

FIGURE 19-4. Skinfold calipers are used to estimate body fat percentage. (Barbara Hilyer)

women, they are commonly made at the back of the arm, top of the hip, and front of the thigh. A mathematical formula is used to convert the skinfold measures into an estimate of body fat percentage. Different formulas are used, depending upon the age and sex of the subject. Skinfold measures are best utilized when they are done at repetitive intervals of time by the same technician. The experience level of the skinfold technician is an important factor in the reliability and validity of the measurements.

The FIT✓CHECK program described earlier uses the chart in Table 19-3 to assign a point value to the body fat percentage. Similar charts can be constructed for any group of emergency responders based upon their job descriptions.

To reduce excess body fat:

- Consume fewer calories
- Cut down on portion size

TABLE 19-3. Points Assigned for Body Composition, Based on Percentage of Body Fat

	Men (Age)				
Points	20–29	30–39	40–49	50–59	60+
100	<13.1	<14.1	<15.1	<16.1	<17.1
95	13.1–14.5	14.1–15.5	15.1–16.5	16.1–17.5	17.1–18.5
90	14.6–16.0	15.6–17.0	16.6–18.0	17.6–19.0	18.6–20.0
85	16.1–17.5	17.1–18.5	18.1–19.5	19.1–20.5	20.1–21.5
80	17.6–19.0	18.6–20.0	19.6–21.0	20.6–22.0	21.6–23.0
75	19.1–22.0	20.1–23.0	21.1–24.0	22.1–25.0	23.1–26.0
70	22.1–23.5	23.1–24.5	24.1–25.5	25.1–26.5	26.1–27.5
60	23.6–25.0	24.6–26.0	25.6–27.0	26.6–28.0	27.6–29.0
50	25.1–26.5	26.1–27.5	27.1–28.5	28.1–29.5	29.1–30.5
40	26.6–28.0	27.6–29.0	28.6–30.0	29.6–31.0	30.6–32.0
30	28.1–29.5	29.1–30.5	30.1–31.5	31.1–32.5	32.1–33.5
20	29.6–31.0	30.6–32.0	31.6–33.0	32.6–34.0	33.6–35.0
10	31.1–33.0	32.1–34.0	33.1–35.0	34.1–36.0	35.1–37.0
0	>33	>34	>35	<36	<37

	Women (Age)				
Points	20–29	30–39	40–49	50–59	60+
100	<18.1	<19.1	<20.1	<21.1	<22.1
95	18.1–19.5	19.1–20.5	20.1–21.5	21.1–22.5	22.1–23.5
90	19.6–21.0	20.6–22.0	21.6–23.0	22.6–24.0	23.6–25.0
85	21.1–22.5	22.1–23.5	23.1–24.5	24.1–25.5	25.1–26.5
80	22.6–24.0	23.6–25.0	24.6–26.0	25.6–27.0	26.6–28.0
75	24.1–27.0	25.1–28.0	26.1–29.0	27.1–30.0	28.1–31.0
70	27.1–28.5	28.1–29.5	29.1–30.5	30.1–31.5	31.1–32.5
60	28.6–30.0	29.6–31.0	30.6–32.0	31.6–33.0	32.6–34.0
50	30.1–31.0	31.1–32.0	32.1–33.0	33.1–34.0	34.1–35.0
40	31.1–32.0	32.1–33.0	33.1–34.0	34.1–35.0	35.1–36.0
30	32.1–33.0	33.1–34.0	34.1–35.0	35.1–36.0	36.1–37.0
20	33.1–34.0	34.1–35.0	35.1–36.0	36.1–37.0	37.1–38.0
10	34.1–35.0	35.1–36.0	36.1–37.0	37.1–38.0	38.1–39.0
0	>35	>36	>37	>38	>39

- Reduce the calories derived from fat to less than 30 percent of the total caloric intake
- Increase aerobic exercise
- Increase muscle mass by weight training

For people who are clinically obese, physician-supervised weight control programs are usually the most successful. For people who are overly fat, increasing physical activity and reducing calorie intake, especially calories derived from fat, usually gives the best results. The American College of Sports Medicine recommends losing no more than 2.2 pounds each week, and never reducing daily calorie intake below 1,200. It is difficult to maintain a balanced nutritional intake while consuming fewer than 1,200 calories a day.

Joint Flexibility

Joint flexibility describes the ability of joints in the skeletal system to move. Human physical activity has evolved around the movement capabilities of the joints. The range of motion in a joint or sequence of joints allows the joint to be flexible. When joints are not properly exercised and conditioned, connective tissue in the tendons, ligaments, muscles, and joint capsules becomes dense and shortened; any attempt to regain the lost range of motion in the joint is resisted by the tight tissue. This accounts for much of the limitation in range of motion of most joints in the body.

Having good flexibility

- Reduces the risk of back injury and disability
- Makes joint injuries less likely to occur and helps one avoid sprained ankles, torn shoulders, and injured knees
- May keep a person from developing arthritis later in life

Emergency responders with good flexibility can also gain

- Improved competence at pushing, pulling, bending, crawling, climbing, and reaching
- Better balance

Having good flexibility reduces not only the risk of muscle and joint injuries, but also the severity of sprains and strains. Those who are injured recover more quickly if they are participating in a flexibility training program. A review of medical, physical therapy, and sports medicine literature suggests that flexibility training may be the appropriate intervention to reduce overall injury rates.

Poor flexibility correlates with a high incidence of back injury in particular. In a program to increase the health and physical fitness of firefighters in Los Angeles, it was found that back protection was associated with better than average physical fitness evidenced by increased flexibility, strength, or work capacity. Firefighters with better than average physical fitness and flexibility had fewer back injuries than those who were less physically fit.

Tests used to measure flexibility include

- The sit-and-reach test for lower back flexibility and hamstring extension
- The twist-and-touch test for trunk rotation
- Flexion and extension of the knee as determined by a goniometer, a mechanical device that measures angle of flexion at a joint
- Flexion and extension of the shoulder as measured by a Leighton flexometer, a device similar to the goniometer

The most commonly used test of flexibility is the sit-and-reach test (Figure 19-5). The test is performed from a sitting position with the legs extended. The subject's feet are placed against a metal box that has a sliding ruler extended toward the subject. The subject extends the arms and hands forward and places one middle finger over the other on the ruler. Then, keeping the knees straight, the subject bends at the waist and pushes the sliding ruler forward as far as possible along a measuring chart. A measurement of zero means that the subject can touch the toes. Positive measures indicate flexing beyond the toes, and negative measures indicate not reaching the toes. The relative distance from the toes is recorded to the nearest ¼ inch.

FIGURE 19-5. The sit and reach test provides a quantitative measure of joint flexibility. (Barbara Hilyer).

The sit-and-reach test can be used efficiently with large groups. The test provides a very important flexibility measure because poor back flexibility is related to low back injury. The FIT✔CHECK system uses the chart in Table 19-4 to assign points based on flexibility, as measured by the sit-and-reach test.

Improving flexibility and staying flexible can be accomplished by

- Staying physically active
- Stretching daily
- Holding stretches for 15–30 seconds
- Stretching before and after any exercise
- Reaching and maintaining an optimal percentage of body fat

When the effects of warming up, massage, and stretching on range of motion in the lower extremity were compared, researchers found that stretching significantly increased the range of motion. Massage and warming up, used separately or in combination, gave no increase in range of motion.

TABLE 19-4. Points Assigned For Flexibility, Based on Distance Achieved in the Sit-and-Reach Test

Points	Inches Beyond Toes
100	+7″ or more
95	6–6¾
90	5–5¾
85	4–4¾
80	2–3¾
75	1–1¾
70	0–¾
65	−¼ to −1
60	−1¼ to −2
50	−2¼ to −3
40	−3¼ to −4
30	−4¼ to −5
20	−5¼ to −6
10	−6¼ to −7
0	Less than −7

Muscle Endurance

Muscle endurance is defined as the ability of a muscle or muscle group to work repetitively and without fatigue for a determined period of time. Adequate muscle endurance enables a person to carry out the normal tasks of living without undue stress. Muscle endurance is closely linked to cardiorespiratory endurance and muscle strength. Poor cardiorespiratory endurance will hasten the onset of muscle fatigue. Poor muscle strength creates a demand for a greater exertion of effort for each repetitive move and thus hastens fatigue. When muscles fatigue, they are unable to continue contracting effectively, and work becomes more difficult.

The physically harder the task, usually the more rapid the onset of fatigue. Fatigue leads to injury when the responder tries to continue to work and is unable to do so efficiently. Adequate muscle endurance is extremely important for emergency responders, since much emergency response work involves doing repetitive tasks over a period of time. Emergency responders with inadequate muscle endurance will not be capable of working at tasks that require effort over a period of time.

Having good muscle endurance

- Facilitates everything one does—standing, sitting, walking, running, lifting, pulling, pushing
- Prevents fatigue
- Protects the internal organs
- Reduces the risk of back injury and disability
- Improves physical appearance

Emergency responders with good muscle endurance are also

- Better able to perform all physical tasks
- Less likely to become fatigued on the job

Measurement of muscle endurance usually requires the subject to do a task for a given period of time, and is quantified by the number of completed successful repetitions in that interval. Another method of evaluating muscle endurance, although not as efficient, is to allow the subject to continue the repetitive task until fatigue causes failure. The number of completed repetitions is reported as the measure of muscle endurance. The latter method is usually not the most desirable, since the subject is physically exhausted and more susceptible to injury. The time required for testing by the second method is usually excessive if large groups are being tested.

A commonly used test of muscle endurance is the sit-up test for 1 minute. In this test, the subject performs assisted sit-ups for 1 minute, and the number completed becomes the measure of muscle endurance. The subject lies on the back with arms crossed over the chest, knees bent at right angles, and feet held firmly to the floor by an assistant. When bent-knee sit-ups are used to measure abdominal strength and endurance, the FIT✔CHECK chart shown in Table 19-5 can be used to convert the number of sit-ups into points. Although the sit-up test specifically measures the endurance of the abdominal muscles, it is also a good measure of

TABLE 19-5. Points Assigned For Abdominal Strength and Muscle Endurance, Based on the Number of Sit-Ups Completed in 1 Minute

			Men		
Points	20–29	30–39	40–49	50–59	60+
100	≥50	≥45	≥40	≥35	≥30
95	47–49	42–44	37–39	33–34	28–29
90	44–46	39–41	34–36	31–32	26–27
85	41–43	36–38	31–33	29–30	24–25
80	37–40	33–35	28–30	27–28	22–23
75	34–36	30–32	25–27	25–26	20–21
70	30–33	27–29	22–24	21–24	18–19
65	27–29	24–26	19–21	18–20	16–17
60	24–26	21–23	16–18	15–17	14–15
50	20–23	18–20	14–15	12–14	11–13
40	17–19	15–17	12–13	9–11	8–10
30	14–16	12–14	10–11	6–8	5–7
20	11–13	9–11	8–9	4–5	3–4
10	8–10	6–8	5–7	2–3	1–2
0	<8	<6	<5	<2	<1
			Women		
Points	20–29	30–39	40–49	50–59	60+
100	≥45	≥40	≥35	≥30	≥25
95	42–44	37–39	32–34	27–29	23–24
90	39–41	34–36	29–31	24–26	21–22
85	36–38	31–33	26–28	21–23	19–20
80	33–35	29–30	23–25	19–20	17–18
75	30–32	26–28	20–22	17–18	14–16
70	27–29	23–25	17–19	14–16	12–13
65	24–26	20–22	15–16	12–13	10–11
60	21–23	17–19	13–14	10–11	8–9
50	18–20	14–16	11–12	8–9	6–7
40	15–17	11–13	9–10	6–7	5
30	12–14	8–10	7–8	4–5	4
20	9–11	5–7	4–6	3	3
10	6–8	3–4	3	2	2
0	<6	<3	<3	<2	<2

overall muscle endurance, or the ability to work hard for an extended period of time without tiring.

Muscle endurance can usually be improved by repeating in multiple sets the contraction of the muscles used in the task one wishes to perform. For example, increasing one's ability to do sit-ups can be accomplished by determining the maximum number of sit-ups one can perform in 1 minute and then repeating one-third of that maximum number. Rest for 3 minutes, then repeat the number. Do at least four sets, three or four times each week. Every 2 weeks, repeat the sit-up test to determine a new maximum.

Muscle Strength

Muscle strength is defined as the ability to exert force to push or pull an object. Muscle strength is measured not by repetitively doing a task, as is done in measuring muscle endurance, but instead by how much force can be exerted in one repetition. Adequate muscle strength is defined as the strength needed to do the task one wishes to complete. Obviously, this varies among individuals and is different in different jobs.

Adequate muscle strength is very important in emergency response. There are often tasks at the emergency scene that require bursts of force, such as are needed to move heavy equipment. An emergency responder with inadequate muscle strength will find many tasks very stressful, and the weak responder is at high risk for injury. Muscle strength in the upper body is especially important.

Having good body strength

- Makes many daily tasks easier, like lifting children, carrying groceries, and pushing a lawn mower, and helps to prevent stress and fatigue from ordinary tasks
- Reduces the risk of injuries, arthritis, and disability
- Improves physical appearance

Emergency responders with good body strength are also better able to

- Lift, push, and pull heavy objects, such as vehicles, equipment, diking materials, and repair kits
- Climb over a fence, pull themselves onto a ledge, or handle other tasks which may be part of the response to a hazmat incident

The measurement of muscle strength is usually accomplished by exerting force against an object and determining how much force is needed to move it. A common method of measuring upper body muscle strength is the one-repetition maximum bench press test (Figure 19-6). It is usually best to use a weight machine to test large groups. Using free weights entails more risk, as balance and coordination are also required. Using a stationary weight machine minimizes those factors.

In strength testing, body weight is an important variable. Usually the greater the body weight, the more the force that can be exerted by an untrained person. Most strength testing is scored by measuring the percentage of body weight that can be moved. The bench press test measures upper body strength—the strength of the shoulders, arms, chest, and back. Table 19-6 shows the FIT✓CHECK chart for scoring the bench press test.

Body strength can be improved by

- Doing push-ups and/or pull-ups three or four times a week
- Training with weights or exercise machines three or four times a week (Figure 19-7)

FIGURE 19-6. The maximum bench press lift can be used as a measure of muscle strength. (Barbara Hilyer).

TABLE 19-6. Points Assigned For Upper Body Strength, Based on the Percentage of Body Weight Bench Pressed

Points	Men	Women
100	≥120%	≥68%
95	110–119	64–67
90	99–109	61–63
85	90–98	56–60
80	79–89	50–55
75	74–78	45–49
70	69–73	41–44
65	64–68	37–40
60	59–63	34–36
50	54–58	31–33
40	49–53	29–30
30	44–48	26–28
20	39–43	24–25
10	29–38	12–23
0	≤28	≤11

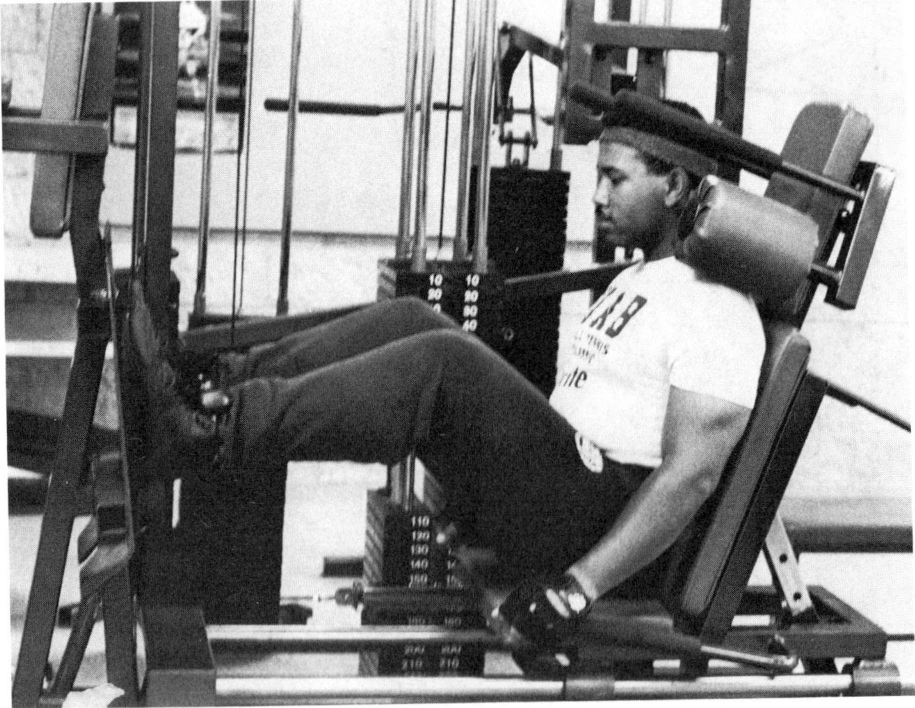

FIGURE 19-7. Training on a leg press machine can be part of a program to improve muscle strength. (Barbara Hilyer).

PHYSICAL FITNESS TRAINING

Specificity of Training

The best results in physical fitness training are usually obtained when there is specificity of movement in training which approximates the skill to be improved. For example, the best way to increase the number of sit-ups one can do in 1 minute is to train by doing sit-ups. The best way to improve a maximum bench press lift is to train by doing bench presses.

The Overload Principle

Adaptation is the process by which an organism makes adjustments to meet the demands of the stresses placed upon it. Adaptation in physical activity is called the "overload principle." Systematic adaptation is the process that allows a person to improve physical functioning so that tasks that once were too difficult can be completed safely. By gradually increasing the resistance to be overcome, increasing the length of time an activity is to continue, or shortening the time frame within which the activity is completed, an individual can improve functioning until the goals and objectives of training are met.

Adaptation in physical fitness training is accomplished by following a planned, systematic training program. Manipulation of the frequency, duration, and intensity of exercise is planned so that an individual can estimate the time period needed to complete the adaptation process. Adaptation planning for predicted outcomes assumes that the individual is free of organic disease, will regulate food intake to meet the demands of the training or the objectives to be reached, and will get sufficient rest and sleep. If any of these assumptions is untrue, adaptation is not predictable.

An example of this process is given to enhance understanding. Mike Smith wants to be assigned to the hazmat team. The selection standard that must be met before appointment includes successfully doing one repetition of the bench press on a Universal machine or a similar piece of equipment with 75 percent of body weight pressed. In experimenting, Mike finds that he can bench press only 50 percent of his body weight on the Universal. If he weighs 200 pounds, Mike must bench press 150 pounds to meet his training objective. If he has 6 weeks to train, a progressive, systematic training program can be planned that should enable Mike to reach his goal. If Mike is to reach his goal following the planned training program, the following assumptions must be correct:

- Mike is free of organic disease or functional disability that will interfere with the physical activity needed to accomplish the prescribed training.

- Mike understands proper nutrition and will maintain adequate nutrient intake during the training period.
- Mike is able to get adequate rest and sleep for adaptation to occur.
- Mike has the time to train, the motivation to train, and the perseverance to overcome obstacles that probably will occur during this process.
- Mike has a training site, the necessary training equipment, and an experienced, knowledgeable training coach or training partner to assist if needed.
- Mike is untrained in this activity. Further adaptation occurs more slowly if he has already been doing bench presses regularly.

The absence of any of these conditions may interfere with Mike's ability to reach his objective through training for adaptation. Once these preliminary assumptions are reasonably accurate, a systematic program can be planned for his 6 weeks of training. By breaking his training time into segments, a determined number of sessions at each level can be planned. There are 42 days in 6 weeks, so there are 21 training opportunities, assuming a day of rest between each period. This rest day is important to untrained individuals.

In strength training, four progressive stages of training are as follows:

- Establishing an endurance base that allows safe progress to a higher intensity level. This is accomplished by working at 65 percent of maximum resistance for four sets, each including 8 to 10 repetitions.
- Retesting, then decreasing repetitions to 5 to 7 and increasing resistance to 75 percent.
- Retesting, then decreasing repetitions to 2 to 4 and increasing resistance to 85 percent.
- Retesting, then doing 1 lift with 90–100 percent of maximum resistance, six to seven times with rest between each, per workout.

In setting up Mike's training program, the following schedule will help him reach his goal of a one-repetition 150-pound bench press:

- Workouts 1–7: 4 × 8–10 repetitions (four sets of 8 to 10 repetitions per set)—65 percent of max (lifting 65 percent of his maximum single repetition bench press weight)
- Retest
- Workouts 8–13: 4 × 5–7 repetitions—75 percent of max
- Retest

- Workouts 14–18: 4 × 2–4 repetitions—85 percent of max
- Retest
- Workouts 18–21: 7 × 1 repetition—90–100 percent of max

If Mike does not reach his goal by the end of this training program, then a reevaluation and in-depth analysis of the preliminary assumptions must be made.

Usually physical fitness training does not focus on a single activity. Even when the objective of the training is to improve only one activity, this is usually accomplished more successfully by a basic training program that promotes adaptation in all of the health-related fitness areas.

Planning a Good Workout

A good training program, and even each individual workout session, should include activities that will improve all areas of health-related fitness. The critical preliminary factors to be considered are

- Amount of time available for training, both in time for each exercise session and in frequency per week
- Availability of a training site and equipment
- Motivation to train

The essential features of a good workout program include

- A warmup period
- Muscle strength and muscle endurance training
- Cardiorespiratory endurance training (aerobic activity)
- A cool-down period

The Warmup

The best warmup sessions include a preliminary low-intensity aerobic exercise for 3 to 5 minutes, followed by 7 to 10 minutes of flexibility training or stretching. Walking at a reasonable pace for ¼ mile or riding a stationary bicycle for 3 to 4 minutes is a good way to lead into the stretching routine. The stretching routine should be primarily static stretching, where each stretch is held for 15 to 30 seconds. Static stretching is different from dynamic stretching in that no stretch is forced and movement after the maximum stretch is reached is kept to an absolute minimum. Examples of static stretching are pulling the knees to the chest while lying on the back and reaching for the toes while in a sitting position with the knees slightly bent. All major muscle groups should be stretched, with emphasis on the lower back, shoulders, and legs.

Muscle Strength and Endurance Training

Muscle strength and endurance training is best accomplished with resistance training equipment. Free weights and resistance machines offer the most convenient methods for this type of training. Equipment is more important in this phase of training than in the other phases, but even here it is not absolutely necessary. Very successful training programs have been implemented using a person's own body weight for resistance. However, training with resistance equipment usually produces the quickest and most significant increases in muscle strength and endurance. Push-ups, pull-ups, and sit-ups are usually included in such programs whether equipment is utilized or not.

Cardiorespiratory Endurance (Aerobic) Training

The most significant component in an exercise session is cardiorespiratory endurance or aerobic training. If time does not permit complete training efforts on certain days, the aerobic component should be the last to be shortened. Aerobic training should be approached as the most important component of an exercise program, and should be invested with the most energy and enthusiasm.

The mode of aerobic training utilized can be selected from a variety of effective activities. Running and rapid walking provide the most benefits for the time invested. Other popular modes of aerobic exercise are mobile or stationary bicycles, stair climbing machines, swimming, and aerobic dance. Aerobic exercise sessions should be at least 30 minutes in duration, done at an intensity of at least 65 percent of maximum heart rate.

The Cool-Down

Conclude aerobic training with 2 to 3 minutes of low-intensity activity, like walking, followed by 2 to 3 minutes of static flexibility training or stretching. A cooling-down period allows a slowing-down period for circulation and muscles, and reduces the risk of sore muscles and joints after a hard exercise session.

Time Allotment for Exercise

It is important to the success of a fitness program to make realistic time allotments for exercise sessions. Planning exercise with unrealistic time expectations usually leads to disappointment and a feeling of failure. This often leads to noncompliance, and eventually to dropping out of the exercise program. Plan exercise sessions at times most convenient to you, and treat the time schedule as a business appointment. When something unexpected threatens your alloted time for exercise, learn to respond with "I'm sorry, but I have an appointment at that time that I cannot break or rearrange." This is hard at first, but when it is done faithfully for a period of time, it will become an acceptable behavior.

If you have 1.5 hours to commit to each exercise session, a good time schedule would be:

- Warm up for 15 minutes
- Strength training for 35 minutes
- Aerobic training for 35 minutes
- Cool down for 5 minutes
- Total workout 90 minutes

PHYSICAL ABILITIES TESTING AS A SELECTION METHOD FOR EMERGENCY RESPONDERS

The selection of employees for the emergency response team should include testing of their physical abilities. Job descriptions and task analyses should identify how much of what kinds of abilities are needed to safely perform the specific tasks of the job. Job tasks and measures of physical abilities can be used to determine if an individual can physically perform the necessary tasks, but if they are used as screening tests, they must be thorough; they may face challenges on the basis of discrimination. In this situation, it may be necessary for a qualified professional to do the task analysis. Physical abilities which have been identified as useful in selection processes are

- Static strength: the ability to use muscle force to lift, push, pull, or carry objects
- Explosive strength: the ability to use short bursts of muscle force to propel oneself; this requires effort over a very short period of time
- Dynamic strength: the degree to which individual muscles remain unfatigued when exerted in repeated or continuous movements
- Trunk strength: the degree to which the abdominal and lower back muscles can support part of the body repeatedly or continuously over time
- Stamina: the ability to exert oneself physically over a period of time without fatigue

- Effort: the degree of physical exertion experienced in performing either a simple task or a series of tasks
- Extent flexibility: the ability to bend, stretch, twist, or reach out with the body, arms, and/or legs
- Dynamic flexibility: the ability to bend, stretch, twist, or reach out with the body, arms, and/or legs both quickly and repeatedly
- Mobility: the capacity to move one's body from place to place
- Speed of limb movement: the speed with which a single movement of the arms or legs can be made
- Gross body coordination: the ability to coordinate the movement of the arms, legs, and torso in activities where the whole body is in motion
- Gross body equilibrium: the ability to keep or regain one's body balance or to stay upright when in an unstable position
- Arm-hand steadiness: the ability to keep the hand and arm steady
- Manual dexterity: the ability to make skillful, coordinated movements of the hand, a hand and an arm, or two hands
- Finger dexterity: the ability to make skillful, coordinated movements of the fingers of one or both hands and to move small objects

A COMPREHENSIVE HEALTH AND FITNESS PROMOTION PROGRAM FOR EMERGENCY RESPONDERS

A comprehensive health and fitness promotion program for emergency responders should include the following components:

- Physical abilities and physical fitness testing prior to selection
- Medical screening before selection, including tests of cholesterol and lipid levels, respiratory functioning, organic illness, and other parameters selected by medical and health and safety personnel
- Annual fitness testing and job task testing
- Annual medical screening and health risk assessment, including behavioral lifestyle assessment and nutritional intake analysis
- Physical fitness training
- Other interventions to reduce the risk of injury and illness

Equipment for Physical Fitness Training

The implementation of a physical fitness training program does not require special equipment; however, if it is possible to purchase equipment, a more sophisticated and effective program can be developed. There are many different types, brands, and prices of equipment from which to choose. Several equipment manufacturers have developed relatively inexpensive, multistationed strength training units that offer multiple exercise options and allow several people to exercise simultaneously.

In addition to a strength training unit, useful equipment includes a stationary bicycle, a treadmill, a stair climbing unit, and a chinning bar. Free weights are less expensive but require strict supervision for safe use.

SUMMARY

Emergency responders have special needs for acquiring and maintaining job-specific physical abilities. These needs should be high priorities in preparing responders to meet conditions that may be present at the emergency scene safely and effectively. Employers must take the lead in promoting good health and physical fitness among the hazmat team members. This often requires administrative decisions and implementation.

Physical abilities testing should be a requirement before assignment to the hazmat team, and annual fitness testing should be required of all team members. Physical abilities testing should be based upon careful job and task analyses. Programs to improve physical fitness should be based on sound principles. Effective physical fitness programs address all five areas of health-related fitness: cardiorespiratory endurance, body composition, joint flexibility, muscle endurance, and muscle strength. To be most effective, physical fitness programs should be part of on-the-job schedules.

Physical fitness is rapidly moving into a dominant role in health promotion and injury prevention programs for emergency responders. Physical fitness is an important feature of good health and a high quality of life for responders, both on and off the job.

REFERENCES

Allsen, P.E., J.M. Harrison, and V. Barbara. 1975. *Fitness for Life*. Dubuque, IA: Wm. C. Brown Publishers.

Cooper, Kenneth H. 1977. *The Aerobics Way*. New York: M. Evans and Co., Inc.

Fleishman, E.A. 1964. *The Structure and Measurement of Physical Fitness*. Englewood Cliffs, NJ: Prentice-Hall, Inc.

Hilyer, J.C., and L.M. Artz. 1987. *FIT✓CHECK Handbook*. City of Birmingham, AL.

National Committee for Injury Prevention and Control (U.S.). 1989. *Injury Prevention: Meeting the Challenge*. New York: Oxford University Press. Supplement to the *American Journal of Preventive Medicine*, 5:3.

Nylander, S.W., and G. Carmean. 1984. *Medical Standards Project: Final Report*, Vol. 1. County of San Bernadino, CA.

Sharkey, B.J. 1979. *Physiology of Fitness*. Champaign, IL: Human Kinetics Publishers.

Sinacore, J.S., and A.C. Sinacore. 1982. *Health: A Quality of Life*. New York: Macmillan Publishing Co.

Williams, M.H. 1985. *Lifetime Physical Fitness*. Dubuque, IA: Wm. C. Brown Publishers.

20

Exercising the Emergency Response Plan

Kenneth W. Oldfield, M.S.P.H., and
Lori P. Andrews, P.E.

The Center Point, Alabama, fire department and the Jefferson County LEPC conducted a simulated emergency response to a chemical fire at an industrial facility. The exercise included community volunteers, local businesses and industries, emergency fire and medical personnel, hospitals, relief agencies, news media, and regulatory agencies. Special effects such as real fires, small detonating devices (for exploding drums), and bones from a butcher shop (for victim wounds) provided a sense of realism for all persons involved. The response went very well, and the mood during the post-incident analysis was good. During the analysis, it was discovered that no provisions had been made for decontamination of personnel, equipment, or victims.

Each of the previous chapters of this text has dealt with a single topic. However, a successful response to a hazardous materials incident will require aspects of most or all of these topics at one time. How can the emergency response team confirm that it will be able to function as a whole to conduct response operations successfully? This chapter will focus on the concept of practicing response skills in a training-course environment and as a part of the emergency response plan (ERP).

OBJECTIVES

- Understand the benefits of conducting exercises to evaluate and practice emergency response skills
- Know the types or levels of exercises that can be used

- Understand the process of preparing, conducting, and evaluating exercises at various levels
- Understand the importance of follow-up to exercises

BENEFITS OF CONDUCTING EXERCISES

The development of a facility's ERP and the training of emergency responders are processes. In both, the participants will benefit from the opportunity to apply what is learned or developed. The potential benefits of exercises include

- The opportunity to assess the capabilities of an individual or a group
- Hands-on application of the ERP and/or hazardous materials response training to improve responders' confidence
- The opportunity to detect deficiencies in the plan or training before an actual emergency
- The opportunity for departments, organizations, and agencies to work together before the tension of an actual emergency

It is hoped that an industrial facility will not have many hazardous materials emergencies. Of those that occur, no two will be exactly the same. There will be enough unknowns and uncertainties during an actual incident without having to wonder if the members of a team can work together or use equipment properly. An actor can learn his part in a play without the other members of the cast, but he will not perform well if he does not rehearse with the other performers before opening night. Likewise, response team members should rehearse the ERP before an actual emergency.

LEVELS OF EXERCISE

Consider a football team. When they hit the field on game day, it is the culmination of a continuous process of learning and practicing. The players all learn the functions of their positions and the plays the team will run. They get together and listen as the coaches map out a strategy for the game. Then the players get out on the field and practice the individual components of the game such as throwing passes, blocking, and rushing the passer. Finally, they have scrimmages and practice games. By the day of the game, they don't know exactly what will happen, but they know they're as ready as they can be. When the game is over, they start the process over again the next day, beginning with an evaluation of the previous game.

In the same way, a hazardous materials team must rehearse the ERP in stages. Four levels of exercise are described by the National Response Teams (*NRT-2* 1990) and by Kelly (1989) in discussions of emergency preparedness:

- Orientation seminar
- Tabletop exercise
- Functional exercise
- Full-scale exercise

These exercises follow a pattern that is similar to the example of the football team. They start with the basic individual components of the plan and build on these to produce a whole team effort. Table 20-1 compares these levels.

The first step in the process of exercising the components of an ERP is determining the need for such exercises. Figure 20-1 is a sample checklist of the status of the ERP components. The checklist should be reviewed regularly and updated as conditions or personnel change or as exercises are conducted.

In the following sections, examples of the exercises will refer to the hypothetical metal fabrication plant shown in Figure 20-2. The major chemical or physical concerns in each area are indicated, though it is recognized that smaller quantities of other hazardous materials would be present. Methylene chloride is stored in a 2,500-gallon aboveground tank, xylene in drums in the storage area, and oxygen and acetylene in 150-pound steel cylinders. The plant is a subsidiary of a corporation which has similar facilities in other states.

Orientation Seminar

The orientation seminar is typically used as an opportunity to introduce a new or recently modified ERP to individuals who may have a part in a response. This is the most basic type of exercise and typically involves an informal discussion of the ERP and the general actions to be taken during an emergency response. Each of the various groups that may be involved in a response to a hazmat incident has the opportunity to see how its role fits into the overall plan and become familiar with the roles and capabilities of the other groups.

TABLE 20-1. Comparison of Exercise Levels

	Orientation	Tabletop	Functional	Full Scale
USES	Introduce new or revised plan; acquaint new personnel with plan	Practice group problem-solving; study specific scenario; observe information sharing	Evaluate any function; reinforce established procedures; test seldom used resources	Demonstrate cooperation; test procedure & equipment capability; media attention; evaluate overall response capabilities
HAZARDS	High Profile	Any priority	To highlight function	Highest priority
NUMBERS OF ON-GOING ACTIVITIES	Single functions	One or two functions	Few to several disparate functions	All disparate functions for a particular scenario
TYPES OF ACTIVITY	Walk through; identify roles and responsibility	Problem solving; brain storming; resource allocation task coordination	Hands-on decision making; coordination; communication	Actual field operations; field command coordination
DEGREE OF REALISM	None	Scene setting with scenario narrative and low key messages	Limited; some simulated messages	Intense, full transmission of simulated messages

Adapted from Birmingham Jefferson County EMA (1991)

Plan Component	New	Updated	Exercised	Used In Emergency	Dormant
Emergency Operations Plan					
Plan Annex(es)					
Appendices					
Standard Operating Procedures					
Resource List					
Site Maps					
Reporting Requirements					
Notification/Warning Procedures					
Mutual Aid Pacts					
Fire Brigade					
Security					
Off-Site Coordination					
Air Surveillance Procedures					
Command Center					
Communications Systems					
Evacuation Procedures					
Emergency Shut-Down					
Damage Assessment Techniques					

FIGURE 20-1. The need for exercises of particular areas can be determined using a checklist.

In a broad sense, the lectures and discussions during training courses serve as the orientation seminar (Figure 20-3). The course is intended to provide the emergency responder with an opportunity to discuss and learn how to safely conduct emergency responses to hazmat incidents. A specific orientation discussion may be used to introduce higher-level exercises, such as when specific instructions in how to patch drums are given.

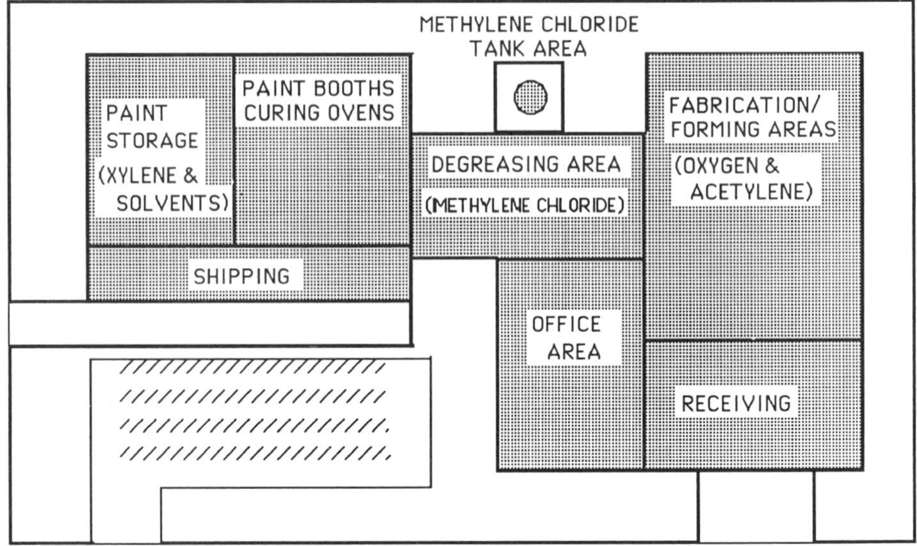

FIGURE 20-2. Layout of the metal fabrication plant used in the examples.

Tabletop Exercise

Tabletop exercises are used to practice group problem-solving skills, and they may take a variety of forms (Figure 20-4). During training, this type of exercise focuses on particular problems or decision-making processes. For instance, the participants may be given a chemical name and asked to find specific information about its hazards from various reference materials. Or they may be given a scenario and asked to write a safety plan that would require them to plan for safety procedures and personal protective equipment (PPE). The exercise may take the form of an ongoing scenario in which the trainees respond to new information as the scenario unfolds.

Emergency planners can use tabletop exercises to bring together various groups that may work together during a response. The purpose is to work out specific roles and responses to various potential incidents. These exercises should explore the potential involvement of different departments of the facility or plant without the stress of an actual response. During this discussion, the needs or deficiencies of the ERP can be revealed in a nonthreatening environment.

In the metal fabrication plant, the emergency planning committee may conduct a tabletop exercise to discuss the specific procedures to be followed in the event of a spill and ignition of xylene in the paint storage

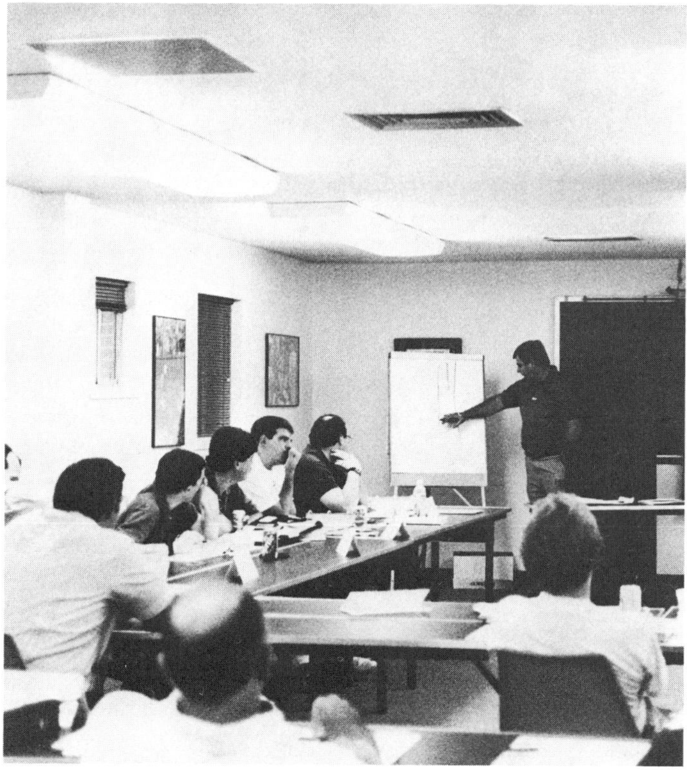

FIGURE 20-3. Orientation exercises communicate information, often in a discussion or lecture setting.

room. The exercise could involve personnel from the groups described in the section on development in this chapter. The tabletop exercise provides an opportunity for these groups to plan a consolidated response to such a spill or fire before one actually occurs.

In designing tabletop exercises, it is important to select certain objectives to be met and to design into the exercise information, clues, or messages that will trigger the desired response. This does not mean that the controllers or instructors necessarily lead the discussion or that written exercises are of the simple fill-in-the-blanks type. Rather, the exercise should be introduced by clearly stating the focus to the participants and then left to take its own course toward the objectives. Occasional direction from instructors may be necessary when the discussion gets off course or when an important response is missed. The computer software

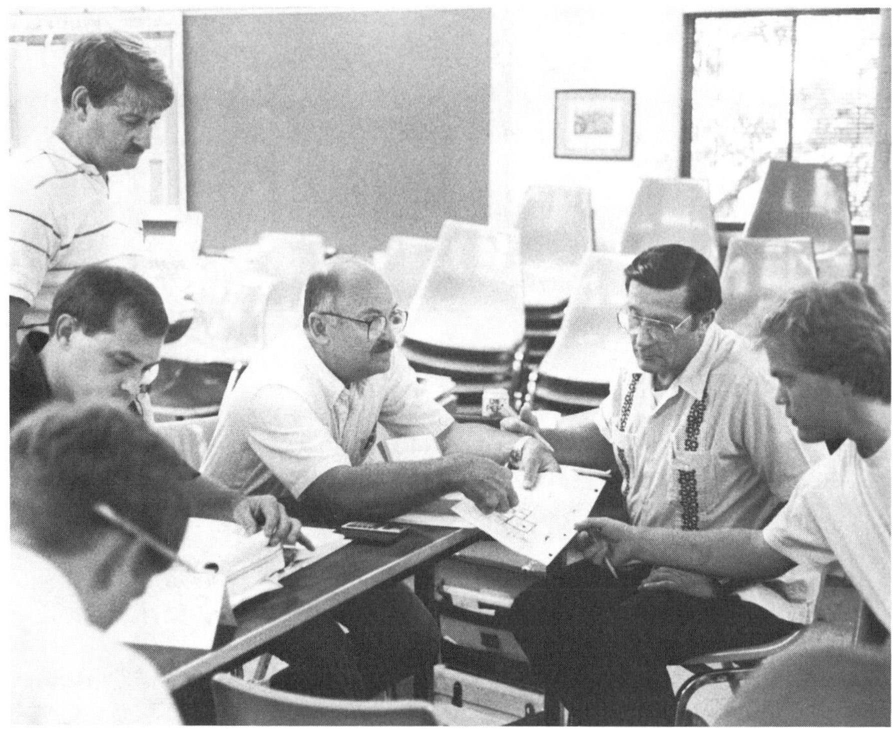

FIGURE 20-4. Tabletop exercise practice the problem-solving skills of groups.

described in Chapter 16 can provide an excellent format for tabletop exercises.

Functional Exercise

Some emergency response procedures or activities cannot be adequately demonstrated in a tabletop or discussion exercise. In order to practice these, one must do them. For instance, donning respirators and other PPE, plugging or patching leaking containers, or operating decontamination lines are all procedures which should not be performed for the first time during an actual emergency. In training courses, these functional exercises are often referred to as "hands-on" activities.

In practicing the ERP at the fabrication plant, the exercise may focus on specific actions such as initiating the warning system, establishing on-scene incident command, or evacuating areas of the plant. The selection of functions to be tested will be based on the response capabilities and

FIGURE 20-5. Functional exercises practice individual activities that may be part of an emergency response.

potential hazards at the facility. For instance, if the plant has no fire brigade and plant personnel are not expected to take offensive actions to fight fires, exercises in procedures for entering the hazard area would not be appropriate. An exercise in the replacement of broken valves on the methylene chloride tank system while wearing appropriate PPE is a possible functional exercise.

The functional exercise may include more than one focal activity, such as having participants patch drums while wearing PPE (Figure 20-5). However, care must be taken to avoid diminishing the effectiveness of the exercise by including too many activities. Functional exercise should not become full-scale scenarios. Having several exercises with well-chosen objectives is preferable to having one in which participants are overwhelmed or rushed through important points.

Examples of hands-on activities which may be used in emergency response training include

- Inspecting and donning an SCBA and performing ordinary tasks which involve some physical exertion, such as playing frisbee

- Simulating chemical spills with water and practicing confinement procedures
- Practicing various containment procedures and going through decontamination in full Level A PPE
- Calibration of and measurement with DRIs

Full-Scale Exercise

A full-scale exercise offers the opportunity to practice the full implementation of the ERP for specific likely scenarios. Training courses can use the simulated incident response to pull together and apply all of the concepts presented in the course. The exercise necessitates the cooperation and integration of the various response groups and adds the stress of time. The degree of realism involved depends on the exercise organizers and on such factors as time, budget, and personnel availability. A full-scale exercise may be as large or as small as desired. The important feature is that it involves all the personnel and procedures required to respond to an emergency situation.

To the extent that budget and time constraints allow, the actual equipment, materials, and procedures involved in the response should be incorporated into the exercise to enhance realism. However, some activities may be simulated if they (1) are part of the normal daily operations of the personnel (e.g., traffic control by police officers or normal structural firefighting techniques for firefighters), (2) involve the use and disposal of expensive consumable items, or (3) would normally take a longer time to complete than can be incorporated into the time allotted for the exercise.

An important way to enhance the realism of the exercise is to use visual or audio props where possible. These props, such as dry ice or smoke bombs, can provide visible clues of a hazard that is being simulated (Figure 20-6). It is important that the props be tied to a specific hazard that is designed into the scenario and not simply thrown in for effect. Intentionally sending participants on a wild goose chase with misleading clues or information does not serve the purpose of the exercise.

Some specific areas that should be tested by the exercise include

- Implementation of the Incident Command System
- Site control and zoning procedures
- Coordination of emergency services from off the site
- Hazard recognition and assessment, including the use of reference documents and monitoring equipment that will be available to the responder

FIGURE 20-6. The use of props, such as smoke bombs, helps make the exercise realistic.

- Selection and use of proper PPE
- Communication, both primary and backup
- Logistical and administrative functions
- Decontamination, including control of any fluids generated for proper disposal
- Containment and confinement procedures that will likely be performed by the responders
- Termination procedures

A scenario that may serve as the basis of a full-scale exercise at the metal fabrication plant is the spillage of a drum of xylene in the paint storage area. A spark causes ignition of the vapors, and the building becomes engulfed in flames. The exercise could practice the response to this emergency, from the initial warning and notification of the fire department through the termination of the incident and the reporting requirements for regulatory agencies. The actions could be triggered by planned messages carried over the emergency communications system, the known hazards in the area and the surrounding area, and the normal actions called for in the SOPs of the ERP as the incident unfolds. The concerns for potential hazards from the degreasing and fabrication areas would have to be addressed, as would such concerns as toxic vapors produced during the fire. The entire response as simulated in the exercise could then be evaluated at its conclusion.

THE EXERCISE PROCESS

Even an apparently simple written or tabletop exercise must be developed, conducted, evaluated, and followed up to be effective as a testing or learning tool. Each part of this process takes time and effort; how much will depend on the complexity of the exercise.

Development

The first step in the process of developing an exercise is to establish the objective(s) to be met. The objective of training exercises may be to provide a practical application of a subject that has been discussed or presented in a lecture. It may be to test the trainee's ability to perform a particular task or function. Or it may ultimately be to give the trainee an opportunity to practice all areas covered in a course during a simulated incident, thereby fulfilling multiple objectives.

A facility's emergency planning group may desire to evaluate the level of understanding of the ERP by the various groups that may be called upon to participate in an emergency response. For fixed facilities, these groups may include

- Health and safety department
- Production supervisors
- Maintenance
- Spill control team
- Security
- Equipment operators

Groups from outside the plant may be involved also, such as

- Local fire department
- Emergency medical services
- Corporate health and safety personnel
- Hospital emergency room personnel
- Law enforcement agencies
- The LEPC
- State or federal environmental protection agencies
- State or local politicians

Representatives from these groups should participate in planning the exercises that will involve them. This will ensure that the exercise activities provide a realistic test of their functions during the response. Practicing the ERP also allows the chance to look for flaws or gaps in its implementation.

An exercise may fail because the participants are not given the necessary information and direction to respond effectively. When the objectives are clearly stated first, the exercise can be designed to accomplish them. For instance, if the objective of a written training exercise is to select the proper direct-reading instrument for a scenario, the description should not emphasize information important to spill confinement techniques.

Another danger is that an exercise may be too ambitious, expecting too much from the participants. For example, participants may be asked to write a safety plan based on a scenario that describes too many or unrealistic hazards. The likely end is that the participants will not take the exercise seriously, and little or nothing will be accomplished. For an exercise to be effective, the participants must feel that it simulates a task or activity they may actually participate in or accomplish during a response.

It is important not to present a complex exercise without preceding it with simpler exercises that deal with its components. For instance, if a training course does not conduct functional exercises before the day when a full-scale incident simulation is conducted, this will result in a confused exercise where participants are unable to perform individual tasks efficiently. Instead of demonstrating an imperfect but successful response to an incident, the exercise frustrates and discourages the participants.

Once the objectives are established, the specific content of the exercise can be created. For functional or full-scale exercises requiring equipment and supplies or integration of several groups, planning should start early. The whole exercise must be well thought out. Possible response actions other than those specifically intended by the planner should be predicted so that exercise controllers are not caught off-guard during the exercise itself.

Conducting the Exercise

The way the exercise is presented and managed will directly affect its success. Three types of people are involved in conducting an exercise: controllers, simulators, and evaluators (Kelly 1989). In tabletop exercises conducted during training courses, a single instructor may be able to fill all these roles. Obviously, larger exercises will require more people. All people involved in conducting the exercise must be familiar with their roles and with the exercise as a whole.

The controller keeps the exercise on course by monitoring activities and intervening when necessary to maintain professionalism or prevent distractions. The simulator is an "actor" who provides realism by portraying groups that may be involved during a response but are not included in the exercise. Simulators will be interacting with the participants throughout the exercise to provide information and may play such roles as a chemical manufacturer's representative, an injured person, or a media representative. The evaluator's role is to observe the exercise and provide an objective assessment of what areas went well and what areas need improvement (Figure 20-7). The evaluator usually will not intervene in the exercise while it is underway, but will observe and then report after the exercise has been terminated.

If the exercise is well prepared, the participants can work through the activities and/or respond to information as it is presented, with little assistance from controllers or simulators. During tabletop exercises, discussion, whether in large or small groups, is the primary activity. The controller should allow participants to discuss the activity or scenario openly, with minimal interference. However, controllers should be very

FIGURE 20-7. Evaluators should observe exercise activities without intervening.

attentive to the progress of the incident in order to intervene quickly to prevent distraction, provide additional information, or simply answer questions.

Conducting functional and full-scale exercises typically requires several people. The activity may be spread out over a wide area or in several locations. Functional activities may involve one task, such as inspecting and donning respiratory equipment, but may have many people participating. Full-scale incidents may involve many people performing many activities. In both instances, sufficient personnel should be available to act as controllers, simulators, and evaluators.

Full-scale and functional exercises should be realistic simulations of activities to be conducted during an emergency response. Therefore, the exercises should be conducted at locations which may be real incident sites, under appropriate time constraints, and with equipment that will be used. Response activities may be triggered by the mere presence of labeled hazards (e.g., a drum labeled "acetone" to initiate concern for flammability hazards), by the simulation of events or hazards (e.g., water leaking from a container labeled as a hazardous chemical), or by messages provided at specific times by exercise controllers (e.g., "An explosion has been reported in the supply warehouse"). The controllers and simulators should ensure that the pace of the response is realistic and that participants maintain a professional attitude. However, the actual direction and involvement of the controllers should be kept to a minimum to allow the participants to find their own solutions.

The exercise should be closely monitored by the controllers to ensure that at no time are the health and safety of the participants, controllers, or general public endangered. Signals or code words that are different from the communications used in the exercise should be established that will stop the exercise if unsafe conditions arise or if a real emergency occurs. Dangerous response activities that might be considered during a real incident should be simulated during the exercise. Exercise controllers should be instructed to observe participants for signs of trouble, such as symptoms of heat stress.

Evaluation

If the exercise activities are important to practicing a response, the evaluation is equally important to learning from the exercise. The evaluation of exercise activities can take place at two levels. First, the participants may conduct a post-incident analysis or critique as a part of the exercise, if time permits. This would be an extension of the incident simulation, in which the response action as a whole is reviewed to learn from mistakes and to reinforce successful actions. The exercise controller should monitor the critique to ensure that it does not become a "blame" session or that personality conflicts do not deter constructive criticism. This level of evaluation causes participants to practice critical evaluation of the performance of the response team.

Second, the exercise should be evaluated from the perspective of the exercise staff (controllers and evaluators). This level of evaluation should assess how well the objectives set forth for the exercise were met, as well as the performance of individuals or groups involved in the response. The exercise staff should be able to compare the decisions and actions observed during the response to those expected according to the ERP. For exercises conducted during training courses, the exercise staff can use this opportunity to evaluate the effectiveness of the training course in preparing trainees for emergency responses.

In addition to evaluating the participants and the ERP in the exercise, the exercise itself should be evaluated. Lessons learned during one exercise can be used to improve future exercises. Exercises should not be one-time events but should be conducted regularly at all levels (orientation through full-scale). The frequency should be determined by the emergency response planning committee based on an assessment of previous exercises. Therefore, an exercise evaluation should be conducted incorporating the comments of participants and exercise staff alike, as well as any nonparticipating observers who may have been invited. A written report should be prepared for distribution to all parties involved which describes the exercise, a summary of needs (planning, training, equipment, and/or personnel) disclosed during the exercise, and recommendations for follow-up activities.

Follow-Up

It is unlikely that a simulated incident will demonstrate a perfect hazmat response. The value of an exercise is lost if follow-up activities are not conducted to correct problems or deficiencies uncovered during the exercise. The process begins with the compilation of the exercise's final report. For exercises intended to test a facility's ERP, the report should summarize

- The objectives of the exercise
- The pertinent points raised during the post-incident analysis
- The lessons learned
- Deficiencies in planning, training, or equipment
- Specific recommendations for improving the response operations or ERP

Once the report and recommendations have been distributed, the emergency response planning committee should monitor the progress of implementation of the recommendations. Also, any changes in the ERP that result from the exercise must be distributed to the appropriate personnel. Recommendations for improvements in the exercise process itself should be incorporated into planning for future exercises.

Exercises conducted during a training course, with participants from various companies, will also involve follow-up activities. The follow-up will be largely the responsibility of the individual trainee, though an observant exercise staff may detect areas where trainees need additional work. The trainee should be encouraged to reflect on the exercise and to note honestly areas where he or she did not understand what was going on. These areas should be studied further. Instructors and exercise staff should be able to provide additional assistance or references to assist the trainee even after the course has been completed.

SUMMARY

A well-planned and executed exercise need not result in a flawless response to a simulated hazardous materials incident to be a success. Exercises can be conducted at four levels: orientation seminar, tabletop, functional, or full-scale exercises. The exercise will involve development, execution, evaluation, and follow-up activities. The success of an exercise is determined by the performance of the participants and the degree to which deficiencies and problems are resolved following the exercise. But failing to conduct exercises of an ERP can result in unnecessary confusion during an actual emergency.

REFERENCES

Birmingham-Jefferson County Emergency Management Agency. 1991. *Emergency Operations Procedures*. Birmingham, AL: Birmingham-Jefferson County EMA.

FEMA. 1989. *Hazardous Materials Exercise Evaluation Methodology (HM-EEM) and Manual*. Washington, DC: Federal Emergency Management Agency.

Kelly, R.B. 1989. *Industrial Emergency Preparedness*. New York: Van Nostrand Reinhold.

National Response Team. 1990. *NRT-2 Developing a Hazardous Materials Exercise Program—A Handbook for State and Local Officials*. Washington, DC: U.S. Government Printing Office.

Index

Access control points
in personnel decontamination line, 229–230
in site control, 112
Acclimatization, 432
Acetylene, 329
handling and storage procedures for, 299
Acetyl peroxide, 331
Activated charcoal, 391
Active sampling systems, in air sampling, 389–390
Activity-related fitness, 446
Acute effects studies, 342–343
Acute stress, 421
Adhesives, 281–282
Adrenocorticotropic hormone, 422
Aerobic capacity, 448
Aging, and heat stress, 432–433
Air monitoring, in hazard and risk assessment, 67
Air-purifying respirators (APRs), 160–161
advantages of, 165
considerations for using, 167–168
disadvantages of, 165–167
inspecting, 214
role of, in emergency response, 169
storage of, 215, 216
Air-reactive materials, 330–331
Air releases, 253
controlling, 264–268
Air sampling, 388–389
sample period, 389
sampling systems, 389
active samplers, 389–390
data, 392–393
laboratory analysis, 392
passive samplers, 390–391
sampling media, 391–392
using, 392–393
Air supply consumption, and safe work mission duration, 211–213
Air surveillance, 366–397
colorimetric indicators tubes, 386
limitations of, 387–388
operation of, 386–387
direct-reading instruments, 374
calibration of, 383–385
commonly used, 376–381
components of, 374–376
gas chromatography, 381–382
inherent safety, 385–386
relative response, 383
selecting equipment, 393
budget, 393–394
hazards, 394
operating range, 394

Air surveillance (*continued*)
operation, 395
planning purchases, 395–397
selectivity, 394
standard operating procedures for airborne hazards, 367–368
surveillance strategies, 368–369
air monitoring versus air sampling, 373–374
initial entry, 370–372
perimeter and background survey, 369–370
periodic monitoring, 372–373
termination monitoring, 373
ALARA, 349
Alcohol consumption
as cause of hypohydration, 432
and heat stress, 433
Alcohol-resistant firefighting foam, 267
ALOHA (Areal Locations of Hazardous Atmospheres), 405, 408
Alpha radiation, 336
Ambient temperature extremes, and safe work mission duration, 212 Ambulance transport, and decontamination, 239
American College of Sports Medicine, 445, 457
American Conference of Governmental Industrial Hygienists (ACGIH), human exposure limits set by, 348
American National Standards Institute (ANSI), and the respirator protection factor, 178–179
Ammonia, 329, 357
Ammonium nitrate, 331
Ammonium persulfate, 331
Angle roll technique, 310–311
Animals, laboratory studies of toxicity using, 341–344
Aqueous film-forming foam (AFFF), 267
ARCHIE (Automated Resource for Chemical Hazard Incident Evaluation), 99, 408–409
Aromatic rings, 321–322
Assessment
in containment operation control, 275–278
in hazardous materials control, 252–253
ASTM standards for personal protective equipment
ASTM F 1001, 205, 209
ASTM F 1052, 205
Atmosphere-supplying respirators, 162
Atoms
combinations of, 314–316
structure of, 313–314

Back injury, 445
Benzene, 322, 356
Benzo(a)pyrene, 362
Benzoyl peroxide, 331
Beryllium, 356
Beta radiation, 336
Bhopal tragedy, 135
Biofeedback, in stress management, 424
Blood-brain barrier, 359
Body fat measurements, 454
Body mass index (BMI), 454
Boiling point of chemicals, 324–325
Bonding procedures, 303
Booms
in controlling land releases, 259
in controlling water releases, 261
sealed, 264
Boot, removal of in decontamination line, 231–232, 233
Boot covers, disposable, 190, 192
Breakthrough time, 180
BRP, site-specific, 276
Buddies, in standard operating procedures, 121
Bulk transportation containers, 68–70
Butane, 329

Calibration, 383–385
Calipers, 454–455
CAMEO 3.0 (Computer-Aided Management of Emergency Operations) Chemical Database, 399, 403, 404–408, 409, 412
Canada, regulation of dangerous goods in, 45
Cancer, 8
chemicals causing, 9
Carbon chain, 320, 321–322
Carbon tetrachloride, 323, 362

Carcinogens, 354
Cardiorespiratory endurance, 449
Cardiorespiratory endurance training, 468
CAS (Chemical Abstracts Service) number, 399
Catch basins, 261
C (ceiling), 349
Cell poisoning, 7–8
Center for Labor Education and Research (CLEAR), i-ii, 1
Centigrade scale, 324
CHEMASYST, 400
CHEMBANK, 399, 412
Chemical Abstract Service (CAS) number, 319
Chemical behavior, 312–338
Chemical combinations, 8–9
Chemical Emergency Transportation Center (CHEMTREC), 54
Chemical information, sources of, 337–338
Chemical Information System, 399, 412
Chemical Manufacturers Association (CMA), Community Awareness and Emergency Response program of, 135
Chemical protective clothing, 179, 190
  attacks on, 180
  information on performance characteristics of, 182–189
  inspecting, 213–214
  materials and technologies, 181–182
  problems with information available on, 183, 190
  and resistance to chemical attacks, 180–181
  reuse of, 216
  selection of, 179–180
  types of, 190–192
Chemical reactions, 330
  air-reactive materials, 330–331
  oxidizers, 331–332
    incompatible materials, 332–333
    toxic combustion products, 333–336
    unstable materials, 332
  water-reactive materials, 330
Chemicals
  appearance of, 327
  boiling point of, 324–325
  in environment, 10
  explosive limits and range of, 327
  families of, 316–318
  and fire, 9–10
  flash point of, 326
  and human health, 5–9
  hydrogen ion concentration of, 327
  ingested, 359–363
  names of, 318–319
  odor of, 327
  organic, 320–323
  physical state of, 327–330
  prefixes of, 319, 320
  properties of, 323–327
  solubility of, 325–326
  specific gravity of, 326
  suffixes of, 319–320
  symbols of, 319
  vapor density of, 325
  vapor pressure of, 325
Chemical solutions, for decontamination, 225–227
Chemical spills response procedures for, 125
Chemical treatment processes, for water releases, 264 CHEMLINE, 399
Chemrel suit by Chemron, Inc., 182
Chemtrec, 40, 92
Chime, 290
Chlorine, 317, 318, 329, 358
Chlorites, 331
Chloroform, 362
CHRIS (Chemical Hazard Response Information System) Manual, 92, 154, 399
CHRIS PLUS, 399, 412
Chromic acid, 331
Chromosorb, 391
Chronic effects studies, 343
Chronic stress, 421
Closure failure, in small containers, 290
Cold injury, prevention of, 440–441
Cold stress, 436
  conditions leading to injury, 436–437
  harmful effects of, 439–440
  and thermoregulation, 437–439
Cold zone, in site control, 111–112
Colorimetric indicator tubes, 366–367, 373, 386
  costs of, 396

Colorimetric indicator tubes (*continued*)
  limitations of, 387–388
  operation of, 386–387
Combustible gas indicator (CGI), 366, 374, 376–377
  calibration of, 383–385
  commonly used, 376–381
  components, 374–376
  gas chromatography, 381–382
  inherent safety, 385–386
  relative response, 383
Command section, in incident command system, 30–31, 32, 34–38
Communication
  in incident command system, 31
  and site control, 113–115
Compounds, 315–316
  naming, 319
  organic, 321
Comprehensive resource management, in incident command system, 30
Compressed gases, 328–329
Computer use, 398–414
  in emergency response training, 403–409
  in employee safety, 402–403
  in tracking and recording, 403
  networks and data bases in, 92–93
  in operations, data management, and reporting, 400–403
  in process tracking, 403
  in risk assessment, 99, 399–400
  sources of software, 412–414
  technical considerations of hardware, 410–411
Confinement procedures, 250, 251
Confinement systems, in pre-incident hazard assessment, 64–65
Consolidated action plan, in incident command system, 31
Contact lenses, 357
Containers
  grounding and bonding procedures, 286, 301–305
  labels and markings on, 49–50
  observing, in hazard and risk assessment, 67–70
  overpack, 285, 286, 306–311
Containment kits, 284
Containment procedures, 251, 285–286
  assessment and decision making for, 275–278
  equipment, supplies and tools used in, 278–285
  hazards to personnel in, 277–278
  for large containers, 291–293
  for plumbing leaks, 293–297
  for pressurized cylinders, 297–299
  for small containers, 286–291
Contaminated surfaces, touching, 239–240
Contamination reduction corridor (CRC), 111
Contamination reduction zone (CRZ), 111
Contingency plans, in pre-incident hazard assessment, 64
Continuous-flow respirators, 164
Cool-down, 468
Cooling garments, 193, 218
Coronary heart disease, 444
CPC base, 183, 402
Cracking, 321
Creosote, 322
Cryogenic gases, 329
Cumene hydroperoxide, 331

Damaged containers, procedures for placing, in overpack containers, 285, 286, 307–311
Dams, 264
  underflow, 262–263
Dangerous goods, regulation of, in Canada, 45
DDT, 362
Debriefing, 122
DECIDE process, 60–63, 120, 250, 252, 276
Decision making
  for controlling containment operations, 275–278
  in hazardous materials control, 252–253
Decontamination, 220–221
  definition and justification for, 221
  management of area, 239
    containment of liquids, 241–242
    orderly cleaning and doffing, 239–240
    protection of decontamination personnel, 240

methods of, 221
chemical solutions, 225–227
level of decontamination required, 227–228
physical, 221–225
performance of, in contamination reduction corridor, 111
post-incident management, 242
setting up area for
emergency medicine and decontamination, 236–239
location, 228–229
personnel decontamination line, 229–235, 244–249
tools, equipment, and vehicles, 235–236
Decontamination personnel, protection of, 240
Deep South Educational Resource Center, ii
Degradation, and chemical protective clothing, 180
Designated safety official, in emergency response chain of command, 18–19
Detergent solutions, in decontamination, 225–226
Detoxification mechanisms, 364
Diborane, 330–331
Digestive system, and ingestion of chemicals, 359–363
Diking, and controlling land releases, 256
Dilution, 272
Direct-reading instruments (DRIs), 373, 374
Dispersal, predicting, for hazardous materials, 93–99
Dispersants, 271–272
Disposable overgarments, 190, 192
Diversion, and controlling land releases, 256
Dose-response relationship, 340
Dosimeters, 390
DOT. *See* Transportation, U.S. Department of (DOT)
Downwind monitoring, 370
Drug consumption, and heat stress, 433
Dry brushes, in physical decontamination, 223
Dry ice, 328
Dust, inhalation of, 6

Ear protection, 195
Earth moving, in controlling land releases, 259
Electrochemical sensors, 377–378
Emergency Action Guides, 92, 155, 183, 265–266
Emergency Handling of Hazardous Materials in Surface Transportation, 154
Emergency medicine, and decontamination, 236–239
Emergency operation codes, 126–127
Emergency Planning and Community Right-to-Know Act, 135. *See also* SARA Title III legislation
Emergency responders
physical abilities testing for, 469–470
physical fitness for, 443–471
stress in, 416–421
Emergency response
chain of command and personnel roles, 18–19
computer use in, 398–414
employee safety, 402–403
event tracking and recording, 403
hardware consideration, 410–411
operations, data management, and reporting, 400–403
process tracking, 403
reporting, 401–402
risk assessment, 399–400
software for, 409–410, 412–414
procedures in, 20
role of air-purifying respirators in, 169
role of self-contained breathing apparatus respirators in,173–174
role of supplied-air respirators in, 171
Emergency Response Guidebook, 86–91
Emergency response plan (ERP), 17–18
Emergency response plan exercise, 473–489
benefits of conducting, 474
levels of, 474–475, 476
full-scale, 482–484
functional, 480–482
orientation, 475, 477
tabletop, 478–480
process, 484
conducting, 485–487
development, 484–485

Emergency response plan exercise (*continued*)
 evaluation, 487–488
 follow-up, 488–489
Emergency response training, computer use in, 403–409
Employee safety, computer use in, 402–403
Environment, chemicals in, 10
Environmental heat, 428
EnviroTIPS: Environmental and Technical Information for Problem Spills Manuals, 155
Environmental Protection Agency (EPA)
 Chemical Emergency Preparedness Program of, 135
 classification of hazardous materials by, 45, 46
 criteria in defining hazardous materials emergency, 4–5
 decontamination lines suggestions of, 230, 244–249
 levels of protection for PPE ensembles, 192, 196–203
 on exposure limits, 351
 regulations requiring use of ICS at hazardous materials incidents, 29
Enzymes, 361–362
Epidemiology studies of toxicity, 344–346
Equipment
 access to, and setting up decontamination area, 229
 in containment, 278–285
 decontamination of, 235–236
Ergometer, calibrated, 449
Evaluation in site control, 105–106
Event tracking and recording, computer use in, 403
Exercise, in stress management, 423
Experiments of nature, 346–347
Explosive limits and range, 327
Extremely hazardous substances (EHSs), 117, 351
Eyes
 damage to, 7
 protection for, 356–357

Facepiece
 fit of, 177
 for respirators, 160
Faceshield, 194
Fahrenheit scale, 324
Fat-soluble chemicals, 362–363
Federal Emergency Management Agency (FEMA), on exposure limits, 350
Federal government, OSHA coverage of employees, 14
Federal Insecticide, Fungicide, and Rodenticide Act (FIFRA), 52
Fetal damage, 8
Fiber counting, 392
Fight or flight response, 422
Filter fences, 263
Finance section, in incident command system, 33, 41–42
Fire, and chemicals, 9–10
Firefighter's Handbook of Hazardous Materials, The, 155–156
Firefighting foams, 267–268
Fire Protection Guide on Hazardous Materials, 153
Firesoft EPA Hazmat, 399–400, 412
First responder, training for, 21
FIT CHECK system, 451, 455, 459, 461, 463
Fixed facility, 136
Flame analyzer, 235
Flame ionization detector (FID), 366, 378–379
Flammability, as issue in personal protective clothing, 202–203
Flammable liquid
 grounding and bonding of containers, 301–305
 safe transfer of, 285, 301–305
Flash cover, 202–203
Flash point, 326
Flotation gear, 193
Fluorine, 329
Foams, firefighting, 267–268
Forsberg's Chemical Protective Clothing Performance Index, 402
Foxboro Century OVA, 383, 384
Freelancing, 30
Freezing, in physical decontamination, 224–225
Freon, 329
Frostbite, 439
Fully encapsulating suits, inspecting, 213
Fumes, inhalation of, 6
Functional physical fitness, 446

Gamma radiation, 336
Gas chromatography, 381–382
  analysis on site by, 391
Gases, 328–329
  decontamination by, 228
  dispersal of, 96, 98–99
  inhalation of, 6
Gaskets, 281–282
Geiger counter, 235, 380
Geiger-Mueller, 396
Gelation, 271
General Bulletin Board Systems, 409
Gloves
  disposable, 190, 192
  inspecting, 214
  removal of, in decontamination line, 233
  in preventing cold injury, 441
GlovES+, 183, 402
Goggles, 194
Goniometer, 458
Grounding and bonding flammable liquid containers, 301–305
Grounding clamps and cables, 301–302
Grounding electrodes, 302–303
Guidelines for the Selection of Chemical-Protective Clothing, 182–183
Guide to the Safe Handling of Hazardous Materials Accidents, A, 155

Halogen, 317, 331
Halogenated hydrocarbon, 321
HAMIN, 402
Handbook of Hazardous Materials: Fire/Safety/Health, 154–155
HAZARD, 402, 403, 412
Hazard and risk assessment, 59–60
  decision making in emergencies, 60–63
  entry and on-site survey, 67–71
  incident perimeter survey, 66–67
  observing released material in, 71
  predicting dispersal of hazardous materials, 93–99
  pre-incident, 63–65
  researching identified materials, 71
    CHRIS Manual, 92, 154, 399
    computer networks and data bases, 92–93
    Emergency Action Guides, 92, 155, 183, 265–266
    Emergency Response Guidebook, 86–91
    material safety data sheets, 71, 74–81
    printed materials, 71, 74–86
    telephone hotlines, 92
  steps of, 119
Hazard class labels, 49
Hazardous chemical(s), 44
Hazardous chemical operations
  liquid splash-protective suit for, 208–209
  support function protective garments for, 209
  vapor-protective suits for, 205–208
Hazardous material foams, 267
Hazardous materials
  definition of, 5, 44–45
  detecting presence of, 60–61
  predicting dispersal for, 93–99
Hazardous materials control
  advanced, 274–311
  assessment and decision making for containment operations, 275–278
  basic, 250–272
  collection techniques, 268–271
  containment procedures, 285–286
    grounding and bonding flammable liquid containers, 301–305
    for large containers, 291–293
    overpacking damaged containers, 306–311
    for plumbing leaks, 293–297
    for pressurized cylinders, 297–299
    for small containers, 286–291
  equipment, supplies, and tools used in containment, 278–279
    adhesives, sealants, and gaskets, 281–282
    containment related items, 281, 284, 285
    plugs, patches, and related items, 279–280
    tools, 281, 283
  role of hazard and risk assessment and decision making in, 252–268
Hazardous materials emergency, defining, 4–5
Hazardous Materials Emergency Response Guidebook, 153–154

Hazardous materials specialist, training for, 22–23
Hazardous materials technician, training for, 22
Hazardous materials terminology, 43–44
- to communicate hazard information, 46, 49–57
- to describe hazard classification, 45–46
- for materials which pose hazards, 44–45

Hazardous Materials Transportation Act (1975), 44–45
Hazardous Materials Transportation Uniform Standards Act (1990), 45
Hazardous substances, definition of, 45
Hazardous waste operations and emergency response (HAZWOPER)
- applicability to emergency response operations, 16–17
- background and development, 15–16
- OSHA regulations pertaining to, 15–17
- purpose and applicability, 16

Hazardous wastes
- definition of, 45
- human health effects of, 339–364

Hazardous Waste Tracker, 403
Hazardous Waste Worker Training Project, 1
Hazmat emergencies, dangers in, 5–10
HAZ-MAT II, 402, 412
Hazmat Manager, 402, 403, 412
HAZMIN, 412
HAZWASTE, 403, 412
Haz Waste Tracker, 403, 412
Head phones, 195
Health-related fitness, 448
Heart rate, as indication of heart strain, 436
Heat cramps, 430
Heat exhaustion, 430–431
Heat-induced illness, 429–430
- prevention of, 431–436

Heat injury, prevention of, 218
Heat rash, 430
Heat strain, monitoring, 434–436
Heat stress, 428
- illnesses induced by, 429–431
- and personal protective equipment, 217–218
- reducing individual's response to, 431–434
- sources of, 428

Heat stroke, 431
Heat syncope, 430
Heat tolerance, screening for, 434
Helicopter transport, and decontamination, 239
Helium, 329
Hepatotoxins, 353
Hexachlorobenzene, 346–347, 362
Hexane, 356
HMIX (Hazardous Materials Information Exchange) bulletin board system, 409
H-Nu P1–101, 384
Hole repair procedures
- for large containers, 291–293
- for small containers, 287–290

Hood, 194
Hotline telephone numbers, 152–153
Hot zone
- setting boundary for, 221
- in site control, 108–111

Human exposure, assessing, to chemicals, 5–9, 339–364
Human exposure limits
- routes of entry into the body, 355–356
  - contact with the body surface, 356–357
  - ingestion into the digestive system, 359–363
- setting safe, 341
  - epidemiology studies, 344–346
  - experiments of nature, 346–347
  - exposure limit terminology, 349–355
  - extrapolating study data to human exposure limits, 347–348
  - laboratory studies using animals, 341–344
  - sources of information, 354–355
  - workplace versus emergency response exposure limit, 350–351

Hydration, 431–432
Hydrocarbon, 321
Hydrogen, 329
Hydrogen chloride, 357
Hydrogen ion concentration, 327
Hyper CPC, 402, 412
Hyper CPC Stacks, 183
Hypochlorites, 331
Hypohydration, 432
Hypothalamus, 429
Hypothermia, 440

Immersion foot, 439–440
IMO hazard class number, 49
  comparison of hazard classes with DOT, 48
  for hazard class labels, 49
Incident commander (IC), 32
Incident Command System (ICS), 18, 28–29
  basics of, 29–34
  command of
    single and unified structures, 34–36
    staff responsibilities, 36–38
  components of, 30–32
  finance section, 32, 41–42
  functional areas of, 32–33
  interaction of the functional areas of, 33–34
  logistics section, 32, 41
  objectives, 29
  operations section, 32, 38–40
  planning section, 32, 40–41
  utilization of preestablished, in site control, 104
Incident information summary sheet, 146
Incident perimeter survey, 66–67
Incompatible materials, 332–333
Infrared spectrophotometer, 379
Ingestion injury, 7,
Inhalation injury, 6, 357–358
Inlet blockage, and controlling land releases, 256–257
Insoluble materials, controlling, 260–263
International Maritime Dangerous Goods (IMDG) Code, 46
International Maritime Organization (IMO) regulations, for transport of hazardous materials, 46
International Union of Pure and Applied Chemists, 319
Inventory control, computer use for, 400
Inverted overpack technique, 309–310
Isolation, in site control, 104–106

Job-related fitness, 446–448
Job segregation, 419
Joint flexibility, 457–459

Knife, 195

Labels
  for hazardous containers, 49–53
  for pesticides, 52–53
Laboratory analysis in air sampling, 392
Laboratory studies of toxicity using animals, 341–344
Land, dispersal of liquids released on, 94–95
Land releases, 252–253
  controlling, 254–259
LC_{10}, 349
LD_{50}, studies of, 341–342
LD_{10}, 349
Lead, as teratogen, 353
Leighton flexometer, 458
Levels of concern (LOCs), 351
Liaison officer, command staff responsibilities of, 37–38
Liquids, 328
  containment of contaminated, 241–242
  decontamination by, 228
  dispersal of, 93–96, 97
  volatile, 328
Liquid splash-protective suit, for hazardous chemical emergencies, 208–209
Local Emergency Planning Committee (LEPC), 117, 135–137. *See also* SARA Title III legislation
Local Emergency Response Plans (LERPs). *See also* SARA Title III legislation
  components of, 144–152
  emergency telephone roster in, 149–152
  governing principles, 148
  implementation procedures, 146–147
  incident information summary sheet in, 146
  instructions on plan use, 149
  organizational rules and responsibilities, 148
  promulgation document in, 146
  relationship to other plans, 149
Logistics section, in incident command system, 32, 41
LOTUS, 99
Lower explosive limit (LEL), 327

Manual for Spill of Hazardous Materials, 156
Maps, in pre-incident hazard assessment, 65
Material Safety Data Sheets, 40, 44, 54
  in pre-incident hazard assessment, 64

Material Safety Data Sheets (*continued*)
in researching identified materials, 71, 74–81
Maximum oxygen uptake (max Vo_2), measuring, 449
Medical surveillance and monitoring, 24–25, 363–364
in assessing human exposure to chemicals, 363–364
Meditation in stress management, 423
MEDLINE, 355
Mental stress, 415–416. *See also* Stress
Metabolic heat, 428–429
Metabolites, 364
Meteorological conditions in establishing hot zone, 110–111
Methane, 329, 367, 374
Microbes as dispersants, 271–272
MicroCHRIS, 399, 412
Mist, inhalation of, 6
Monomers, 332
MSA Model 260, 384
Muscle endurance, 460–462
Muscle strength, 462–464
and endurance training, 468
Mutagens, 352

Naphthalene, 328
National Fire Protection Association (NFPA), 1
certification standards for chemical protective garments, 203–209
National Electrical Code of, 385
professional qualifications standards, 443, 447–448
704 system of, 51–52
National Institute of Occupational Safety and Health (NIOSH)
human exposure limits set by, 348
and screening for heat tolerance, 434
on skin notations, 356
National Research Council's Committee on Toxicology, 350
National Response Team (NRT), establishment of, 137
Negative-pressure respirators, 164
Negative-pressure tests, 174
Nephrotoxins, 353
Neurotoxins, 353
Neutralizing solutions, in decontamination, 227
Nickel, 356
NIOSH Pocket Guide to Chemical Hazards, 81–86, 156
Nitrogen, 329
Nonchemical hazards, protective clothing for, 191, 192
Nonencapsulating chemical protective (NECP), 190

Obesity, 453–457
Occupational Health Guidelines for Chemical Hazards, 154
Occupational safety and health, coverage under state plans, 14–15
Occupational Safety and Health Act, purpose of, 12–13
Occupational Safety and Health Association (OSHA)
background and creation of, 12–13
human exposure limits set by, 348
provisions of 29 CFR 1910.120 applicable to emergency response operations
chain of command and personnel roles, 18–19
emergency response plan, 17–18
emergency response procedures, 20
medical surveillance and consultation, 24–25
personal protective equipment, 25–26
training of emergency responders, 20–24
regulations pertaining to hazardous waste operations and emergency response, 15–17
regulations requiring use of ICS at hazardous materials incidents, 29
requirements for training in PPE use, 211
responsibilities of, 13
safety and health standards, 13–14
scope of authority, 14–15
selection and use of PPE, 25–26
setting permissible exposure limits (PELS), 7–8
on skin notations, 356
specific requirements for PPE, 26–27

on standards for toxic chemicals, 363–364

OHM/TADS (Oil and Hazardous Materials/Technical Assistance Database), 399

On-scene incident commander, training for, 23

Operations section, in incident command system, 32, 38–40

Organic chemicals, 320–323

Organic compounds, 321

Overflow dams, 260–261

Overload principle, 465–467

Overpack container, 286
- procedures for placing damaged containers in, 307–311
- safe transfer of, 285

Overpacking, 291
- purpose and basic concept of, 306
- safety considerations in, 306–307

Overweight, 445

Oxidizers, 9–10, 331–332

Oxygen, 329

Oxygen meters, 377–378

Ozone, 358

Paradichlorobenzene, 328

Passive samplers, in air sampling, 390–391

Patches, 279–280

PEL (permissible exposure level), 7, 349

Penetration, 212
- and chemical protective clothing, 180

Peracetic acid, 331

Perchlorates, 331

Perchloroethylene, 369

Permeation, 212
- and chemical protective clothing, 180
- rate of, 181

Permissible exposure limits (PELs), 7, 349

Peroxides, 331

Personal dosimeter, 195

Personal locator beacon, 195

Personal protective equipment (PPE), 25–26, 158
- accessories, 192, 193–197
- chemical protective clothing and accessories
  - attacks on, 180
  - information on performance characteristics, 182–189
  - inspecting, 213–214
  - materials and technologies, 181–182
  - problems with information available on CPC, 183, 190
  - resistance to chemical attacks, 180–181
  - reuse of, 216
  - selection of, 179–181
  - types of, 190–192
- classification of protective ensembles, 192
  - EPA levels in, 192, 196–203
  - NFPA certification standards in, 203–209
- maintenance of, 216
- permeation and penetration of, 212
- in preventing cold injury, 440–441
- respirators
  - classification of respiratory protective equipment,159–165
  - importance of respirator fit and fit testing, 174–177
  - respirator protection factor, 178–179
  - respiratory protection requirements, 159
  - selection of, 165–174
- safe use of, 209–210
  - doffing, 215
  - donning, 213
  - factors limiting safe work mission duration, 211–212
  - heat stress, 217–218
  - inspection of, 213–214
  - personal factors affecting respirator use, 212–213
  - program requirements, 210–211
  - storage of, 215–216
  - training in use of, 211
  - in-use monitoring of, 215
- and safe work mission duration, 212
- selection and use of, 25–26, 330, 393
- specific requirements for, 26
- stress in wearing, 418

Personnel, hazards to, in containment operations, 277–278

Personnel decontamination line
- access control points, 229–230
- stations in line, 230–235

Pesticides, labels for, 52–53

Petrochemical industry, 321

Phenobarbital, 362
Phenol, 322
Phosgene, 358
Photoanalyzer, 235
Photoionization detector (PID), 378
Physical abilities testing for emergency responders, 469–470
Physical fitness
  body composition in, 453–457
  cardiorespiratory endurance in, 448–453
  defining, 446–448
  health benefits of, 444–445
  and heat stress, 433
  joint flexibility in, 457–459
  muscle endurance in, 460–462
  muscle strength in, 462–464
Physical fitness training
  equipment for, 470–471
  specificity of, 465
    cardiorespiratory endurance training, 468
    cool-down, 468
    muscle strength and endurance training, 468
    overload principle, 465–467
    planning workout, 467–469
    time allotment for, 468–469
    warmup, 467–468
Physical methods of decontamination, 221–225
Physical state, changes in, 329–330
Picric acid, 322
Pipe connections, procedures for leaking, 293
Pipe repair clamp, 294–295
Pipe saver, 294–295
Piping, 294–295
Placards, 54–57
Planning section, in incident command system, 32, 40–41
Plugs, 279–280
Plumbing leaks, procedures for, 293–297
Polybrominated biphenyls (PBBs), 322
Polychlorinated biphenyls (PCBs), 322
Polyethylene, 332
Polymerization, 330
Polypropylene, 332
Polystyrene, 330, 332
Polyvinyl chloride (PVC), 332
Positive-pressure respirators, 164–165
Positive-pressure tests for respirator fit, 175
Postemergency response operations, 26–27
Potassium dichromate, 331
Potassium permanganate, 331
Potassium peroxide, 331
Potentiation, 362
Powered air-purifying respirators (PAPRs), 164
Pressure-demand respirators, 164
Pressurized cylinders, procedures for, 297–299
Private industry, OSHA coverage of employees in, 14
Process tracking, computer use in, 403
Promulgation document, 146
Propane, 329
Propylene, 329
Protection-in-place procedures in site control, 106
Public information officer, command staff responsibilities of, 36–37
Pumping, in collecting spills, 268
Puncture procedures
  for large containers, 291
  for small containers, 286–287

Qualitative fit tests, for respirator fit, 174
Quantitative fit tests, for respirator fit, 177
Quick Selection Guide to Chemical Protective Clothing, 183

Radiation, 336
  detectors of, 380–381, 396
Radioactivity, 336
  detecting, 396
  recognition, 336–337
  response, 337
Relative response, 383
Relaxation, in stress management, 422–423
Released material, observing, 71
Releases, controlling
  air, 253, 264–268
  land, 252–253, 254–259
  water, 253, 260–264
Reproductive damage, 8
Resource Conservation and recovery Act (RCRA) (1976), 45
Respirator protection factor, 178–179

Respirators
  classification of equipment, 159–165
  importance of fit and fit testing, 174–177
  inspecting, 214
  personal factors affecting use of, 212–213
  respirator protection factor, 178–179
  respiratory protection requirements, 159
  selection of, 165–174
Respiratory protective equipment, 159
  examples of, 165
  facepiece type, 160
  method of protection, 160–163
  mode of operation, 164–165
  selection of, 165–174
Respiratory system, inhalation of chemicals into, 357–358
Responder by Life Guard, Inc., 182
Roller technique, 310
RTECS (Registry of the Toxic Effects of Chemical Substances), 399, 354

Safety glasses, 194
Safety helmet, 193–194
Safety officer, command staff responsibilities of, 37
Safe work mission duration, factors limiting, 211–212
SARA!, 402, 412
SARA Title 1 legislation, 16
SARA Title III legislation, 134–135. *See also* State Emergency Response Plans (SERPs)
  chemical lists and their purposes, 142–143
  focus of, 398
  history, 135–137
  reference sources, 152–156
  relevant provisions of, 137–144
  state and local ERPs, 144–151
SARATRAX, 400, 402, 412
Scientific management theories, 425
Scrapers, in physical decontamination, 223
Scrubbers, in physical decontamination, 223
SDMS Chemical Management System, 412
SDMS Injury/Illness Surveillance System, 402
Sealants, 281–282
Sealed booms, 264
Self-contained breathing apparatus (SCBA), 26, 162, 377
  advantages and disadvantages of, 171
  considerations for using, 171–173
  inspecting, 214
  removal of, in decontamination line, 234
  role of, in emergency response, 173–174
  storage of, 215, 216
Shipping papers, 53–54
Shivering, 439
Short-term public exposure guidance level (SPEGL), 393
Showers, in physical decontamination, 222
Silica gel, 391
Sit-and-reach test, 458–459
Site control, 101–102
  communication, 113–115
  enforcement of, 103–104
  isolation, 104
    access control, 104–105
    evacuation and protection in place, 105–106
  objectives of, 102–103
  zoning, 106–113
    access control points, 112
    cold zone, 111–112
    hot zone, 108–111
    warm zone, 111
Site location maps, computer use for, 400–401
Sit-up test, 460–461
Skilled support personnel, in in emergency response chain of command, 19
Skin
  chemical contact with, 356–357
  damage to, 7
Sodium chloride, 317
Sodium nitrate, 331–332
Sodium peroxide, 331
Solidification, 271
Solids, 328
  decontamination by, 227–228
  dispersal of, 93
Solubility, 325–326
Sorbent collection, 268–271
Sorbents, 391
Span of control, in incident command system, 31
Specialist employees, in emergency response chain of command, 19

Specialty kits, 278
Specific gravity, 326
Splash hood, 194
Stamina, 448
Standard operating procedures (SOPs), 116, 117
  for airborne hazards, 367–368
  chemical spill response procedures, 125
  considerations in, 117–119
  emergency operation codes, 126–127
  and enforcement of site control, 104
  establishing and need for communication in site control, 114
  generic operating guides, 119–120
  for HAZ-1, 128–133
  response activities, 119
  typical, 120–122
State Emergency Response Commissions (SERCs), 135
State Emergency Response Committees (SERCs), 117
State Emergency Response Plans (SERPs). *See also* SARA Title III legislation
  components of, 144–152
  emergency telephone roster in, 149–152
  governing principles, 148
  implementation procedures, 146–147
  incident information summary sheet in, 146
  instructions on plan use, 149
  organizational rules and responsibilities, 148
  promulgation document in, 146
  relationship to other plans, 149
Steam, in physical decontamination, 224
STEL (short-term exposure limit), 349
Strap iron bandaid patch, 292
Stress
  mental, 415–416
  reaction to, 422
  reducing on the job, 424–425
  shaping up to manage, 422
    biofeedback, 424
    exercise, 423
    meditation, 423
    relaxation, 422–423
    structure, 423
  symptoms of, 421
  temperature, 427
    cold, 436–441
    heat, 428–436
  work-related, 416–417
    acute and chronic stressors, 421
    political change, 419–420
    social change, 418–419
    technological change, 417–418
Structure, in stress management, 423
Sublimation, 328
Submaximal testing, 449
Sulfur dioxide, 358
Superfund Amendment and Reauthorization Act (SARA) (1986), 11. *See also* SARA Title I legislation; SARA Title III legislation
Supplied-air respirators (SARs), 162, 377
  advantages and disadvantages of, 169
  considerations for using, 169–170
  inspecting, 214
  role of, in emergency response, 171
  storage of, 215
Supplies in containment, 278–285
Support function protective garments, for hazardous chemical operations, 209
Sweating, 429
Swipe tests, 227
Synergism, 362

TC_{Lo}, 349
TD_{Lo}, 349
Telephone hotlines, 92, 152–153
Temperature. *See also* Cold stress; Heat stress
  ambient extremes in, and safe work mission duration, 212
  impact on spilled material, 66
Tenax, 391
Teratogens, 352–353
Termination procedures, 116, 122–124
  emergency operation codes, 126–127
  job descriptions, 124
Thermoregulation, 429
313 Advisor, 402, 412
3M Select Software, 303, 412
Threshold planning quantities (TPQs), 117
Time allotment, 468–469
Time-weighted average permissible exposure limit (TWA-PEL), 379
Tissue, impact of chemicals on, 8
TLV (threshold limit value), 349
Toggle patches, 287
TOMES Plus System, 402, 412

Tools, 281, 283
in containment, 278–285
decontamination of, 235–236
Topography
and controlling land releases, 254–255
impact on spilled material, 66
in setting hot zone, 108–109
and setting up decontamination area, 229
Totally encapsulating chemical protective suits (TECP) suits, 190, 192
inspecting, 213
Toxic combustion products, 333–336
Toxic exposure data, 379–380
Toxicity, 340
Toxicological profiles, 355
TOXLINE, 399
TOXNET, 355
T patches, 287
Training
in air surveillance techniques, 368
of emergency responders, 20–24
in personal protective equipment, 211
and stress management, 424–425
Transportation, U.S. Department of (DOT)
classification of hazardous materials by, 45–46, 47
comparison of hazard classes with IMO, 48
four-digit identification numbers for hazardous materials, 50
Transportation Safety Act (1974), 44
Trench foot, 439–440
Trichloroethane, 317–318, 321, 356
Trimethylaluminum, 331
Trinitrophenol, 322
TWA (time-weighted average), 349
Two-way radio, 195
Tyvek, 182

Underflow dams, 262–263
United Nations system, 46
U.S. v. *Johnson Controls* (1991), 353
Unstable materials, 332
Upper explosive limit (UEL), 327
Upwind monitoring, 370

Vacuum, in collecting spills, 268
Valves, procedures for broken, 295, 297
Vapor(s)
density of, 325
dispersal of, 96, 98–99
inhalation of, 6
knockdown or dispersion, 265–266
Vapor pressure, 325
Vapor-protective suits, for hazardous chemical emergencies, 205–208
Vapor suppression, 266–268
foams in, 267–268
Vasoconstriction, 437–439
Vasodilation, 429
Vehicles, decontamination of, 235–236
Viscosity of a liquid, 328
Volatile chemicals, 10
Volatile liquid, 328

Warm zone in site control, 111
Water
access to, and setting up decontamination area, 229
dispersal of liquids released into, 95–96
solubility in, 326
Water-reactive materials, 330
Water releases, 253
controlling, 260–264
Water-soluble chemical, 362
Water-soluble materials, controlling, 263–264
Water streams, in physical decontamination, 222
What to do before the Ambulance Arrives, 408, 412
White phosphorus, 331
Wind, and setting up decontamination area, 229
Work
and rest cycles, in preventing cold injury, 440
stress at, 416–420

Xylene, 374

Zoning in site control, 106–113
cold zone, 111–112
hot zone, 108–111
warm zone, 111

Richard R. Paul Oct 94

Labels supplied courtesy of Labelmaster, Div. of American Labelmark Co., Inc.